Studies in History and Philosophy of Science

Volume 56

Clemency Montelle
School of Mathematics & Statistics
University of Canterbury
Christchurch, New Zealand

Ofer Gal
Unit for History and Philosophy of Science
University of Sydney
Sydney, Australia

Studies in History and Philosophy of Science is a peer-reviewed book series, dedicated to the history of science and historically informed philosophy of science. The series publishes original scholarship in various related areas, including new directions in epistemology and the history of knowledge within global and colonial contexts. It includes monographs, edited collections, and translations of primary sources in the English language. These cover a broad temporal spectrum, from antiquity to modernity, and all regions of the world.

More information about this series at http://www.springer.com/series/5671

Lukas M. Verburgt

John Venn: Unpublished Writings and Selected Correspondence

Springer

Lukas M. Verburgt
Freudenthal Institute, History and Philosophy of Science
Utrecht University
Utrecht, The Netherlands

ISSN 0929-6425 ISSN 2215-1958 (electronic)
Studies in History and Philosophy of Science
ISBN 978-3-030-79828-4 ISBN 978-3-030-79829-1 (eBook)
https://doi.org/10.1007/978-3-030-79829-1

This Springer imprint is published by the registered company Springer Nature Switzerland AG
The registered company address is: Gewerbestrasse 11, 6330 Cham, Switzerland

Introduction

John Venn, eminent English logician, President of Gonville and Caius College, Cambridge, Fellow of the Royal Society, and inventor of the famous diagram that bears his name, was born in 1834 and died in 1923. His long and remarkable life spanned several major intellectual developments which his wide-ranging activities as a scholar and as a teacher helped shape, to a greater or lesser degree – ranging from the professionalization of British philosophy and the reception of *Essays and Reviews* (1860) *and The Origin of Species* (1859) to the development of the Moral Sciences Tripos at Cambridge and the creation of modern logic.

The present volume contains unpublished writings and correspondence documenting Venn's transition from an Evangelical son to a Cambridge don. Venn has been very little studied compared to Moral Sciences colleagues like Henry Sidgwick and Alfred Marshall and scientific collaborators such as Francis Galton and William Stanley Jevons.[1] In lieu of something that might be called 'Venn scholarship', Venn's writings and correspondence offer a rich source of insight into his life, work and thought and, as such, can add new dimensions to understanding the world to which he belonged. This world was essentially that of what has become known as – and what Venn himself also called – the mid-Victorian generation of reading men at Cambridge for whom John Stuart Mill was the 'dominant teacher who expounds for all' and Alfred Tennyson the 'dominant poet who sings for all' (see letter **134**). Like so many of his contemporaries, including Henry Sidgwick (1838–1900) and his cousin Leslie Stephen (1823–1904), Venn struggled deeply and intensely with the relation between science and religion, between reason and

[1] My intellectual biography of Venn, entitled *John Venn: A Life in Logic*, is forthcoming at The University Chicago Press. On the other figures mentioned see, for instance, Bart Schultz, *Henry Sidgwick – Eye of the Universe: An Intellectual Biography* (Cambridge: Cambridge University Press, 2004); Peter Groenewegen, *Alfred Marshall: Economics, 1842–1924* (Basingstoke: Palgrave Macmillan, 2007); Simon J. Cook, *The Intellectual Foundations of Alfred Marshall's Economic Science. A Rounded Globe of Knowledge* (Cambridge: Cambridge University Press, 2009); Nicholas Wright Gillham, *A Life of Sir Francis Galton. From African Exploration to the Birth of Eugenics* (Oxford: Oxford University Press, 2001); and Harro Maas, *William Stanley Jevons and the Making of Modern Economics* (Cambridge: Cambridge University Press, 2005).

faith and, perhaps most accurately said, between intellectual honesty and emotional attachment to tradition. Unlike John Henry Newman, for instance, Venn never experienced a crisis of faith followed by the emergence of a new certainty. Instead, in the period between 1862, when he returned to Cambridge after a series of curacies, and 1883, when he resigned Holy Orders, Venn moved from a hereditary Evangelicalism to a liberal Broad Church position. Rather than following those around him into agnosticism and antagonism to the Church of England, Venn for the rest of his life retained an undogmatic form of faith as well as a keen affection for clerical work (see for instance letters **124** and **134**). Both were inspired in great part by reverence for five generations of Evangelical Venns, including his own father, Henry Venn, the influential honorary secretary of the Church Missionary Society (C.M.S.) (see letter **121**). From around the turn of the century, Venn keenly felt that the mid-Victorian world to which he had belonged was gone; tellingly, the years between 1900 and 1923 were spent in looking back, upon his personal history as well as that of his family, College and University. One outcome was the monumental *Alumni Cantabrigienses* (11 vols., 1922–54) which is still used today as a standard reference source. Originally a project of his son, John Archibald Venn (1883–1958), who would become President of Queens' College, Cambridge, the drudgery of plodding through long lists of names and dates of matriculations and degrees helped Venn fill up all his time and occupied his thoughts during the First World War (see letter **137**).

Venn's change of religious identity, which he analysed in detail in *Annals: Autobiographical Sketch* – reproduced here for the first time (Part I) – was paralleled by the acquisition of an academic identity. In fact, both informed each other on several levels. When Venn entered Gonville and Caius College in 1853, Cambridge was still largely a classical and mathematical seminary where religious tests kept all degrees, fellowships and offices as Anglican monopolies, though university reform had been brought nearer with the appointment of the Graham Commission in 1850. Venn's undergraduate experience as a pensioner and scholar not only provide a point of contrast for the reforms that followed, and in which he himself was a participant; in their recorded form, they are also the chief documents for the mid-nineteenth-century history of his College. About a decade later, in 1862, when he resigned his curacy in Mortlake to take up the role of Catechist, teaching Logic and Political Economy to Indian Civil Service candidates, Venn threw in his lot with the 'ardent young' reformers at Cambridge (see letters **117** and **123**). Facilitated by recent institutional changes that had created opportunities for an academic career independent from religious expectations, Venn came to belong to a small circle of religiously unorthodox men committed to bringing the new course of study called the Moral Sciences to maturity. Many of them gathered around John Grote (1813–1866), who in 1855 had succeeded William Whewell (1794–1866) as Knightbridge Professor of Moral Philosophy, forming the 'Grote Club', where Venn spent some of his 'happiest hours' (see letter **40**) freely discussing all kinds of philosophical topics. During that same period, Venn continued to occasionally take clerical duty in churches near Cambridge in response to requests from prominent Evangelicals, emerged onto the public Evangelical platform – for instance attending meetings of missionary and philanthropic societies – and became a regular contributor to the *Christian Observer*, the Evangelical outlet of which his father was editor between 1868 and 1872. Much to his father's delight, in 1869 Venn was elected to give the Hulsean lectures before the University, to be delivered from the pulpit of Great St. Mary's Church. Venn's chosen subject – the nature of belief, both religious and scientific, and the evidences for it – was an attempt to make the intellectual demands of academia (rigour, method and honesty) bear upon one of the major religious themes of the day (church party differences). His central point was that, from a purely logical standpoint, there was no essential difference between religious and scientific belief. Both were mental states in which we are prepared to act upon the truth of a proposition, and both were true in terms of their beneficial consequences for our daily interaction with the world. It was no surprise that there was far less unanimity on religious matters than on scientific subjects, as religion embodied truths intended for all humankind and for all human experiences. Venn's definition of belief and truth – boiling down to a hitherto forgotten early formulation of Cambridge pragmatism[2] – went against the grain of positivism and secularism by showing that religious faith was entirely rational. But it was not nearly orthodox enough to satisfy his father, who insisted that his son referred to the Bible as the origin of the truth of religion (see letters **49–50**).

At Cambridge, the Moral Sciences Tripos had been established in 1848, but initially only led to honours: a degree had first to be earned in Classics or Mathematics. It was not until 1860 that a degree could be earned in moral sciences alone, with philosophy, logic and ethics, on the one hand, and history, political philosophy, jurisprudence and political economy forming two separate groups of the curriculum. A rather exciting opportunity for Venn, Sidgwick, J.B. Mayor and other young teachers, who managed the moral sciences – in which none of them had been educated – more or less by themselves, the Tripos itself was nonetheless held in low repute until the 1870s, when gifted graduates like William Cunningham and F.W. Maitland moved into desirable positions. Venn captured the main issue in a letter to his father from October 1863: 'Those who could really understand it [i.e. moral sciences topics] are already engrossed in Classics or Mathematics, so for the most part the poor remainder fall to my share' (letter **27**). Other problems would surface in the 1870s–80s: new Triposes were founded (e.g. the History Tripos in 1875), which further reduced the number of potential moral sciences students and it also turned out that there were almost no employment opportunities at Cambridge for Moral Sciences Tripos graduates. As Sidgwick wrote to Venn a few months before his death, in 1900: 'As you say our work in Moral Sciences at Cambridge does not afford a triumphant retrospect' (letter **118**). Sidgwick, of course, could not predict the great successes of two of his own pupils: G.E. Moore and Bertrand Russell.

Among his colleagues at Cambridge, Venn emerged as the most dedicated follower of John Stuart Mill's work – with which he had become acquainted around

[2] See Lukas M. Verburgt, 'Pragmatism at Cambridge, England before 1900', *British Journal for the History of Philosophy* 29:1 (2021): pp. 84–105.

1860 – and the undisputed authority in the field of logic. As well as being a dedicated teacher, he became a productive scholar, publishing three books and two dozen papers and reviews, especially in *Mind* – the world-leading journal on philosophy founded in 1876 of which Venn became a co-editor in 1892, when G.F. Stout, another Moral Sciences graduate, was made the editor. There was a direct link between Venn's teaching and writing as most of the ideas found in his logical trilogy, *Logic of Chance* (1866), *Symbolic Logic* (1881) and the *Principles of Empirical or Inductive Logic* (1889), had originally been developed in the lecture-room. Venn, in other words, became an expert in the fields he himself introduced into Cambridge: probability theory, formal (algebraic or symbolic) logic and inductive logic.

All his contributions to these fields bore the stamp of Mill's influence in one way or another. Indeed, the most helpful way of structuring Venn's academic career as a logician is to see it as falling into three periods of engagement with Mill's groundbreaking *System of Logic* (1843): support (1860s), criticism (1870s) and disillusion (1880s). During the 1860s, Venn, with the help of his cousin James Fitzjames Stephen (see letter **15**), published one article ('Science of History', 1862) – a short commentary on the discussion between Auguste Comte, Henry Thomas Buckle and Mill on the possibility of a social science – and one book (1866). This book, the *Logic of Chance*, was written as a systematization of Mill's definition of probability as the frequency of like events in the long run. Mill himself praised it as the 'best and most philosophical I have met with' (letter **42**). The five years from 1876 to 1880 was the period in which Venn published almost all his journal articles. After a lengthy review of his main rival William Stanley Jevons's *Principles of Science* (1874), there followed a steady stream of increasingly critical discussions of the so-called material outlook on logic, originating in Mill and further elaborated by Alexander Bain, Herbert Spencer, Carveth Read and John Neville Keynes. Venn never deserted to the enemy's camp – that of the conceptualists, such as George Boole, Augustus De Morgan and Jevons – who thought that logic dealt with laws of the mind – but he did abandon the ideal of a purely objective treatment of logic, one in which logic deals with inferences from and about 'matters of fact'. Instead, Venn developed a pragmatic, proto-conventionalist alternative, which said that logic's subject matter was the world as it was constructed out of 'assumptions' and 'conventions'.[3] This alternative would be fleshed out in *Principles of Empirical or Inductive Logic*, whose first five chapters discussed all the subjective and objective postulates needed to construct 'the universe as the logician regards it'.[4] The rest of this almost 600-page book provided an account of all parts of, what Venn and his peers took to belong to, logic – traditional deductive logic (term, proposition and argument) and its formal-algebraic extension, inductive logic, and methodology (e.g. scientific units, standards, explanations) – viewed from this new perspective. Although the book itself was considered a failure by the reviewers, the admittedly

[3] John Venn, 'The Difficulties of Material Logic', *Mind* 4:13 (1879): pp. 35-47, on p. 47.

[4] John Venn, *The Principles of Empirical or Inductive Logic* (London: Macmillan & Co., 1889), on p. 1.

half-hearted break with the material view of Mill, whom Venn, in Jevons's opinion, simply admired too much (see letter **65**), had opened up a theoretical door for Venn. Most importantly, it allowed him to engage with the recent developments in formal logic associated with George Boole (1815–1864), whose *Laws of Thought* (1854) Venn first read in the late 1850s upon the recommendation of Isaac Todhunter, his former mathematics coach at Cambridge. It took him several periods of study to make sense of it and, especially, to see how Boole's new algebraic logic, that is, his calculus of deductive reasoning, could be reconciled with Mill's material system of inductive logic. This task, which he shared with Jevons – one of his frequent correspondents (see letters **62–66**, **68**, **70**, **72** and **74**) – was by no means an easy one, especially not for someone like Venn who – unlike Jevons – wished to remain loyal to both Mill and Boole, who were almost complete opposites. For example, Mill not only equated deductive logic with syllogistic logic but also held that there is no inference of a non-deductive kind; Boole, for his part, initiated the move that took logic far beyond the syllogistic and argued that what Mill called induction did not belong to logic at all. What Venn tried to do, in brief, was to come up with a compromise based on a criticism of Mill and an opinionated reading of Boole. Venn believed that Mill's claim that there is no process of inference involved in deductive arguments was due to his 'over-objectifying' of logic. There was a crucial difference between the *subjective* recognition of facts and what is given in the *objective* facts themselves: to go from the premises to the conclusion always requires a mental step, even in such a simple case as inferring Socrates's mortality from the mortality of mankind. For this reason, Venn argued, any tool that can aid us in recognizing what conclusions can be drawn from given premises are welcome, especially when dealing with cases far more complex than those considered by Mill. For over two thousand years, this tool had been Aristotle's syllogism. What Venn took Boole's algebraic logic to have done was nothing less, but also nothing more than this: the generalization of Aristotelian logic in such a way that it can deal with much more complex cases of reasoning. Unlike Boole, however, Venn remained committed to the view that deductive logic is just a formal tool: the premises from which we reason are drawn from, and the conclusions at which arrive ultimately concern, the observable world. This was, roughly put, the position elaborated in *Symbolic Logic*, which also contained Venn's famous diagrams.

Upon finishing *Principles*, Venn quit logic, donating his private collection of 1100 logical books to Cambridge University Library, where it is still held today. He longed for more applied work, which he found, for instance, in the collaboration with Francis Galton in the Cambridge Anthropometric Laboratory (see letters **78**, **80–83**, **85**, **88**, **94–97**, **99**, **103–104**), aimed at testing the physical characteristics of students, comparing these with the results from the laboratory at South Kensington and, thereby, investigating their relation to mental abilities. Venn's task was to apply the kind of statistical tools given pride of place in the third edition of the *Logic of Chance* (1888) – means, medians, averages, correlation, etc. – to the data on, among other things, eye-sight, head-size and strength of pull, obtained with Galton's instruments at the Library of the Philosophical Society in Cambridge. By that time, Venn had already campaigned with his colleague and former pupil James Ward (1843–1925) for the establishment of a laboratory in psychophysics, on which subject Venn also claimed expertise in his 1889 application for a chair in Oxford (see letter **90**). Venn's practical involvement in anthropometry and psychophysics incidentally led him to give the first-ever lecture course on statistics in England in 1890 (see letters **100** and **104**).

It is rather difficult to assess Venn's influence and legacy as a logician. On the one hand, he received academic recognition for his work through election as a Fellow of the Royal Society in 1883, the award of the degree of Doctor of Science in 1884, and the offer of an honorary degree at Princeton in 1896 (see letter **111**). On the other hand, his books and articles may be said to have stood at the apogee of developments in philosophy that soon went out of date. For example, at the beginning of the twentieth century, the tradition of algebraic logic – which had started with Boole and was climaxed by Ernst Schröder – was superseded by the mathematical logic of Gottlob Frege, Giuseppe Peano and others. One way of approaching the issue, and of highlighting its complexity, is to focus on the notion of a local tradition of Cambridge philosophy. Due to the short-lived rise of British Idealism, associated mainly with J.M.E. McTaggart, G.E. Moore and Bertrand Russell came to regard the work of Sidgwick, Venn and other men from that generation as hopelessly old fashioned. At the same time, it were Moore and Russell who, through their own work as well as through their teaching of pupils like C.D. Broad, W.E. Johnson, J.M. Keynes and Frank Ramsey, made logic at Cambridge return to its formal, inductivist and probabilist course. Indeed, Venn was actively read and discussed by Broad, Johnson and Keynes, who recognized him as the 'father' of probability theory at Cambridge (see letter **149**).

Venn's career was not altogether a successful one, in terms of academic preferment. Despite several attempts (1866, 1872, 1890 and 1897), Venn never became a professor, and in 1903 he was passed over for the position of Master of his College, instead being elected President, more or less by way of consolation. Upon losing the 1897 election to the newly erected Chair of Mental Philosophy and Logic at Cambridge to Ward, Venn, bitterly disappointed, resigned his teaching position and even talked of leaving Cambridge (see letters **112–115** for this episode). The appointment of Ward, as painful as it was to Venn on a personal level, was also a clear sign that the times were changing. Like Cunningham, Maitland, Stout, and J. Neville Keynes, Ward, about ten years Venn's junior, was a home-grown product of the Moral Sciences Tripos who had continued as one of its teachers and who stood at the vanguard of a rapidly developing new academic discipline, in his case psychology. Venn, for his part, belonged to an older, mathematically trained generation whose research was less discipline-based. This is borne out perhaps most clearly by the freedom Venn felt – and which was permitted to him by the income from his fellowship, teaching and examining – to abandon logic as a topic of research and start exploring other kinds of intellectual pursuits. Although he continued to teach logic until the late 1890s, from around 1889 onwards, until his death in 1923, Venn was *de facto* an historian and antiquarian. During these three decades, he compiled a three-volume biographical history of Gonville and Caius College, worked on the first part of the *Alumni Cantabrigienses*, edited historical records of the University of Cambridge, contributed numerous articles to the College magazine, *The Caian*, and published a history of his family.

The year 1900 marked another turning point. While Venn had already shown a strong interest in the history of his family, College and University, the past also came to play a central role in Venn's own life. When Sidgwick died in 1900, Leslie Stephen wrote to Venn that 'the old world to which we belonged is vanishing very rapidly' (letter **120**). Venn's correspondence with Albert Venn Dicey (see letters **139–141**, **143**, **145–148** and **150**), the surviving male family relation with whom he retained the closest contact, and with Laura M. Forster (see letters **121–124**, **126**, **134–137** and **144**), a descendant of the Clapham Thorntons and, like Venn, a friend of the Cambridge Darwins, was centred around sharing and reflecting on memories of past events, friends and relatives. Venn's feeling of a 'tie of gratitude to the past' also came with a sense of 'obligation to the future'.[5] Once a dedicated reformer who believed that all that had gone on at Cambridge 'for more than 30 or 40 years must be bad' (letter **123**), with old age Venn became a political and cultural conservative. In 1912, he wrote to Dicey that much strength was needed to 'stamp the life out of' such developments as female suffrage (letter **139**). He also made no secret of the underlying ambition of his many historical volumes: to strengthen a 'sense of historic unity and continuity' which he believed was urgently needed in a modern society where lack of respect for national traditions was leading to 'anti-patriotism' (letter **136**).[6]

Structure of the Book

The present volume is divided into two parts. Part I contains Venn's *Annals: Autobiographical Sketch*. This text, originally written in 1887, was revised in 1903 and offered to his wife, Susanna Venn (née Edmonstone) to correct in 1917. Although never published, it was privately circulated among close family members: Venn's brother Henry Venn read it in 1910, his son Archibald in 1917 and Albert Venn Dicey in 1919. The *Annals* cover the period of Venn's upbringing, early years at Cambridge, curacies and his academic life up to 1866. It is both selective and incomplete. Venn's main goal was to figure out the emotional logic, so to speak, behind his gradual abandonment of the Evangelicalism inherited from his father. His careful analysis of the intellectual inadequacy of a religious orientation founded largely on emotion leaves out key elements, such as his writing for the *Christian Observer* in the 1860s–70s. What is also left out is an explanation of the long period between his rejection of Evangelicalism and his eventual resignation of Holy Orders in 1883. The *Annals* do nonetheless offer many interesting materials, especially on College life in the 1850s and the early days of the Moral Sciences Tripos at Cambridge. Part I also includes a short, unpublished lecture delivered in 1889 at

[5] John Venn, 'Address Delivered in the College Chapel, After the Commemoration Service, Dec. 21, 1894', *The Caian* 4:3 (Lent Term 1895): pp. 188–198, on p. x.

[6] John Venn, 'A Chapter in College History', *The Caian* 1:1 (Easter Term, 1891): pp. 13–39, on pp. 36-38.

Caius House, Battersea, entitled 'Science and Common Thought'. Although addressed to a non-expert public, it arguably offers a valuable insight into Venn's mature philosophical thinking in the late–1880s. More specifically, its emphasis on science's continuity with everyday life re-emphasizes the pragmatic aspect of Venn's work, already present in his 1869 Hulsean lectures. As such, it adds to the recent interest in the connections between the British and American roots of the pragmatism of William James and Charles Sanders Peirce.[7]

Part II contains annotated transcriptions of a selection of some 150 letters from and to Venn, divided into four periods starting in his undergraduate years at Cambridge. Venn's surviving correspondence is highly uneven in interest and value. On the one hand, most likely because of their close proximity, no letters appear to have been exchanged between Venn and his Moral Sciences colleagues in the 1860s–90s. These letters would have contributed much to the picture of the moral sciences at Cambridge in this fascinating period. On the other hand, a large number of letters were sent only for practical matters, such as those from his father Henry Venn of C.M.S., or for information, with men contacting Venn at Caius to inquire about their pedigrees and with Venn asking Caians to consult local records to fill gaps in College information. Some of the letters from and to Venn have unfortunately not survived. For example, there is no trace of his long correspondence with the Scottish logician Hugh MacColl, mentioned in a letter to Jevons (see letter **74**).

The primary aim of the selection is to present a representative picture of Venn's life and work in the period between 1854 – his first year at Cambridge – and 1923 – the year of his death – the focus being on his academic career, which began in 1862 and officially ended around the turn of the twentieth century. These editorial choices have several implications, of which the following are the most important. First, this selection covers about one-tenth of the total amount of surviving letters. Second, the early letters exchanged between Venn and his father, Henry Venn of C.M.S. – which give a rather vivid impression of an Evangelical childhood in the early Victorian era – have not been included. The same goes for many of their written exchanges during Venn's time as a regular curate. Third, whereas only a few letters from Venn's extensive correspondence with Albert Venn Dicey, which took place in the first two decades of the twentieth century, have been included, his correspondence with Jevons and Galton, from the 1870s–90s, is reproduced almost in full. Fourth, only those letters from after 1900 have been included which are particularly exemplary of that which characterizes the last twenty years or so of Venn's life: the fact that, with his academic career behind him, and most of his personal links to Cambridge snapped by the death of relatives, friends, and colleagues, his was a life lived 'mostly in the past' (letter **121**). (Many of these letters have been abridged, in so far as some of their more miscellaneous details have been omitted.) This also means that those letters pertaining, for instance, to Venn's antiquarian pursuits have not been included.[8]

---

[7] See Cheryl Misak, *Cambridge Pragmatism: From Peirce and James to Ramsey and Wittgenstein* (Oxford & New York: Oxford University Press). See also Verburgt, 'Pragmatism at Cambridge', and Alexander Klein, 'How American is Pragmatism?' forthcoming in *Philosophy of Science*.

[8] This aspect of Venn's career is explored in detail in Verburgt, *John Venn*, chapter 12, and Michelle Clewlow, 'Intersecting Sets: John Venn, Church and University', unpublished PhD-thesis, The

Any selection of writings and correspondence is unavoidably arbitrary. It goes without saying that other editorial decisions could have been made. For example, it would have been possible to omit and condense certain passages in Venn's *Annals*, especially the long descriptions of his childhood surroundings and country rambles. However, given the autobiographical nature of the text, the objective in preparing it has been to maintain its integrity, representing it as Venn himself (re-)wrote it. To give another example: an alternative volume could have been made with a focus on Venn's extensive correspondence with his father in the 1850s–70s, seeking to highlight the way in which Venn parted ways with the religion of his youth and contrasting it with that of his cousins, Leslie and Fitzjames Stephen, and his second cousin, Albert Venn Dicey. Likewise, the reproduction of the letters, notes, notebooks, and printed material used by Venn for his books on the history of Gonville and Caius College would give an interesting account of the work of an 'amateur' or 'gentleman' historian.[9] Instead, the present volume seeks to provide a picture of Venn as a university don, a moral sciences teacher and a professional logician, offering the source material for a further exploration of the mid-Victorian generation of Cambridge thinkers to which he belonged.

It may here be mentioned that the bulk of the material reproduced in this volume comes from three archival collections held, in order of size, at the Church Missionary Society Archive, University of Birmingham; Gonville and Caius College, Cambridge; and the Society of Genealogists, London. These collections were originally part of one larger collection held by Venn's son John Archibald Venn, who was the last member of his branch of the family. After his death in 1958, the Venn family papers were divided and portions were sent to the three aforementioned locations.

The 'Biographical List of Names', found at the end of the book, contains short biographical sketches of the individuals whose names occur frequently (at least more than twice) in the texts in Part I and, in Part II, of the recipients and senders of more than one letter. When an individual is included on the list, his or her name appears in bold in the annotated text. When a name or individual occurs only once or twice, biographical information – limited to dates and a one-phrase characterization – is provided in footnotes. Further details of editorial practice are given in the 'Notes to the Reader' preceding Part I and Part II.

Freudenthal Institute,
History and Philosophy of Science
Utrecht University,
Utrecht, The Netherlands

Lukas M. Verburgt
l.m.verburgt@uu.nl

Open University, 2007, chapter 7.

[9]This interesting category is discussed in Philippa Levine, *The Amateur and the Professional. Antiquarians, Historians and Archaeologists in Victorian England, 1838–1886* (Cambridge: Cambridge University Press, 1986).

Contents

Abbreviations

CMS	Church Missionary Society Unofficial Papers Accession 81: Venn Manuscripts, Cadbury Research Library: Special Collections, University of Birmingham
CUL	Cambridge University Library
GCA	Gonville and Caius College Archive (Venn Papers)
GL	University of Glasgow Library, Archives & Special Collections
JRUL	John Rylands University Library of Manchester (Jevons Papers)
SGL	Society of Genealogists Library, London
TCL	Trinity College Library, Cambridge
UCL	University College London Library (Francis Galton Papers)

Part I
Unpublished Writings

Note to the Reader

Part I contains two documents: *Annals: Autobiographical Sketch* (CMS/ACC81 F27), written in 1887 and revised in 1903 and 1917, and the lecture 'Science and Common Thought' (GCA D1), delivered in October 1889 at Caius House, Battersea.

All editorial interventions, including the indication of the original pagination in the main text and information in footnotes, appear within square brackets. Venn's original footnotes – which were inserted according to the old-fashioned system of beginning with '1' on each page – appear at the bottom of the page, in a smaller font; additions and corrections made in the original text in 1903 and 1917 are rendered as follows: [*del*: deleted word] and //added word in main text or footnote//. Editorial annotation in Part I has been kept to a minimum and is appended only to identify names and where further clarification of meaning or allusion was deemed essential to the readability and usefulness of the material. All biographical information has been taken from the *Alumni Cantabrigienses*, 11 vols. (1922–1954) unless otherwise stated. As explained in the Introduction, names appear in bold in the footnotes when included in the 'Biographical List of Names'.

L. M. Verburgt, *John Venn: Unpublished Writings and Selected Correspondence*, Studies in History and Philosophy of Science 56,
https://doi.org/10.1007/978-3-030-79829-1_1

Annals: Autobiographical Sketch

John Venn

[1] My grandfather[1] kept a careful diary, which, though very brief, indicates most of the events of his life, from the data of his being admitted at College in 1777, till near the time of his death.[2] My father[3] always thought it desirable to do the same, but though he made repeated attempts to keep it up during parts of his early and middle life, he only very partially succeeded. Towards the close of his life however he used often of an evening to dictate to my sister[4] the principal details of his day's work. These details have been drawn upon for his published Life.[5] As to myself unfortunately, I never at any time attempted to note down the events of the passing day. And now, in late life, (//June, 1887//, June 28, 1902) wishing to leave some record which shall give information to those who come after me, I find that the only contemporary records consist in letters, account books, pocket Diaries, etc. I have however a good memory for past events, so far as I am personally concerned, and with the help of such records as those just mentioned, composed by my father and myself, I have now put the following brief narrative, or rather notes, together.

1834

I was born at Drypool, Hull, of which parish my father was then the incumbent, on the 4th of August.

[2] 1839

This is the year from which my earliest recollections date. In the autumn my mother first developed symptoms of consumption, and for her sake, - as well as that of my father, who was only just recovering from a dangerous heart-attack –, we went to spend the winter at Torquay. Of our life there I have just a few fragmentary glimmerings of recollection, but nothing worth record.

---

[1] [John Venn of Clapham (1759–1813).]

[2] [As Venn wrote in *Annals of a Clerical Family*: 'To my grandfather, John Venn, we owe a large part of our family reminiscences. He was the first of our line to take any interest in genealogical inquiries. The *Parentalia* were begun by him, and carried down to his father's time; and without his industry, and his care in noting what he heard and remembered, many early facts would have been hopelessly lost. Of the main incidents in his life we have a complete account, for he kept a brief diary, or at least made notes of current events, from the time he entered college'. John Venn, *Annals of a Clerical Family* (London: Macmillan and Co., 1904), on p. 12.]

[3] [**Henry Venn of C.M.S.** (1796–1873).]

[4] [Venn had one sister, Henrietta Venn (1832–1902).]

[5] [William Knight, ed. *Memoir of Henry Venn, B.D. Prebendary of St. Pauls, and Honorary Secretary of the Church Missionary Society,* with an introductory biographical chapter and a notice of West African commerce by his sons, the Rev. John Venn, M.A., and the Rev. Henry Venn, M.A. (London: Longmans, Green, and Co., 1880).]

1840

My mother[6] died March 21, 1840. I remember my father coming into the room and telling us that she was gone: my sister bursting into tears (she was two years older than myself); and my doing the same, rather from sympathy with her than from appreciation of what it meant. She was buried at St John's Holloway. My father went by sea, with the body, from Plymouth to London, a journey of 48 hours. Shortly afterwards my father returned home to the house in Hornsey Lane, Highgate, where he had lived since about 1838. He had driven the whole way to Torquay in his two-horse phaeton, the same vehicle, I suppose, in which he had been accustomed to take so many long drives about England with my mother. He returned in the same way as far as Basingstoke, where he took the railway. I have just one vivid recollection of this homeward journey. This was as we were crossing the vast open plain of Salisbury, to which I remember his calling my attention as we came to it. I remember well, too, our stepping at Stonehenge, and my nurse (old "Burder" as we called her, – though she was not then old) perching me on [3] one of the smaller stones to get a sight of that wonderful monument.

For the next six years, till we went to school, (1840-6) my brother[7] and I lived under the charge of governesses. As our father found it essential to his health to go to Switzerland as frequently as he could, – not every year, and never for more than a month –, we used generally to be taken in at Hereford by our devoted uncle and aunt, John and Emelia Venn[8]: sometimes we went with the governess and nurse to Brighton, to join our cousins the Stephens or the Elliotts.[9]

1841

Thus, June 6, 1841, we paid one of these visits to Brighton. I have just the common recollections of the sea-side, which might apply to any other time: - the chain-pier, the bathing, the camera on the cliff above the pier, and so forth.

1843

July 5, 1843, my father took us down to Hereford, left us there, and went to Switzerland. We had been at Hereford before, but this is probably the visit from which all my earliest recollections of our delightful stays there date. It was then that we made our first acquaintance with several neighbours and friends, the Bulmer children, and others of our own age. At the end of August my father returned to us at Hereford, and took us for a 4 or 5 days trip into Wales. We saw Llangollen, the Devil's Bridge, and one or two other places. This is my earliest recollection of those Welsh mountains which I got to love so in after days.

---

6 [Martha Venn (née Sykes) (1800–1840).]

7 [**Henry Venn of Walmer** (1838–1923).]

8 [**Henry Venn of C.M.S.**'s brother John Venn of Hereford (1802–1890) and sister Emelia Venn (1794–1881).]

9 [The children, respectively, of (Sir) James Stephen (1789–1859) and (Lady) Jane Catherine (née Venn) Stephen (1793–1875), and Henry Venn Elliott (1792–1865) and Julia Marshall (1816/17–1841).]

[4] 1844

In May 1844 we went to Brighton, and in June into France. This was not actually my first visit abroad, for I had been taken to Dieppe during my mother's life, and again in 1842, but it is the earliest I can remember. We went to Havre, Honfleur, and Caen. There was then much of the old-world life of France still to be seen. I remember, – as a boy would –, a church or two near the coast crammed with the small ships put up by sailors as votive offerings for their escape, and which naturally seemed to me to be sadly wasted there. The country women still often wore astonishingly lofty and varied caps of lined and lace: - our cousin Emelia Gurney was one of the party, and already something of an artist; we used to be set to pick out in the market-place the finest of these caps, for her to sketch them and their wearers. My first recollection of wild flowers dates from this visit. In the chalky soil of Normandy orchids are very plentiful. I remember being startled by my first sight of the bee-orchids, which for a moment I really took for the live insect.

We went on this occasion, I believe, partly with a view to meet our Aunt Emelia, who had been spending the winter in Italy with her sister Caroline and niece Emelia, and was returning to England with them.[10]

1845

My first visit to Cambridge was in May of this year. My father took us three children to the Lodge at Queens',[11] staying [5] there for a few days from May 16. He drove us down, in the day, in his phaeton, having sent his own horse on, half-way, and hiring one for the first part of the journey. I remember "the Wash at Edmonton" passed by John Gilpin,[12] and which then lay across the road. My father told us, as we approached Cambridge, that if it had been earlier in the afternoon we should have seen numbers of men in caps and gowns out for their walk. We of course, as children, lived mostly in the nursery, with the King[13] children, but we had our early dinner in the President's dining room. He was already completely paralysed, but was put in his seat and fed by his footman. One little incident dwells in my mind connected with this visit. It was the sight of the Boat Procession. Our nurse

---

[10] [**Henry Venn of C.M.S.**'s sisters Emelia and Caroline Venn (1798–1870), the mother of **Emelia Russell Gurney**; and Caroline Emelia Stephen (1834–1909), daughter of (Sir) James and (Lady) Catherine Stephen.]

[11] [**Henry Venn of C.M.S.** studied at Queens' College, Cambridge, graduating as Nineteenth Wrangler in 1818. He was elected Fellow in 1819.]

[12] [John Gilpin was the main character in a well-known comic ballad by William Cowper, written in 1782. The ballad describes John Gilpin riding a runaway horse. Venn's reference is to the stanza "Thus all through merry Islington / These gambols he did play, / Until he came unto the Wash / of Edmonton so gay".]

[13] [Joshua King (1798–1857), Senior Wrangler in 1819 and Lucasian Professor of Mathematics between 1839 and 1849, served as the President of Queens' College from 1832 to his death. King had been appointed to the professorship despite his physical abilities. He was succeeded in 1849 by G.G. Stokes. See A.A.D. Craik. *Mr Hopkins' Men: Cambridge Reform and British Mathematics in the 19th Century* (London: Springer-Verlag, 2008), on p. 101.]

came running in to us, – I suppose we had gone to our bedroom –, telling us to come and look out of the window on the river below. It was the well-known "boat-procession" night.[14]

In the early summer we went again to Hereford. Most of my early recollections of riding on a coach are connected with these visits. The Great Western Railway did not for many years get nearer to Hereford than Gloucester, – I can remember still earlier when we had to start driving from "Cirencester Road", as the station was called from whence we went by coach through Ross. It was a charming drive, and at one point (as we got to know in later days, in 1878) the road commanded a superb view over the Welsh mountains. In the course of this same summer my father took us for a short trip into N. Wales. With his usual liberality where missionaries were concerned, he extended his hospitality during part of this holiday to [6] most of a family of six boys, named Adley, whose father was in India, and who from my remembrance, proved rather a handful. They were schoolboys, away for their holiday from King William's College, Isle of Man. We went in the usual way travelling in open cars and driving from place to place. We stayed amongst other places, at Llanberis, Beaumaris, etc. Later on we paid our first visit to our Sykes cousins at Raywell near Hull.[15] We went by steam-boat from Beaumaris to Liverpool, – the Holyhead Line was not then quite finished, I think –, stopped a night at Liverpool, and thence by train to Hull. There was a mechanical invention then on show at Liverpool, (I suppose in some public garden) which has been since revived under the name of switch-back, but instead of only descending and ascending, the cars made one, or perhaps two, complete revolutions. I have a vivid recollection of the sight of an old gentleman in the car, whose hat flew off whilst he was upside down. The 'cousinhood' at Raywell was then very large, for, of the 13 in the family, most were alive and at home. They were very friendly and hospitable, and I wish we could have seen more of them in other years, as this was my only introduction to the family life of a country squire.

Our general life at home as children was not, I think, a joyous one, though it was far from being actually unhappy. Our father, who was now beginning to recover his health, was naturally much absorbed in his parish, and he soon began to give more and more of his time to his work at the Missionary House, Salisbury [7] Square.[16] We were therefore mostly in the care of the governess for the time being, who was a changeable person, and the nurse, who was a fixture. The latter deserves some notice. She was a vigorous Yorkshirewoman who had come as nursemaid soon after my sister's birth in 1832, and who never left the family until she died at Mortlake in the summer of 1868. We did not think of criticizing her then; but looking back at her

---

[14] [The annual procession of the University boats upon the river Cam.]

[15] [Venn's mother, Martha, was the daughter of Nicolas Sykes. Her sister, Frances (Fanny) Sykes (1797/8–1878), married Matthew Babington (1792–1836). They had six children.]

[16] [The home of the Church Missionary Society (C.M.S.), and the nucleus of the Evangelical party in the Church of England. Henry Venn had been appointed honorary secretary of the C.M.S. in 1841. For his work as one of the foremost Evangelical mission strategists see Wilbert R. Shenk. *Henry Venn Missionary Statesman* (Eugene, OR: Wipf & Stock Publishers, 1983).]

now I should say that she was a perfectly honest, faithful, and devoted woman, with scanty education and not much manner for her position, and with a decidedly hot temper. This latter she kept, like a dog, for those whom she regarded as trespassers; and many and violent were the altercations that went on, – upstairs, between her and the governess, and downstairs between her and the other servants, especially the footman or whatever maid he might be supposed to be attached to. The dominant governess of our childhood, – a Miss Lemon, – was no mean antagonist, and on the whole held her ground. Now and then she got routed by a combination of allied forces. This was mostly when some aunt, of whom we had two or three who were most affectionate, came and took our part against the common foe: – our Aunt Fanny (Babington)[17] our mother's younger sister, was the chief of these. It used to be her delight to come in from Kentish Town where she then lived with her young family, and make her way into the nursery and stir the nurse on to resistance and rebellion. The governess however mostly held her own in the long run. Another aunt, calling one day, related how she found us [8] three children distributed: – my sister was on the mat outside the room (being a girl she could be trusted not to run away): my brother was standing on the balcony outside the window (being too small, he could not swarm down the posts and get into the garden, as I should have done): I was in the corner of the room, in similar but more obvious disgrace; whilst Miss Lemon sat at the table in the middle of the room in solitary state.

The person to whom we really owed the most at the time in question was, I think, our Aunt Emelia. She was my father's third sister, and at the time in question (1846) would be about 52. She did much to supply our mother's place, though as she lived at Hereford with our uncle John, and had not only to look after him and her own old aunt, but also to take all the clergy-wife's share of a large parish as well, she was not able to be with us nearly so often or so long as she and we could have wished. Her coming was always a bright and happy incident in our lives. It was she who selected the governesses, with a labour and a patience which only those can appreciate who remember what sort of work it was, in the days before women's education and the rise of examinations, to ascertain what a teacher knew and whether she could teach. It was our aunt again who mainly chose our books for us. Here too she must have taken infinity of pains, and considering the strictness of her views it was an astonishing how various and interesting a selection she managed to make. She used also to spend some time during her stay with us in reading such books [9] aloud to us. Miss Martineau's "Feats on the Fiords"[18] was one book of her choice. She was a remarkable woman. Full of life and interest and cheerful humour which rather blended with, than set aside, her strong evangelical convictions, one never had a dull moment in her society. It was not so much that she was full of tales of the past, like many old people; her interests in the present whether near or remote, never seemed to flag to the last. She had paid several visits to Rome in which [*added in pencil*: city] not even her hereditary horror of Popery could prevent her from taking the keenest interest. To her last day she would delight in comparing notes with any

[17] [See note **15**.]

[18] [Harriet Martineau's "Feats on the Fjord", first published in 1841.]

friend or visitor who had recently been in Italy. She was one of the very few old people I have known who never ceased to the last [*del*: to] deliberately //to// cultivate new friendships and call on new comers. On our last visit to Hereford in her life time, – when she was about 86, – she took [*del*: Susie] //my wife//[19] quite a round of visits to new neighbours and friends on Aylestone Hill.

She remains the centre of some of my brightest and happiest childish recollections. We saw most of her and of her brother, - my uncle John, - on our summer holiday visits to Hereford. For some time these visits took place about every other year, and were keenly enjoyed by us. Not only did we get a welcome such as that from indulgent parents, and a most enjoyable house and garden on the top of the hill overlooking the town, but we also had opportunities of making friends which we could not secure at home. These were the sons and daughters of clergymen and gentlemen in the neighbour-[10]hood, young people with whose parents our uncle and aunt were on familiar terms, and who therefore supplied an element which was greatly lacking in our own surroundings at home. Our holidays at Hereford were the pleasantest we had.

1846

In the autumn of this year the governess regime came to an end for my brother and myself, for we were sent to the Cholmondely School on Highgate Hill.[20] This was about the middle of November. Whether my previous notice had been given to us, as is tolerably certain, I do not remember, but when my father told us, after breakfast, that he was about to take us to school; it came as [*del*: such] a surprize, [*del*: that] //and// we raised so tearful a commotion that he was obliged to defer the proceeding to the next day. The fact was that we had heard such vivid accounts of Mr Dyne,[21] the master, from my cousin Frank Babington,[22] who had been at the school before going to Eton, – he used to declare that he was one of the only two boys who had not been flogged, and gave us [*del*: vivid] //graphic// descriptions of the weapons used, – that we were fairly terrified. Dyne certainly was of the old type; and the birch for serious offences (and his interpretation of 'serious' covered a very wide range) and the cane for lesser matters, were constantly resorted to. The school was, I apprehend, pretty much in the condition of most grammar schools of the day. I may have a better memory than many persons for early details, but I seem to recall hardly anything [11] except the sound of the floggings heard overhead in the headmaster's room, the fights

19 [**Susanna Carnegie Venn (née) Edmonstone** (1844–1931).]

20 [A free grammar school founded in 1565 by an endowment from Roger Cholmondeley (or Cholmeley) (c. 1485–1565). It still exists today under the name of Highgate School, London.]

21 [Rev. John Bradley Dyne (1809–1898), Fellow of Wadham College, Oxford 1832–38 and Dean and Lecturer in Divinity, 1837–38, was headmaster from 1838 to 1873. Under his direction, the school grew from a small institution of nineteen boys to more than 200 boys. See J.S. Cockburn et.al., eds. "School: Highgate School," in *A History of the County of Middlesex*, vol. 1 (London: University of London Press), pp. 302–304. Dyne was generally remembered as a proficient classics teacher but a cruel headmaster.]

22 [Francis Evans Babington (1831–1920), Trinity 1849; B.A. 1853; subsequently manager of a bank at Norwich.]

of the boys amongst themselves, the bullying of the small by the big, and the habitually filthy and blasphemous language of almost every boy, at least among the lower forms. From subsequent report I gather that the boys in the upper forms were well grounded in classics, and that they had some respect for Dyne. J.M. Fletcher,[23] afterwards fellow of Caius, was near the head of the school at that time; and one distinguished man, – Skeat,[24] afterwards Professor of Angl0-Saxon at Cambridge –, must have joined the school soon after I left. After a very short experience however we got to like the school life; and, in fact, there was almost as much of a commotion when we had to quit it as there had been when we were first called on to join it.

1847

After the Christmas Holidays, when the weather, I remember, was very severe, we returned to school. In the spring I must have broken down in health, and was kept at home for a time. I was certainly away on Oak-apple day (May 29). It was a custom of the school that every boy, who did not bring a piece of oak in some form or other with him that morning, should be severely punished, and I remember my brother coming back and telling me about the goings on. I was in the 1st (or lowest) form of the six,[25] – below this there were 'senior grammar' and 'junior grammar boys' – and got a prize for classics, the only subject in which we were [12] examined. The prize was however, not awarded, as I had left the school before Speech Day. I think that, if I had remained at the school, I should have done fairly well, though I am sure that I had not in me the making of an accurate, or least graceful 'scholar' in the classical sense.

1847 [*sic*]

In the autumn of this same year my father took us by Folkestone to Boulogne, where we left with the governess whilst he went for a short trip to the Rhine, returning to take us home in September.

During the winter of this year and the spring of the next we remained at home, under the charge of tutors. My father may have come to the conclusion that the Highgate school was not a good one, but in my case he had resolved to quit Highgate and move nearer to London, for convenience in attending to his secretarial work at Salisbury Square.

1848

As at midsummer 1848 we quitted Highgate, and moved to Highbury in Islington, I may say a few words about our life in the former place, where my father occupied the house (afterwards known as No 9, Hornsey Lane, and as the home of the Edmonstones) for 10 years.[26] It was on the whole a //fairly// happy time. This, I think,

---

[23] [John Martineau Fletcher (1832–1915), Caius 1850; B.A. (Classical Tripos, First Class) 1854; Fellow, 1854–67.]

[24] [Walter William Skeat (1835–1912), Christ's College 1854; B.A. (14th Wrangler) 1858. Skeat was elected Professor of Anglo-Saxon at Cambridge in 1878.]

[25] [The system was to count downwards from six, making the first class the lowest.]

[26] [This is the house where Venn's wife grew up, being the daughter of the Rev. Charles Welland Edmonstone (1811–1897).]

was largely due to the garden. Only those who had a free run [13] in a big garden when they were children, can realize the delight, and I should say the profit, of such an experience. Our garden was unusually large, and must have been planted long before the neighbouring houses were built, as it extended behind these and hemmed then in on each side. It contained about 4 acres, and much money must have been spent on it by some former occupant. There was a pond artificially made, a small rookery, some magnificent elms, a big lawn suitable for cricket, and a very extensive kitchen garden. The governess and nurse for some reason interfered but little with us here, and the result is that I have such a vivid recollection of every hole and corner of the grounds that I could at this day draw a fairly accurate plan of them and insert the name and place of all the principal trees. I wish I had, at visits of a later date, made notes of the size of some of these trees, as they were really remarkable*.

> // *My brother contests this, & maintains that any 'recollections' of the magnitude of the trees, &c, is due to the usual mirage of old age. I might knock of 6 inches from the diameter of the laurel, but not more. I may say that I paid a visit to the garden, in after years, about the time of first penning these reminiscences. //

There was a Portugal Laurel on the lawn, such as I have never seen or heard of elsewhere, which was at least two-feet diameter in the trunk a little above the ground; an arbutus very little less than this; and the finest cedar I have seen near London. The remark as to size may be applied also to a larch tree, and to two Spanish chestnuts. The garden must have been 'established' some time in the 18th century, and I much wish we could have known something about its history. It so imprinted itself on my childish mind, that, though I never saw it after the age of 14, except on occasional visits after my engagement and marriage, it constantly recurs in dreams. [14] I doubt if a month ever passes, even now, without my nightly fancies carrying me back to wander there amongst those splendid old trees. Our guides in this garden were two. First a wonderful old gardener – [*del*: John] //James// Boston, a Lincoln-shire man – who managed somehow by almost superhuman industry, to keep all this large extent of ground in admirable order. He was enthusiastic in his work, and would have gladly taught us more than we ever had the patience to learn from him. Never did anyone have a more faithful and honest servant. He had come to the garden before our time, and remained on with the Edmonstones after we left, dying of cancer in old age [*del*: about] //in// [*del*: 1868] //1869//. Our other 'tutor', out of doors, was Robert the manservant. He was a thorough-bred country youth having been brought up on the estate of Lady Olivia Sparrow, a well-known Evangelical lady in Huntingdonshire.[27] He was intimately acquainted with the bird and beast life of the place, and all I ever learned as a boy about such things was gained from him. He had got possession of an old gun which my father had bought (when an undergraduate, on the occasion of his long vacation visit to Tenby in S. Wales) and used to take us out into the neighbouring fields. There was more sport then to be had than most believe would now

---

[27] [Lady Olivia (née French Acheson) Bernard Sparrow (c. 1778–1863), a devout Christian committed to education and care of the poor whose houses in London and Huntingdonshire were strongholds of Evangelicalism.]

believe. I remember the flocks of field-fares which used in winter to frequent the fields in Maiden Lane. This lane was then absolutely unoccupied by any building (except one old empty barn) from Highgate Hill, where 'St. Joseph's Retreat'[28] now stands, where there was a [15] turnpike. This lane was our frequent resort, for it was the way to Fort[*del*: r]ess Terrace, where our cousins the Babington's lived till 1845, and was also the route by which my father used to drive from Hornsey Lane to Salisbury Square. Robert (I forget his surname) was also our stand-by in most domestic matters, and I suspect in most matters of mischief. He taught us cricket; he taught us to climb; he helped us in carpentry; he gave us an introduction to bird stuffing, an art of which he had picked up the rudiments; and, it must be admitted, aided and abetted us whenever there was an opening to resist the governess or bring her into ridicule. He was not a very edifying companion; but I don't suppose he did us harm in any way, and we certainly picked up a great deal of miscellaneous information, which we should not have acquired elsewhere, from his suggestions.

Altogether we had a very fair life as children, our main deficiency being that of suitable friends. In this respect we were certainly at a strange disadvantage compared with the generation before us. Of course much of what those at Clapham rectory[29] were, and what they did, was due to their own intrinsic character; but I cannot suppress a slight feeling of envy when I compare the host of eminent and life-long friends [with, *added*] whom they had first [acquired, *del*] [made acquaintance, *added*] as children, with the [very, *added*] few companions of our youth whom, owing to their mental [qualities, *added*] and social position we could regard as permanent acquaintances. Our father, with his old-fashioned courtesy and refinement, must have keenly felt the difference between the friends of his own youth and those of ours. He [16] probably regarded it as the inevitable result of circumstances, and of the life to which he had devoted himself.

I seem to have said but little about our father and our relations with him, and I find it difficult not to give a misleading impression. I can never recall the time when my feelings towards him were not those of the profoundest reverence, in fact of awe. [*Del*: With a boy of a higher moral stamp than the ordinary, – as was indeed pretty much the case with my sister, – I imagine that the feelings entertained would] //It is quite possible that with some very exceptional kind of boy, – of unusual maturity of character & moral strength – the relations between father and son might in consequence// have been those of perfect confidence and affection. As it was, in my own

[28] [Founded in 1858, St. Joseph's Retreat, 107 Highgate Hill, Highgate, London, was the first Catholic church in the area (Islington, London).]

[29] [Venn's grandfather, another John Venn, was the rector of Clapham, near London, from 1792 to his death in 1813. Here he ministered among the families associated with the famous coterie of religious philanthropists later called the Clapham Sect – the Wilberforces, the Thorntons, the Stephens, the Macaulays and the Venns. Both through friendship and inter-marriage, these and other families continued to form a close-knit network in the nineteenth and twentieth centuries. See, for instance, Michael Hennell, *John Venn and the Clapham Sect* (London: Lutterworth Press, 1958) and Adam Kuper, *Incest & Influence. The Private Life of Bourgeois England* (Cambridge, MA: Harvard University Press, 2009).]

case, I must admit that there was a very strong infusion of fear: not the fear as of a master, but rather that of a being of a superior order. On occasion of any [*del*: display of] //detection in// falsehood his tone and attitude of moral condemnation were overwhelming, though he said but little. And yet no one could be more thoughtful and affectionate, nor more prompt in inventing plans of amusement when on a tour. When we ran away //into the wood// to escape him one might really adopt the old explanation of our behavior – "I heard thy voice in the garden, and I was afraid."[30]

In June 1848 my father gave up the house at Highgate, and moved some two miles near two London. One object was to bear near what he thought a better school, viz. the Islington Proprietary[31]: but he was probably more influenced by the desire to be near his work at Salisbury Square. We moved therefore to No. 11, Highbury Crescent, from whence we were only 10 minute walk to the school; and he was within a walk of his office. The change from the garden [17] at Highgate to what seemed to us the yard at Highbury was [*del*: simply] distressing. After all these years I can still remember, as a sort of dismal vision, the hideous contrast between the one place and the other; a contrast emphasized by the fact of the move being made in June. When we went for walks, which we sometimes did with our father, there was only one way by which we could, within the time allotted, get out into anything resembling country, and this was by way of Newington. My father who was always extremely liberal in such matters, wished us to ride, and from as early a time as I can remember there was always a horse for him or the manservant, and two ponies for us. So long as we started from Highgate Hill we used to enjoy the rides, and I got to know pretty thoroughly all the beautiful country to the N.W., N., and East. But from Highbury, two miles nearer town, and with the town itself rapidly spreading, it was not so easy to reach any attractive country. Moreover to boys, especially in the summer months, cricket offered much more attraction. So my father, though he had gone to the expense of building a stable and coachhouse at Highbury, at the bottom of his garden, came to the conclusion to part with his horses and let the stable.

From the middle of August, – for the //summer// holidays at Islington were absurdly early, – we began our second period of school life. We had spent the month of July and first part of August at Hereford, and came straight back from there to school. The school itself was no improvement on the one before. The boys were of much the same class, – the sons of professional and trading [18] men in the City, with a small sprinkling of clergymen's sons; – and there was distinctly less of anything like 'school spirit'.[32] It must surely be a bad sign when no boy has a good word

---

[30] [Genesis 3:10.]

[31] [The Islington Proprietary School was founded in 1830 'for instruction in Latin, Greek, French, German, and Hebrew, and in all knowledge preparatory for a University career, for a professional or commercial life, or for Government Competitive Examinations'. Herbert Fry, ed., *Our Schools and Colleges* (London: Robert Hardwicke, 1868), on p. 145.]

[32] [The public school spirit of a healthy mind in a healthy body. See, for instance, J.A. Mangan, *Athleticism in the Victorian and Edwardian Public School*. With a Foreword by Sheldon Rothblatt and an Introduction by Jeffrey Richards (Abingdon: Routledge, 2000).]

for his own school, either at the time or afterwards, and that is my recollection of the place. There was of course a sprinkling of superior and intelligent boys, with two of whom, – William Berkley, afterwards vicar of Navestock, and Francis Jourdain, afterwards vicar of Ashbourne –, I retained an intimacy until their death.[33] There were two or three others who went to Oxford and did fairly well there, and one very brilliant boy, – Charles Hoole,[34] son of the Secretary to the Wesleyan Missionary Society, //and afterwards// for many years a student of Christ Church –, but these were not enough to leaven the mass. The school had been started by a body of Subscribers and Proprietors, at the head of whom was Daniel Wilson,[35] vicar of Islington, and my father probably thought the character of these directors would act on the character of the school, and that the vices then so rampant in most of the big public schools would be avoided. As a matter of fact I long ago came to the conclusion that if a day school had been chosen, – as I think was best for myself and probably for my brother–, we should have done incomparably better at one of the large new day schools in London, such as King's College School, or the City of London School.[36] But it certainly was not from any lack of anxious care that my father made his choice.

I was just 14 when I entered the school, about the beginning of August, 1848. I was put into the 3rd class, – they measured from [19] 1st downwards there – which was about the place that my age indicated. For one year I remained in that class, being rather idle, and making somewhat bad companions.

1850

About this time my father gave us a turning lathe. It was an old-fashioned wooden lathe, picked up second hand, – at a shop in Cow Cross Street, Smithfield, for £4 – the iron lathes not having come in yet for general use. I got a great deal of work and pleasure out of it, however, and attained some skill in its management. The only misfortune was that no one was found to give me any instruction in the art of turning. I gradually picked up for myself a number of tricks of the trade, but never attained to even amateur skill or the higher order. Amongst other things I made several brass cannon of about 6 or 8 inches in length which were considered a great marvel by my school companions. Also had a try at a steam engine, but the wooden bed of the lathe, and the absence of any slide-rest, made it impossible to construct a cylinder true enough to work successfully.

The boys who get just the sort of youthful training which their later judgment will approve as being that which on the whole made the most of their characters and capacities, must necessary [*del*: be] //comprise// a very small minority. I was not one

---

[33] [Probably William Berkeley (1837/8–?), studied at Oxford; B.A. 1860; Fellow of Trinity College 1862; vicar of Navestock, Essex 1868. Francis Jourdain (1834–1898), studied at Oxford; B.A. 1856; vicar of Ashbourne 1878.]

[34] [Charles Holland Hoole (1837–1902), went up to Oxford in 1854 to study at Magdalen College, graduating in 1859; he became a senior student and tutor of Christ Church, Oxford in 1861.]

[35] [**Daniel Wilson** (1778–1858).]

[36] [These schools were established, respectively, in 1829 and 1834.]

of these fortunate ones. A very considerable sum of money must have been spent upon our education by our father, and some of this in ways which were not [20] then very commonly appreciated, but somehow they failed to catch us at the right time or in the right way, or they were not the sort of developments which fitted in with our own idiosyncrasies. For instance my father felt the importance of some acquaintance with the art of drawing, and we had from time to time [*del*: more] //several// drawing masters [*del*: than I can recall to memory], but it came to nothing in the case of any one of us three. Wood-carving was, I have since seen, a form of art for which I should have had some real aptitude; but it was not recognized for an art then, and really no one who had much else to do than to investigate the [*del*: drawing] //dawning// tastes of a raw boy could have been expected to go so far out of the beaten track as to think of this. Again, my father had felt very acutely in his own case the disadvantage of not being able to speak the language of the country when on a foreign tour. Accordingly he took the greatest pains that we should be better off in this way than he whose own boyhood had been spent in days when Frenchmen were only known as enemies to be fought. I can't remember the coming of the first French master or mistress, it occurred at so early an age. Much the same was it with German, when this was a language hardly heard of in the schools. But such is boys' obstinacy, and so bad were the arrangements, that the outcome of it all was this: – before I went to school I could talk and read French after a fashion, but by the time I went to College I could hardly ask the simplest question or read the easiest book in that language. So with German. A trifle [21] of this tongue had been got into me at the age of about 13, but in my second year at Cambridge, when I first felt how desirable it was to know something of it, I found that I had to begin from the beginning, with the very letters. I had, I think, as a boy, – I am speaking of the time before I began to have a turn for mathematics, which did not take place until I was nearly 16, – two decided tastes. One of these, for lack of encouragement and fitting opportunities of development soon dwindled, the other was sufficiently strong to assert itself for a number of years. The former of these was one for natural history, for collecting and observing insects plants and birds. Very many boys doubtless have a fancy for catching and pinning down butterflies, but I had something more than this. I can remember now the keen interest with which I first distinguished the notes of the yellow hammer, and succeeded in recognizing and naming the bird. So with the large green woodpecker which abounds about Hereford. I can never hear its wild ringing notes now without the revival of strong associations of the woods by the Wye, where I first used to hear it. At Highgate the commoner birds were numerous enough to give one some practise in distinguishing them, but when we went to live down in Islington [*del*: the] //such a// taste had to be cultivated under difficulties which were quite enough to smother it so far as I was concerned.

The other taste was for mechanism of every kind, – an inheritance, possibly, from my grandfather.[37] I had more than the usual taste for this, and greater manual apti-

[37] [In the *Annals of a Clerical Family* Venn wrote that his grandfather was 'a man of strong mechanical and scientific tastes (I still possess instruments which must have belonged to him; e.g. a Dollond telescope, sextant, chronometer watch, dividing compasses). Though not in any way an

tude than most boys possess. [22] My father encouraged me in this, having had somewhat similar tastes himself in his early days, and used to buy me tools. Later on, as I have said above, he bought me a turning lathe, and for some years I was devoted to this pursuit.

During my first year or two at the Islington school I was, I think, rather tending to the bad. I was decidedly low in the school, for my age, and this in a school where the standard of attainment and of diligence was far from high. When I was nearly 16, I was still only in the third class of the school, and had no interest in any subject taught there. I did get a prize certainly but it was for French, and its attainment was entirely due to such remains of the governess-taught French as I had not succeeded in forgetting. Suddenly one day an impulse came over me to spend an hour at home in doing some algebra – I had been for several months painfully and slowly working at this subject, but it had never entered into my head, or //into// that of any boy or master, that anything was to be done at [*del*: it] //a lesson// except in school hours. The impulse came as the whim of the moment, and seemed rather strange to me, I remember, even at the time, so as to seem to call for some sort of explanation to my younger brother. I can remember the place I sat at in the dining room, and the spot from which I took down that volume of Colenso's Algebra.[38] I regard it as the first dawn of intellectual activity or interest, coming, as it did, at a time of life when most of those who ever do anything at college have already got several years of effective mental history behind them. It was in [23] the holidays that this occurred, which, with the standard in vogue at school, made the occurrence the more astonishing. However, having once started my new career I went on with it, and had made some advance before school began again. I remember the master's surprize when I came back, and he found that I had actually moved on from the spot where he had seen me leave off at the end of the previous half-year. The master's name was Harding.[39] He was no mathematician, and had never been at college. In fact he was not supposed to teach anything but Arithmetic and Euclid, and a little elementary Algebra. There was a Cambridge graduate, Coombe[40] by name, who taught Mathematics to the two upper classes //& under whose charge I afterwards came//; but any little taste he may have once had for the subject had apparently been driven out of him by the time he came to teach us. Harding, who knew [*del*: extremely] //but// little had really more interest in what he did know.

---

original investigator, or with much philosophical capacity, he had a decided taste for trying practical experiments'. Venn, *Annals of a Clerical Family*, on p. 121.]

[38] [John William Colenso's *The Elements of Algebra*, first published in 1848 with many later editions, was a widely-used textbook at schools and universities.]

[39] [J.J. Harding (?–?), Fellow of the Royal Geographical Society, was Master in the Hackney Grammar School before coming to Islington Proprietary School.]

[40] [Probably Charles George Coombe (1823–1902), administered pensioner at Peterhouse, Cambridge in 1842; B.A. (19th Wrangler) 1846; Fellow, 1849.]

The teaching all through the school was of a formal and unintelligent kind. Mr Bush,[41] the headmaster, was a somewhat finical Oxford scholar. As composition master at any of the large schools he would probably have done his part well; but set to direct the studies of a hundred boys, mostly middle class and commercial, his traditional maxims and instincts were utterly baffled. By the time [*del*: we] //I and my brother// went there he seemed gradually settling down into the attitude which he afterwards displayed, of a morbid and disappointed drudge. Had he known his day and generation he might have done good work, for the school set out at its foundation [24] with some prestige*

// *It was started, I believe, in 1836. It had the advantage of beginning under a good master, Mr Jackson,[42] afterwards Bishop of Lincoln & of London, who held the post for a number of years. //

and had a good start for developing it according to a reformed standard suitable to the wants of 9 out of 10 of his pupils. As it was he had not a notion of treating the youth who was to go into the counting house at 16 in any different way from he thought best for those who were to enter Oxford at 19. I remember his asking a rather intelligent lad of the former class what he had been reading in the holidays: the boy replied that he had been reading a number of Walter Scott's novels: considering what sort of literature, – if any at all, – most of them resorted to, this seemed to show unusual literary taste. But the master's only reply was, 'Dear, dear, how many books of Homer you might have got through!'. It was a great pity. The mass of the boys suffered from the absence of any attempt to interest them in History, English Literature, Politics, etc.; whilst the few in the upper classes, who were naturally a select body in such a school, got none of the personal encouragement which would have been so much appreciated.

We boys, in the upper classes, were so unsophisticated that, – unlike the generality of those at the great classical schools, – we should have been deeply gratified at being ever asked to the master's house and having a little talk about College life and prospects. It was the same with Coombe and the mathematics. For most part of my time with him I was the only boy who had a glimmering of interest in the subject, but he never gave me the slightest help beyond the perfunctory instruction required during [25] school hours. The supreme weariness and disgust with his position which Bush gradually displayed as time went on, – especially after he had suffered from a severe illness, – at last did its work, and the school simply died out. When new schools were springing up on every side, and old ones extending their studies and developing new branches, the Islington school stood as an almost solitary instance of one which was so obviously drawing towards complete decay that at last the proprietors sold it up, and gave what little money it fetched as a composition to

---

[41] [Robert Wheler Bush (1820–?) had studied at Oxford (Worcester College), where he was a University Theologian Prize Essayist and Select Preacher.]

[42] [John Jackson (1811–1885), educated at Pembroke College, Oxford, was appointed Bishop of Lincoln in 1853 and in 1868 became Bishop of London, being selected by Benjamin Disraeli, then Prime Minister.]

the Head Master who had brought it to that pass. It then started under new management (about 1880) as a middle class school professedly, and for some time had a decided success.*

// * It was afterwards abandoned as a school, & the building used as a shop or warehouse. //

I do not imply the slightest blame to my father. His anxious watchfulness for all our interests was far too sincere for him not to have considered the matter; but it was a very difficult thing for one who was not inside scholastic circles to have known how to choose aright. He thought, and I suppose, with perfect truth, that the tone at the big grammar schools was thoroughly bad, and the training intellectually far from what it should be. Our cousins the Stephens were at Eton; but Leslie, who was nearest my contemporary, was removed at 16 in order to go to King's College. My father was very likely influenced by the consideration that the vicar of Islington, Daniel Wilson, was the chairman of the proprietors of the school, and had sent his own sons there. All things considered, the best available schools for us at that time were [26] probably two of a //kind// somewhat similar [*del*: kind] to what Islington was meant to be, viz., King's College School and the City of London.

At either of these, our training both intellectual and moral, and our opportunities of making suitable friends, would have been different indeed from what they were at Islington.

I remained at this school for five years, from 1848 to 1853. During the last three years of this time I made fair progress in mathematics, and had read further than usual for a schoolboy. But my knowledge was not very accurate; partly owing perhaps to the slovenly character of the teaching, and partly to the fact that I had absolutely no competitors to keep me up to the mark. As regards classics I plodded on with the usual results secured at English schools by the average diligent boys. That is, as an outcome of so many years teaching, I could just make shift to read an ordinary Latin author as one is expected to read a French or German book after a few months study; whilst any Greek author, unless the simplest, was quite unintelligible except by prolonged study with help of a dictionary. It should be added that Bush had given us a slight drilling in Theology after a fashion. This was a result presumably, of the 'Papal Aggression' which threw the Protestant parts of England, – and therefore above all the parish of Islington, – into such a temporary frenzy about 1851.[43] Such was his zeal in this direction, for a short time at least, that he used to keep us in, much to our disgust, after school time, taking notes of the arguments against the Roman Catholic doctrines. I [27] can't say that his teaching in this regard was entirely ineffective, though it did not tell in the way he meant or should have meant. The case against the Romanists was presented as so ridiculously simple and conclusive, that when some eight years afterwards I first happened upon the arguments of a Romanist, presented <u>by</u> him and not <u>for</u> him, – (it was in the Hereford

[43] [Between October 1850 and August 1851, public opinion in Britain was absorbed in an outburst of intense anti-Catholicism in response to the restoration of a hierarchy of bishops announced by Pius IX. See Gerald Parsons, "Victorian Roman Catholicism: Emancipation, Expansion and Achievement," in Gerald Parsons, ed. *Religion in Victorian Britain* (Manchester & New York: Manchester University Press, 1988), pp. 146–183.]

controversy, in which my good uncle had done battle against a priestly champion,)[44] – I well remember the hock of surprise and almost consternation which I felt on finding that these Papist really could say anything even plausible in their defence. It was only one more illustration of the proverbial folly of despising your antagonists.

For English Literature, prose or poetry, for History and for Science, I had then no particular taste; and no attempt was made by anyone in the school to develop or encourage such taste. My father used indeed not unfrequently to read poetry, mostly that of the [*del*: last] //eighteenth// century, to us or with us, and for much of Pope and Dryden he felt a real sympathy, – and this was perhaps the only instigation toward sound literature which we got in those days. But the opportunities were not frequent for this. His own many engagements, – both from home, and at home when missionaries came to spend the night with him, - did not leave many free evenings. Really I think that beyond the school lessons I hardly read anything except Dickens' 'Household Words' which had just then commenced.[45] My father allowed us this periodical, though he [28] certainly did not much like it. His dislike, of course, was a part of the old Evangelical objection to novels, which he fully shared. He never forbade them, but his aversion and contempt for fiction in general was so unmistakable that we never attempted to introduce such books into the house. It may seem incredible to the present generation, but I believe it is the fact, that when I went to College I had only read two novels. One was "Uncle Tom's Cabin" by Mrs Beecher Stowe, which of course stood on a ground apart: the other was Quentin Durward.[46] This //latter// had somehow or other got into the house, probably through my mother or one of her sisters, when it first came out, for it was an early edition. I devoured it with the intensest interest, but such was my conviction that I was doing an illicit act that I always skulked in cupboards or unoccupied bedrooms to peruse it, and carefully hid it between whiles. And I never thought of going on to other volumes of the series.*

// *Among the very few books which found their way into our house during my boyhood & which I can still read, were George Borrow's Bible in Spain, & Gipsies in Spain.[47] It must have been the title of the former, & Borrow's connection with the Bible Society, that franked them through. I devoured them with the greatest delight. //

---

[44] ['The establishment [in 1845] of a Romish monastery at Belmont, on the Wye, near Hereford, was a very severe blow to him [i.e. John Venn of Hereford]. He courageously tried to grapple with the evil by challenging a priest [Rev. James Waterworth] to a public controversy'. Venn, *Annals of a Clerical Family*, on p. 201. The exchange was published in 1845 as John Venn, *Assertions of a Roman Catholic Priest Examined and Exposed; or the Correspondence Between the Rev. John Venn and the Rev. James Waterworth, Respecting Certain Assertions Made by the Latter at the "Hereford Discussion"* (Hereford, 1845).]

[45] [*Household Words* was a weekly literary magazine edited by Charles Dickens in the 1850s. It appeared every Saturday from March 1850 to May 1859. Among other things, it contained Dickens' own *Hard Times*, published in weekly parts between April and August 1854.]

[46] [Harriet Beecher Stowe's *Uncle Tom's Cabin* appeared in 1852. It stood 'on a ground apart' as both the author's views and the book's main theme resonated with one of the Clapham Sect's main missions: abolitionism. *Quentin Durward* is a historical novel by Walter Scott. The first edition of 1823 was published anonymously in three volumes.]

[47] [George Henry Borrow's *The Bible in Spain* (1842) and *The Zincali, Or, An Account of the Gypsies of Spain* (1841). Borrow, a polyglot, was active for the British and Foreign Bible Society in several ways. Among other things, he supervised a translation of the Bible into Manchu.]

As I have implied, my Islington and school life was neither joyous nor painful; it was simply rather dreary and common place. We used at first to ride, – for my father was always most liberal in finding ponies and horses for us, and would have liked us to take a pleasure in the exercise. But the distance, ever increasing, to which one had to go in order to be immersed in the green fields and lanes, and the practical necessity when starting from Holloway of generally going in pretty much the same direction, [29] dulled our enjoyment of the exercise. Then we took to cricket, but in an untrained and inartistic fashion. Those were the pre-athletic days.[48] Not one of our masters took the slightest interest in our pursuits, even if he knew that we played at cricket. We could not afford any professional help, and I cannot recall the presence of any friend or senior person to watch or advise us. We just hired a field for the summer. It was always a vacant plot of ground destined to be almost immediately built over. Four such fields I can remember, all long since densely covered over with rows of small houses. Our last, that of 1851, was the plot adjacent, on the East side, to the then new Holloway Gaol. We once ventured on a match against a private school at St. John's Wood, in which we got ignominiously beaten. It was my only visit, in an active capacity, to Lord's, and I distinguished myself by a double 'nought'. Finally a few of us took to boating on the Lea upon half holidays, but this also a practice carried on according to natural impulses, for we had no one to give us a hint how to hold an oar.

Almost the only really strong taste I remember experiencing in these years, and the only source of intense pleasure, was from wild and especially mountain scenery. I can recall here too the very day and spot when and where this passion first seized me, viz. in an excursion from Penmaenmaur[49] in 1852. I had been repeatedly in North Wales in years before this, – we had been taken thither when children more than once by my father, but I can't recall [30] that we any of us [sic] experienced any particular pleasure in the scenery as such. In 1852 it was very different. We spent that summer at Penmaenmaur, a place then quite unheard of. My father hit upon it almost by accident. He saw from the map that there was just one place where the sea and the mountains came into contact on the north coast of Wales, and where there was also access by railway. So he wrote to the station master of the nearest station, to enquire if there was any inn, and hearing from him a favourable account he took rooms there for our party for some five or six weeks. He was certainly a little surprised when he reached the Inn, a small way-side public house such as is so frequently found on the Welsh roads, and which went by the name of the Jolly Herring. It stood just under the quarries, a few hundred yards short of the spot where the coast road to Bangor reaches its highest point. The place was very quiet and secluded. We had to send to Conway for our meat, and to walk two miles [*added in pencil*: on Sunday] for an afternoon English service. There were no regular lodgings there, and only three houses, of which the Jolly Herring was one, that ever, I believe,

---

[48] [See Mangan, *Athleticism in the Victorian and Edwardian Public School*, for an account of the rise of athleticism – of the merging of the concepts of gentleman and sportsman – in the second half of the nineteenth century.]

[49] [Throughout the *Annals*, Venn often Penmaenmaur instead of Penmaenmawr.]

took in visitors. We enjoyed the place intensely from first to last, and our father was always organizing excursions amongst the mountains behind. The expedition I most vividly remember was the first mountain ramble I ever took by myself. It was nothing much, only extending to the little lake which lies a few miles due south of Penmaenmaur, at the foot of Foel Fras. I had been told of this lake, and discovered it by [31] aid of the Ordnance map. I remember now when I got to the top of the ridge from which I looked down and saw it; how I stood and shouted with delight. With the far grander scenery which lies a little farther back, about Carnedd Llewellyn and Davydd, I did not make acquaintance till several years later.

I recollect one other long country ramble, – also a solitary one, – about the same period, which produced a deep impression upon me. It was to Winchelsea, in the winter of 1851–2. Its aspect of peaceful decay, and the striking view of its ruined gateway over the wide stretch of Romney Marsh, impressed me powerfully. I had doubtless often seen more remarkable places than this, and grander scenery than that which lies within 4 or 5 miles of Penmaenmaur, but the fact of getting to them by myself, made all the difference to me, and I think laid the foundation of most of my subsequent love of wild scenery.

During my last two years at Islington I was in the top class. The upper boys were not //at all// a bad set, but they were far too few in number, – not more than 4 or 5 altogether in the class, – to create anything corresponding to that public feeling, intellectual and moral, which ought to exist in the head class of a good school. Four of these, who were my contemporaries at one time or other (within a couple of years) distinguished themselves fairly at Oxford. F. Jourdain, afterwards vicar of Ashburne, was a couple of years my senior; so was C. Witherby[50] who, I think, also got an open scholarship at Oxford. The two cleverest who came near me were about a couple of years my juniors. These were C.H. Hoole, who [32] was a really brilliant boy and at 16 carried everything before him. He //afterwards// got a studentship at Christ Church, and then subsided into quiet. The other of these was W. Berkley who got a scholarship at Brasenose and afterwards a fellowship at Trinity. He possessed extreme acuteness and mental vivacity, and seemed starting at Oxford, and afterwards as a curate at St Botolph Bishopsgate, with a fine career before him. Soon afterwards however he completely broke down in health. A few more lads of this caliber had they been really my contemporaries, were all that was needed, – given more efficiency on the part of the masters, – to make the school a tolerably good educational training. As it was, I was in the disadvantageous position during the latter part of my stay of being by two years the oldest boy in the school, and of not having a single schoolfellow who was looking forward to entering Cambridge.

Nothing of any significance, so far as I was concerned, happened during these last years of school life. We spent our summer holidays mostly at Hereford, in 1850, 51, and 52, under the cheerful and affectionate charge of our most devoted uncle and aunt alluded to before.

[50] [Cornelius Witherby (1834–?) received an Islington Scholarship in 1852, starting at Lincoln College, Oxford, in March of that same year. He graduated B.A. in 1855 and later became vicar of St. Simon, Bristol (1868–82) and rector of St. Paul Charlton, Kent, 1885.]

1853

The first part of our summer holidays was spent at Hereford. The stay there was marked by the death of our old grand-aunt Jane.[51] She was then 92, and was almost to the last in full possession of her faculties, and a store-house of family information. [33] If only, like her brother, my grandfather, I had developed an early taste for family history our "annals" would be much more complete than they are. As it was, unfortunately, I never cared a bit about such matters until not only she, but nearly every one of the generation following her, had passed away. This was the first time, within my memory, that I had been brought into contact with death. Immediately after she had departed my father took us young people into his room, and read the 15th chapter of I Corinthians to us. It was the occasion of a solemn family gathering; as my uncle James Stephen came down to the funeral.[52]

After this, for part of September, we went to the Isle of Wight. I then first began the practice of long solitary walks. One of these, – it will show my early sympathies and training –, was a devout pilgrimage to the grave of the (1) "Dairyman's Daughter"[53] in Arreton.*

// * Readers of the present generation may need to be told that this was one of the tales of Legh Richmond, another being "The Young Cottager". They were very popular in religious circles & were translated into several foreign languages. //

We were then staying at Cowes. The longest expedition I made, – my first very long walk –, was through Carisbrook to Ventnor, thence to Black Gang Chine, and so home again, about 35 miles. I well remember the splendid burst of sea-view as I came over the ridge which lies just above Ventnor. The pretty church at Bonchurch also much pleased me.

The only external event which I can recall during this stay at Cowes was the great Naval Review, when also I had my only sight of the Queen. The Review was a magnificent spectacle, and was, I suppose, the last of the kind held under the old conditions.[54] [34] The ships were nearly all sailing vessels, and of course there was not a rifled gun in the whole fleet. Most of the cannon were the old the old 24 and 32

[51] [Jane Catherine Venn (1760/61–1853), sister of John Venn of Clapham.]

[52] [Like other Evangelicals in the eighteenth and nineteenth centuries, the Venn family were committed to the narrative of the good death. As Henry Venn of C.M.S. wrote to his children on the occasion of his aunt's death: 'God has brought you into a scene not only of solemn interest, but which will, I trust, through His grace, be very profitable to your own souls'. Knight, *Memoir of Henry Venn*, on p. 300. See also, for instance, Mary Riso, *The Narrative of the Good Death. The Evangelical Deathbed in Victorian England* (London & New York: Routledge, 2015).]

[53] [This popular booklet, written by Rev. Legh Richmond, told the story of the religious conversion experience of Elizabeth Wallbridge. Its success – in 1828 it exceeded four million copies in 19 languages – led many, including Queen Victoria, to make the pilgrimage to Arreton to visit Wallbridge's grave. See Dennis Butts and Pat Garrett, eds., *From The Dairyman's Daughter to Worrals of the WAAF: The Religious Tract Society, Lutterworth Press and Children's Literature* (Cambridge: Lutterworth Press, 2006).]

[54] [For the changing character of naval reviews see Jan Rüger, *The Great Naval Game: Britain and Germany in the Age of Empire* (Cambridge: Cambridge University Press, 2007), 12–49.]

pounders: we noticed what seemed by comparison the terrific crash of the few 64 and 84 pounders on some of the decks. The main review took place several miles out in the channel, – to the dismay of some of the spectators who thought they would not leave the calm waters of the Solent. My father took us all three on a steamboat which accompanied the ships and from which we got an excellent view. I well remember the mysterious passage, as it seemed, of one of the [*del*: few] vessels which had screws, as it shot by us with apparently not motive power to propel it: – most war vessels were still, I think, paddle boats, if they used steam at all. Perhaps the most impressive part of the spectacle was the attack by gun-boats, as it gave one a vivid idea of the old-fashioned "cutting out" expeditions of Nelson's time.[55] A number of small boats were lowered each with a small cannon in the bow, and filled with sailors and marines. These gathered around the ship to be attacked and poured an incessant fire into it for some time.

This Naval Review* took place in the autumn //of 1853// by which time troubles were already anticipated in the East, but it was still sixt months before the declaration of war. [56]

// *Just 50 years afterwards (April 1903) I paid a visit to Portsmouth and went with my son to see the <u>Victory</u>. An old sailor who rowed us thither declared that he was in the fleet at that Review. //

In October 1853 I entered College.[57] As was the case with most freshmen, I had to remain in lodgings during my first years. [35] My rooms were in Peas Hill over a furniture shop kept by a Mr Toft, and looked to the back over a yard belonging to Mr Clayton the fishmonger. I cannot say that the college was then in a very satisfactory state. The name and reputation of Mr Clayton (brother of the fishmonger just mentioned) had made it the Evangelical College in popular estimate.[58] This circumstance attracted a number of very narrow-minded men several of whom had given up business in order to take orders. They formed a strong party which kept entirely to itself, holding aloof from the boat-club and from almost every other form of social amusement; as by natural reaction the opposite party gathered into its ranks all that was idle and noisy. The result was that the College was split into two distinct camps, as it were, which had little communication with each other. As to myself, for example, there were a number of men, not

[55] [Horatio Nelson (1758–1805), British flag officer in the Royal Navy known for his unconventional tactics which often proved decisive, especially during the Napoleonic Wars. A 'cutting-out attack' is one based around the element of surprise.]

[56] [The Crimean War (October 1853–February 1856) in which Britain fought alongside France and the Ottoman Empire against the Russians, chiefly in the Crimean peninsula.]

[57] [Venn was administered as a pensioner at Caius on 25 June 1853 and matriculated in Michaelmas Term, starting in October.]

[58] [The Evangelical **Charles Clayton**, tutor of Gonville and Caius College from 1848 to 1865, had succeeded Charles Simeon in his ministry at Holy Trinity in 1851. He attracted many sons of Evangelical clergy to Caius, making it the leading Evangelical college of the day. See Christopher Brooke, *A History of Gonville and Caius College* (Woodbridge: The Boydell Press, 1985), chapter 5, and Peter Searby, *A History of the University of Cambridge. Volume III: 1750-1870* (Cambridge: Cambridge University Press, 1997), chapters 9–10.]

only in my own College but some even in my own year, with whom, I believe, I never once exchanged a single word during the period of my residence. Of course there was a small body, – of whom my friend J.E. Whiting[59] was one –, who acted as a sort of intermediaries; the misfortune was that thoroughly good men of this type were so few in number. During my first term I fell entirely into the hands of the rigid party, joined their prayer-meetings, and regularly took my walks in an afternoon with one or other of them. My main associated was a certain C.R. Ord,[60] three or four years older than most of us and an extreme Calvinist. He was so narrow and determined in his views [36] that I doubt if I should have got on with him as long as I did, but for his one indulgence, – chess –, in games of which I used to spend one evening in the week with him. Small accidents often have a great influence on our careers, and I have sometimes thought what incalculable benefit it would have been to me had I happened, instead of being alone in lodgings, to have got a room on some staircase in College next to one or two liberal-minded though religious men. I got to know such men afterwards, but my extreme shyness made the process of acquiring new friends a very slow one, and meanwhile much of the short seed-time of youth was lost. A bright young fellow, some years afterwards, when the same state of opinions was still subsisting under Clayton's influence, summed up the situation by declaring that he had no companions, for the good men were too good to have anything to do with him and the bad men too bad for him to wish to have anything to do with them. The fact is that the time was one of transition. A generation or two earlier the Evangelicals were very few, and scattered at various colleges, and were consequently somewhat in the position of a persecuted party.[61] By my time they had become numerous, and (so far as Caius was concerned) almost aggressive. It need not be added that a generation further on the whole character and tone of religious feeling had again profoundly altered.[62]

As I did not belong to the boat club or cricket club, and, besides these, there was only the musical society, in which I could take no part, I was practically cut off from the society [37] of all the men who were not in my own year, – another serious diminution in the number of friends available. Several indeed of the seniors in the College called on me, at Clayton's instigation or for my name's sake, but after the first term or two they hardly found me in sufficient sympathy with them to keep up any intimacy. The saving element

---

[59] [James Edward Whiting (1834–1917), Caius 1853; B.A. (40th Wrangler) 1857; later became chief engineer for irrigation, and controlled schemes for famine relief, in India. He died in Trumpington in 1917.]

[60] [Charles Renshaw Ord (1830–?), Caius 1852; LL.B. 1858; afterwards curate in Sydney, Australia.]

[61] [For example, Venn's grandfather, John Venn of Clapham, Trinity College, Cambridge refused to take him in 1777 because of the Evangelicalism of his father, Henry Venn of Huddersfield, who fostered it at Cambridge through his influential protégé Charles Simeon. See John Gascoigne, *Cambridge in the Age of Enlightenment. Science, Religion and Politics From the Restoration to the French Revolution* (Cambridge: Cambridge University Press, 1988), pp. 252–253. On Evangelicalism at Cambridge see Searby, *A History of the University of Cambridge*, chapters 9-10.]

[62] [For example, **Henry Sidgwick**'s election as Knightbridge Professor of Moral Philosophy in 1883 was a milestone in the emergence of an increasingly secular academic culture at Cambridge, severing the traditional link between the university and the Church of England.]

to me in those early days was the association with three or four men. One of these was my cousin, Henry Babington,[63] a fine specimen of the healthy, moral, shrewd public-school-boy (though as a matter of fact he had never been at school) who did much to keep in in what I should now call the right path. He was so outside the pious world of the University that but for our relationship I should never have made or kept his friendship. As it was, fortunately, we were much together. He was a year my senior, and at the end of his second year, having no vocation to a clerical life, he left Cambridge and secured a place in the Board of Trade, under Captain Fitzroy, in the newly founded Meteorological department.[64] My most intimate friend, by far, was J.P. Garrick,[65] afterwards fellow, and subsequently rector of Blofield. We came up as freshmen together, and from an early date he and I were inseparable. At the end of his second year his eye-sight gave way, and he had to leave College for a year and degrade. He, like myself, had begun his academic life with Ord and the rigid party, but like myself, gradually drifted somewhat away from their path. Another valuable acquaintance was J.E. Whiting, son of the vicar of Royston, a handsome man and very fair cricketer. He had a strong taste [38] for mechanics, and was an excellent geometrician. Had his algebraic capacity been at all commensurate, he would have gained a high place in the Tripos.[66] Ransome,[67] for years one of my most intimate acquaintance, I did not get to know well till towards my third year. Other pleasant and profitable friends were Panton, Ponton, Gedge, Gould, and Holloway, all of my own year.[68]

The outset of a College career seems a suitable occasion to try to form some esti-mate of one's intellectual character and attainments at the time. I think that my case was that of a mind dominated up to a certain stage by [*del*: one] //another mind// of a totally different kind. This superior influence with me was found in my father. He had, as I have elsewhere said, an intensely practical type of mind. He held, with the most unswerving faith the Evangelical opinions he had inherited, and his whole life was an absolutely consistent display of the working of this faith. He was however very tolerant, and would I am certain, have been ready and willing to listen to any

[63] [Thomas Henry Babington (1833–1907), entered Trinity College in 1852 but did not graduate.]

[64] [Thomas Henry Babington became the closest assistant of Robert FitzRoy (1805–1865), captain of the HMS Beagle during Charles Darwin's voyage. In 1854, FitzRoy had been made chief of the new department of the Royal Society which was to deal with the collection of weather data at sea – a forerunner to the Meteorological Office. See Katharine Anderson, *Predicting the Weather: Victorians and the Science of Meteorology* (Chicago & London: The University of Chicago Press, 2005), chapter 3.]

[65] [James Percy Garrick (1834–1919), Caius 1853; B.A. (9th Wrangler) 1858; Fellow, 1858–81; tutor, 1875-6; later rector of Blofield, 1880–1912, and Honorary Canon of Norwich, 1899–1919.]

[66] [Whiting ranked 40th Wrangler in the 1857 Tripos.]

[67] [**Arthur Ransome** (1834–1922).]

[68] [David Brooke Panton (1832–1891), Caius 1852; B.A. 1857; born in Jamaica, he later became Archdeacon of Middlesex, Jamaica, 1884–91. Alexander Campbell Ponton (1835–c. 1867), Caius 1853; B.A. 1858; curate of Bloomsbury, London, 1864–6. William Wilberforce Gedge (1835–1912), Caius 1853; B.A. 1857; headmaster of Cheltenham Proprietary School, 1860–70; rector of Mutford, Suffolk, 1889–93. Joseph Gould (1834–1908), Caius 1853; B.A. 1857; curate of Brewood, 1864–5; rector of Beachampton, 1891–1903. Edward John Holloway (1834–1908), Caius 1853; B.A. 1858; later vicar of Clehonger.]

doubts and difficulties which lay within certain limits. But the awe which I always felt for him, from the earliest days, made it quite impossible for me ever to argue with him except upon matters in which I knew that he had no strong feeling. The only form of belief with which I had had practical acquaintance in others was the Evangelical; neither at home[1], nor in [39] College, did I ever meet with a Highchurchman, or a Broadchurchman.[69] No reading of my own, and no conversation with others, ever brought me into contact with other forms of belief than that which was traditionally familiar to me. It was not so much dogmatic teaching on my father's part as his strong personality, that gave me the initial start in the direction of his own opinions. To me, who wished to be religious, the Evangelical creed seemed the only one in the field. Accordingly I threw in my lot with the strict party in College, and for a time attended their prayer-meetings. In those days, Hall was at 4 or 5, on Sundays and weekdays alike, and Chapel at 6. Those who attended Chapel could not go to the ordinary ser vice at a church, but when their own service was over they were just in time for the sermon. Consequently there was always a large influx of gownsmen at such churches as possesses any popular preacher. There were then practically only two such preachers; Clayton, at Trinity Church (the church rendered famous in former days by Simeon) and Harvey Goodwin at St Edward's.[70] They were respectively 'low' and 'high' but in no extreme degree. Both were good preachers, simple and very earnest; Goodwin had, what Clayton had not, thoughtful matter in his sermons. He was naturally disavowed by our party, and I never remember but once setting foot in his church. So far as one is competent to judge one's own character, I should say that I was naturally inquisitive and sceptical in spirit but profoundly respectful and even reverent towards those who seemed to have a rightful claim to superiority. My father of course occupied this position through life, but I remember starting with [40] the same feeling of adherence and attachment towards my first vicar, Mr Faithfull.[71]

Few men probably enter college now with so little general literary acquaintance. It will sound incredible, but I had, to my almost certain knowledge, only read <u>one</u> //

[1] Of the people whom I met at home, and with whose intimate converse with my father I was most familiar, far the largest number were missionaries. I am bound to say that the missionary as he sometimes appears in fiction never fell under my view. Those whom I met were just ordinary clergymen, with rather more zeal and devotion and something wider experience, than the bulk of their brethren at home.

[69] [The three major parties within the Church of England in the nineteenth century were the High Church, Low Church or Evangelicals and Broad Church parties. There are many ways to define their respective position and influence. One way of putting it, and rather roughly so, is to say that the High Church party emphasized authority, the Low Church emphasized personal vital religion and the Broad Church was liberal in accepting differences of religious opinion.]

[70] [Whereas **Clayton** represented the Evangelical or Low party, the Cambridge mathematician Harvey Goodwin (1818–1891) – who later became Dean of Ely and subsequently Bishop of Carlisle – represented the High Church party.]

[71] [James Grantham Faithfull (1817–1873), first-class cricketer educated at Oxford who later became vicar of Cheshunt, 1858–71.]

real// novel, – Quentin Durward, – under the circumstances already mentioned. In this abstinence I was not chafing against an imposed condition; the food had never been supplied to me, and I had acquired no appetite for it. It is true that we took in, – under toleration rather than with any welcome –, Dickens's Household Words almost from its origin, when I was about 15 or 16, and the short stories therein contained of course supplied a fair amount of the element of fiction. But I suppose the taste had been but little developed, for though I began to get novels occasionally out of the[1] Union Library[72] from my first term, I felt no temptation towards the full feeding, on this provender, which some would expect, of one who had been kept back from such natural nourishment.

I have tried to recall the name of any book which had made any real impression on me before this date, or indeed till a year or more later, and I cannot do so. Outside the school lessons, I can remember Household Words, & many of its stories and articles are fixed in my mind; but beyond this I can remember nothing except a few popular scientific books. There were, of course, [41] scarcely any magazines of the popular kind in my boyish days, and the stream of fiction was shallow indeed compared with the torrent which now flows.

After my first two or three terms, when I got to know Garrick so intimately, who was a great admirer of Byron, and somewhat of a poet himself, I added to my small library, such as it was, but I much doubt if, during my first three terms in lodgings, there was a single book on my shelves, – outside the classical and mathematical stock-in-trade, – except my great-grandfather's Complete Duty of Man.[73]

As to amusements, there were then practically no alternatives except boating and taking walks.[74] Owing to the general character of the college at the time, few or none of the men in my set were members of the boat club, or therefore took their regular exercise on the river. Now and then it is true, an excursion was planned, but more for the purpose of occasional exploration than for regular exercise. Of one such excursion taken in my second term, I have a vivid recollection. A party of 5 or 6* of us resolved upon a trip to Ely.

// * J.P. Garrick, D.B. Panton, J.J.M. Sowell, A.L. Nicholson, E.A. Hillyard, J. Venn.[75] //

[1] In its early days the Union Library strictly excluded novels. By my time this rigour was relaxed, but I think only so far as classical authors and recent works of repute were concerned. The fugitive popular productions of the day were certainly not sought there. The whole establishment was entirely closed on Sunday till some years after this.)

---

[72] [The Cambridge Union Society was established in 1815 as a debating club, with a library, by Henry Bickersteth (1783–1851), Sir Edward Hall Alderson (1787–1857) and Sir J.F. Pollock (1783–1870). It moved into its current home in 1866, housing the Keynes Library.]

[73] [The popular Evangelical handbook written by Henry Venn of Huddersfield and first published in 1763.]

[74] [On 'games for gownsmen' see Searby, *A History of the University of Cambridge*, chapter 18.]

[75] [Joseph James Matthews Sowell (1832–1866), Caius 1853; B.A. 1858; school-master in Georgetown, British Guiana, 1858–59; curate of St Thomas, Trinidad, 1859–61. Armytage Lenox Nicholson (1834–1891), Caius 1853; resided three terms. Edward Augustus Hillyard (1834–1897),

Not one of us was at all used to rowing in Cambridge, though we could all pull a little and keep clear of cutting crabs. I remember the excursion well, as it was my first sight of the fen country. We started early in the day, about 9 or 9:30, and got down to Ely easily enough as the stream was tolerably strong and there was a fresh S.W. wind behind us. We inspected the Cathedral and had our dinner //at the Lamb Inn// there, and started to [42] return home about 4 o'clock. It was nearly dark before we reached Upware and we had some work in distinguishing the stream of the Cam from one or two of the other ditches or canals. I shall never forget the impressiveness of the sight towards dark, when we got out to reconnoiter; and standing on the high canal banks which enclose the river on each side, looked over the vast level, and saw the distant church towers and spired standing cut clear against the light on the horizon. It was six hours work getting back; and with our heavy boat, and four not very expert rowers to work it, we soon found it impossible to make way against the high wind and the stream, to say nothing of the growing darkness. So one of us [*del*: stood] //sat// in the bow with a lantern, another in the stern to steer, and the other[s] [*del*: three] worked with a tow rope. The nine o'clock bell of St Mary's was tolling as we reached Ditton, and it must have been 10 o'clock before we got back to College. This was my first sight, and for sometime my only one, of the fen scenery,[76] and left a very strong impression of its interesting and impressive character.

Otherwise we took our exercise on the high roads.[77] We had at least the sense not to walk alone, and accordingly the country for miles round Cambridge every afternoon was sprinkled over with couples of men. Monotonous and tame as such a pursuit may seem to the modern youth, it had one recommendation, in the opportunities it gave for conversation with selected friends. Some of the best and most profitable hours I have ever spent have been [43] those in which, with some congenial friend by one's side and a stretch of a dozen miles before one, the give and take of conversation could have free play. Sunday, – no one, I believe ever 'worked' on that day –, was specially devoted to such walks, by the quiet and reading men more especially, none but the somewhat extreme having any objection to such an employment of the day. We came out of chapel soon after ten, and Hall was at four, and the whole interval between was sometimes expended in this way. I suppose there is scarcely a single village, within 10 miles of Cambridge with which I have not made acquaintance on such walks.

In the May Term we, (that is, I and my set), indulged in a little mild cricket. Here again the shyness of some of us, and the want of homogeneity in the College, had their drawbacks. Instead of joining the regular cricket club, a few of us constituted,

---

Caius 1853; B.A. 1857; vicar of Christ Church, Belper, 1876–97; a prominent Derbyshire freemason.]

[76] [The area to the north of Cambridge – once a vast and plant-rich region, now reduced through drainage to a few scattered pockets.]

[77] [Until 1885, when the safety bicycle was invented, walking was the common exercise for reading men at Cambridge, often walking in twos to engage in conversation.]

with some outsiders from other Colleges, a small 'scratch' club.[78] We played of course like all other people, – with the single exception of the University Club, which already used 'Fenners' (Fenner, the player then rented it),[79] and Jesus which used their own Green –, on Parker's Piece. This great field was naturally a lively place on a fine May afternoon, and not altogether a very safe one either to play on or to cross.

I am mentioning, in detail, some rather trivial matters, but these, as every historian knows, are just the points which are most difficult to recover afterwards. I should say, speaking from my own recollection, that there was absolutely no football [44] played in those days.[80] With a good memory for trifles I have not the slightest recollection of it, though some friends tell me that, – a few Rugbeians and Etonians used to carry on their old school game. Where they played I cannot imagine, as I never saw them on Parker's Piece. A few men used to ride, – more probably than now –, and there was one tennis court but such pursuits as these were far too expensive for any but the rich. Once or twice within my recollection the old fashioned 4-wheeled 'velocipede'[81] had a short vogue. But from my experience this mode of progression was more tiring than walking, and very little faster. I remember that two of my acquaintances (the present Sir Charles Elliott, and Mr E.E. Bowen, of Harrow)[82] accomplished the feat of getting to London in one day on these machines; but it was thought a remarkable performance, and they would have walked the distance with less fatigue. About my second year the canoe was introduced (or reintroduced) and those who did not regularly row in the eights often took to paddling at the Backs.[83]

In the spring of this year I was for some time unwell, though not in any serious form. I suppose it was what would commonly be called 'want of tone' or 'poor blood', for I suffered from a succession of bad boils. Sir James Clarke, – the same who had been called in 10 years before, when I was ill with inflammation of the lungs, and to whom my father now took me, recommended my giving up reading

[78] [An impromptu team, brought together on a temporary basis and composed of players playing for different sides.]

[79] [Fenner's is the Cambridge University Cricket Club's ground. It was established in 1848 by Francis Phillips Fenner (1811–1896) on land which he had leased.]

[80] [It was only in the 1850s–60s that the two varieties of football, soccer and rugby, began to take on its modern appearance. See, for instance, Adrian Harvey, *Football: The First Hundred Years. The Untold Story* (London: Routledge, 2005), chapter 4.]

[81] [An early version of the bicycle that was ridden by sitting in the saddle and pushing yourself along with your feet.]

[82] [Sir Charles Alfred Elliott (1835–1911), Lieutenant Governor of Bengal, educated at Harrow School, where he won a scholarship at Trinity College, Cambridge. He was the son of Henry Venn Elliott. Edward Ernest Bowen (1836–1901), educated at Trinity College, Cambridge; later became a schoolmaster at Harrow School.]

[83] [The Cambridge Backs are an area of central Cambridge, along the banks of the river Cam, occupied by seven colleges, including Trinity College and King's College.]

for honours.[84] I feel now quite sure that [45] what I was in need of what a somewhat better diet, and earlier hours at night; fortunately I did not follow the advice as to study, and when I went to Wales a month or two later, I found myself quite fit.

As to diet our practice at that day shows the great advance since then in luxury, or at least in ease of living. I can speak with certainly for myself, and with some confidence as to my friends, and I do not think we were peculiar. We had only one real meal in the day. At breakfast a couple of eggs, or a bit of cold bacon, was the only usual supplement to our bread and jam: lunch was effected by going to the cupboard and cutting some bread and cheese, with the addition at most of a glass of beer: dinner was, for the freshmen at 4, and for others at 5. Unless we 'sized', or ordered extras, there was nothing provided but meat, vegetables, and bread. All tarts and puddings were extras. Of course we wanted something later on, but except when a regular supper was in question, this was confined to bread and jam with our tea.

My first May-term differed in no way, as regards work or variety of amusement, from any other term. There were no 'gaieties' whatever, except the customary boat-races, to which few or no friends from a distance ever thought of coming. Even the Flower-show, which for a number of years was the only public amusement provided at that season, was not started, I [46] think, until my second or third year. Our habits were very simple. I went out early in the morning for a bathe on Sheeps' Green, with Panton or some other friend, then to morning Chapel, and, after that used to work all the morning. Breakfast and supper parties were very rare in our set, and lunch and dinner parties absolutely unknown. I can only remember two entertainments of any kind during my first year. One of these was to [*del*: a sort of] //an// evening [*del*: teaparty] //gathering// at one of the doctors in the town, with whom my father had a slight acquaintance, and to which in my ignorance I went in morning costume. The other was to the annual dance which used to be given by my mathematical tutor, Mr Hopkins,[85] – it was, I should think, about the only dance given in the place. I remember that I and one other pupil as raw as myself, went at the hour mentioned on the card of invitation, 9 p.m. The servant, supposing we had come to do a paper, showed us into the 'coaching room' at first: on subsequent introduction into the drawing room we found the Miss Hopkins there, who had not yet gone up to dress. Altogether, as I did not dance, I was considerably out of it that evening.

The 'May' examination, which then took place towards the end of that month, was a very important affair. It was exclusively mathematical, – the classical subjects having been taken before Easter –, but the total result of the two [47] examinations, as deciding the assignment of scholarships, was the first public test of performance.

---

[84] [The B.A. honours degree was then still awarded on the sole basis of performance in the Mathematical Tripos examinations, with honours candidates being divided each year into three categories, called 'Wranglers', 'Senior Optimes' and 'Junior Optimes'. The Senior Wrangler was top of the list. The Tripos was fearfully hard work, both physical and mental.]

[85] [The famous Cambridge private mathematics tutor, and 'Wrangler-maker', **William Hopkins** (1793–1866). See A.D.D. Craik, *Mr Hopkins Men*.]

In these days, early examinations and entrance scholarships fix a man's place from his entry at College, but on the old system there was nothing but common rumour to go by during the first year of residence. Accordingly many a reputation built up on school performances, or the student's own estimation of his powers, was doomed to disappointment.

// (College Examination, 1854.

Classics	Mathematics
Monro 675	Venn 920
Gedge 546	Garrick 750
Holloway 526	Goldson 678
Gould 504	Hyde 633
Garrick 427	Gould 587
Venn 399	Cockson 539
Sowell 49	Ashley 49
(32 classed)	(27 classed) //

In this examination I came out, rather to the surprize of most of my contemporaries, at the top of the list, in a class by myself, which however did not imply very much. I think that I might have done better under more suitable conditions, and that such conditions would have enabled me to get a more thorough grasp of mathematical principles, and probably to take a higher place in the Tripos eventually. During my first term I did not coach: after that for a time I attended William Hopkins' classes. Mr Hopkins had a high reputation, doubtless well deserved for those whom he suited: in fact during many years almost every good mathematician at Cambridge had been under his tuition. Hitherto he had not taken freshmen, but narrowed means (the result of some unsuccessful investments) now induced him to do so; and my father, ever generous where our interests were concerned, thought it would be wise for me to attend him. His charges were very high, – £20 for the term, we went to him every day –, and he used to take us altogether, in one class for each year. [48] His system however did not suit me at all. What I wanted was some one who would talk out every subject with me, attend to my doubts, – often involving speculative points which did not lie quite in the usual curriculum –, and especially in a novel branch like the Differential Calculus, be prepared to explain what it all meant, and was intended to lead to. There was then at least one such man, viz. Percival Frost,[86] and I should have read mathematics to far more profit and pleasure if I had had his assistance all through my time. As so many eminent men have declared that they owed very much to Hopkins' methods of tuition, the fault must have been partly mine; but, as it was, I cannot recall any new or stimulating idea which I owed to him, or any solution of a real difficulty.

---

[86] [Percival Frost (1835–1898), Caius 1835; B.A. (2nd Wrangler and First Smith's Prizeman) 1839; Fellow of St John's, 1839–41; Mathematical Lecturer at Jesus, 1847–59 and at King's College, 1859–89; Fellow of King's College, 1882–98. Like **Isaac Todhunter**, Frost belonged to the second generation of private tutors, succeeding Hopkins, who was then beginning to lose ground as the leading coach.]

There were three or four weeks between the end of the May Term and the commencement of the working Long Vacation. Penmaenmawr had left such delightful reminiscences that I was determined to go to North Wales and get some good walks there. My father, as usual, was at once liberal in providing the means. But very naturally I could find no companion, as none of my friends were able to provide such a luxury. I should have done better and more sociably if I had stayed at home, and gone to see friends and relations, and so forth. But I was bent on getting to the mountains, and accordingly started off one evening and got to Conway about 11 p.m. [49]

1854

I put up at the Caste Inn, and next morning walked out to the Jolly Ferring at Penmaenmawr, where I spent between two and three weeks entirely by myself. I had no books with me, and my only in-door amusement was working out such of the problems set in the late examination as I had not succeeded in solving at the time. The trip seems to me dull enough, looking back at it afterwards, but I certainly enjoyed myself extremely at the time. As common enough in Wales, during June, the weather was very bad during much of the time. I made an excursion of two days away from Penmaenmaur during the time. I walked the first day, round through Conway by the right bank of the river, and Bettws, to Capel Curig, where I put up at the large and rambling old coaching Inn. It was a pity that I had not some older companion with me to point out that, in Wales, instead of thus sticking to high roads I should have avoided them altogether, or at least driving along them, and kept my legs entirely for the hills. In after years I got to know that the direct scramble from Penmaenmawr, over Careddan Llewellyn and Davydd, is simply the most magnificent walk in Wales. I spent a lonely evening in the big hotel, being too shy to speak to the other guests, and started on my way early next morning. I was too ignorant of the ways of Inns to be aware that besides paying the bill the servants had to be tipped, – there was not then the pretense of charging "service" in the account. Before [50] I had gone more than a couple of hundred yards on my way I heard some one furiously pursuing me and found that it was the 'boots' (the waiter had already taken care of himself) who then reminded me that there was still the chambermaid to be satisfied. By the time I reached Lake Ogwen I fell in with a guide who persuaded me to go up and see the Twll Dhu or Devil's Kitchen. It was my first visit to that strange wild place, which I afterwards got to know well. As we were descending on the Llanberis Valley side the guide declared that Snowdon was clearing up, and that it would be well worth crossing the pass, and ascending straight up the other side. This we did, but found, as most often happens, clouds and heavy rain for all the latter part of our way. We reached the summit, saw nothing, and then descended, pretty well soaked, to Llanberis. After my experience at Capel Curig I thought the smaller Inn of the two – the Dolbadarn Arms –, would be pleasanter, so I selected it. There was no one there to speak to, and I beguiled the rest of the evening by such literature as the Hotel afforded. This consisted only of a Methodist Missionary book or two. Next day I walked back through Pentir, to my Jolly Herring, where I stayed out the rest of my time before returning home.

It is a strange experience how, when one looks back on such a trip, one may feel convinced that it must have been extremely dull, and yet has to admit that at the time it was [51] thoroughly enjoyable. My father did not say a word to hinder me from going, though he must have shrewdly suspected what sort of a tour it would prove. Probably he thought that to be thus thrown for a time on my own resources would be a useful piece of worldly experience, and in this I think he may have been right. But it is always disappointing to recall lost opportunities, and I can't but think what such a trip might have done for me, if it had been taken in company with some genial companion who was mentally ahead of me.

About the beginning of July I returned to Cambridge. The Long Vacation was then a most important time of study, as almost all the private tutors gave their men two months, and, with Hopkins we had somewhat more. It was a very enjoyable time, as two or three of my principal friends, – especially Garrick and Whiting –, were up at the same time. It was then that I began to get really intimate with the former. We used to walk almost every afternoon round by Trumpington and Grantchester, generally indulging in a bathe in the then popular "Byron's Pool".[87] About the end of our time, viz. [*del*: early] //late// in [*del*: September] //August//, we wound up with a long walk to Ely, by the towing path: a pleasant excursion which I afterwards repeated several times in subsequent years. The view of the Cathedral growing bigger as one gradually approaches it, the extensive view, all round from the raised towing path, and the wild [52] loneliness of the district, are very impressive. One peculiarity in the ways of the bargees struck us, which now perhaps has been improved away by education. We noticed that there were no gates on the path, but only styles over which we had to climb. The puzzle was, How did the horses get over? This was solved at last when we met a barge, and observed that the men stopped the horses for a few seconds, so as to make the rope slack, then whipped them up into a canter and made them jump the style. The only explanation we could get at a farm house, where we stopped for some refreshment, was that the bargees were such a rough and careless set that if there were gates they could never be trusted to shut them.

In the autumn of this year the cholera was raging in England, especially in London.[88] I got the notion that I had something the matter with my stomach, and was in a panic at the thought of catching the disease. I do not believe that there was anything really the matter with me, but I remember being in a great fright. My father took me with him on his short holiday. We went first for a few days to Brighton (September 9–12) mainly on a visit to his cousin Henry Elliott. A little later we went

[87] [The public bathing-places at the time were on an island called Sheep's Green and Byron's Pool, some two-and-a-half miles from Cambridge. Byron's Pool, which owed its name to Lord Byron's fondness of swimming in that part of the river at Grantchester.]

[88] [The outbreak of cholera known as the Broad Street cholera outbreak, which occurred in 1854 near Broad Street (now Broadwick Street) in Soho, London, during the 1846–1860 worldwide cholera pandemic.]

to the North. Our first journey was to Raywell, to my uncle Joseph Sykes.[89] A great change had taken place since our former visit there. The family had nearly all grown up and dispersed, and my uncle had recently married [53] again, so the household was very small. From there we went to Riley and to Scarborough. The latter place brought back to my father early recollections of his Yorkshire life. I remember his stopping at one point where a path from the sea rose up the cliff, and telling me that it was just there that he first recognized, in my mother's look at him, that she would accept him. From thence we went across Yorkshire, in one very long day, to Lancaster, stopping on the way to see York Minster. That was on the 1st of October, a memorable day, as the (false) news had just arrived that Sebastopol had been taken by a sudden assault.[90] The bells were ringing joyfully over all the city, and for about one day England was what would now (1902) be called "Mafficking."[91] From Lancaster we went on the Lakes of Cumberland and Westmoreland, a district which I afterwards came to know so well, but which I then saw for the first time. I got no mountain walks that time: partly as being with my father, partly as supposing myself somewhat of an invalid. We spent the Sunday immediately after our arrival at the Lowood Inn. It was just 35 years since my father had been there when, as a young graduate reading for his fellowship, he was living with William Wilberforce, and giving some coaching to his son Robert. But every detail of the country that he had then seen seemed to be vividly present to him. We spent about a week in the Lakes, winding up with [54] Keswick, and thence to Penrith where we spent a night. On the 13th October my father went northwards to see a Sykes niece who was stopping at Moffat in Scotland, and I returned home to Highbury. The cholera had by then nearly disappeared from London, and with it my anxieties about my health had disappeared also.

At the usual time in October I went back to College. The result of the examination in May had brought me out as first scholar of the year. This secured me rooms in College, my scholarship being one of the Perse foundation and entitling me to rooms in the Perse Building.[92] This Building, pulled down in 1868, has been pretty

---

[89] [The brother of Venn's mother. The Sykes family, Evangelical business people living in Hull, was a provider of wives and husbands to Clapham for several generations.]

[90] [The Siege of Sevastopol by the French, Sardinians, Ottomans and British, taking place during the Crimean War, would last from October 1854 until September 1855.]

[91] [Venn's reference is to the Siege of Mafeking, the battle for the South African town of Mafeking during the Second Boer War from October 1899 to May 1900. The relief of the siege in May 1900 was experienced as a boost to British morale.]

[92] [In 1615, Stephen Perse (1548–1615), himself a Caius graduate, founded six fellowships and six scholarships, to be elected from the scholars. Perse also left a legacy for building purposes; the construction of the Perse building begun in 1617, and was completed the next year; until 1868 the building, which lay at the north side of what became known as Tree Court, was known as the Perse Building, with Perse scholars being entitled to rooms there. See John Venn, *Biographical History of Gonville and Caius College, 1349–1897. Volume III.* (Cambridge: Cambridge University Press, 1901), pp. 82–83, p. 218 and p. 231.]

fully described in the Bio. Hist. of the College.[93] I need only say now that the chambers, like all old college rooms, had provided no separate sleeping apartments. Accordingly, the bedroom was in every case an after-accommodation. The usual method had been to portion off a part of the sitting room with a lath and plaster partition. This of course gave a very small, and often rather dark, sleeping place, and also made the sitting room comparatively small. In some cases it was almost impossible to have a sponge-bath in the bedroom. In one or two cases the sitting room was left unaltered, and an attic was used for the bedroom, – this attic, of course, having once been the common living place, by day and night, of two //or more// students. This was how my room was [55] arranged. The attic was up two flights of stairs, had no fireplace, and only a small window to the north under the sloping roof.

It was in this term that I first began, I should say, to get any appreciable good out of College life. I had made a few more friends; and, being in College, instead of lonely lodgings, was much better able to see them frequently and at all times of day or night. Moreover I had nearly shaken myself free from the narrow and bigoted set who had been my principal acquaintances at first. During this term I began the study of German, by myself. I had at one time, – long before, in the governess days –, picked up a few words of the language, but had again, whilst at school, entirely forgotten it. I now resolved to know something of it, principally for the use it would prove when on a foreign journey. Accordingly I worked away at the Grammar, and at the translation of 'Undine'[94] for an hour or so every day, finding the time during the evening, after Hall, before setting to work on Mathematics.

About this time I left Hopkins, and, – running into an opposite extreme –, selected a Mr Jameson,[95] a fellow of the College, of whom I heard a good account from Garrick. My object was to get the individual attention and discussion which the famous tutor could not afford. It is perhaps rather characteristic that I did not ask any one's advice in [56] the matter. I obtained by this change the attention that I wanted, but the move was not a success altogether, as Mr Jameson, though fairly well up in his subject, had neither the power nor the originality that was desirable, and, after two terms with him, I made a final change and selected Mr Todhunter.[96]

Towards the end of this term there was something of an epidemic of Small Pox in the town, and three //members// of our College suffered from the disease. It was first introduced by a medical student, who had caught it from a patient at the Hospital, – for all fever cases were then treated at Addenbroke's [*sic*].[97] He, I

[93] [Venn's *Biographical History of Gonville and Caius College, 1349–1897*, published in three volumes between 1897 and 1901.]

[94] [Friedrich de la Motte Fouqué's *Undine*, a popular German fairytale novella first published in 1811.]

[95] [Francis James Jameson (1828–1869), Caius 1846; B.A. (6th Wrangler) 1850; Fellow of Caius, 1852–55; Fellow of St Catherine's, 1855–1862; vicar of Coton, Cambridge, 1862–1869.]

[96] [**Isaac Todhunter** (1820–1884).]

[97] [Addenbrooke's Hospital, the teaching hospital and research centre in Cambridge, founded in 1766.]

suppose, communicated it to a friend of mine, a freshman, Chalmers[98] by name, who afterwards went out to India during the Mutiny as an officer in the Army.[99] I used to go and sit with him every day. Of course I knew there was some risk; but the efficacy of a single vaccination in infancy, was then very widely held, and I did not regard my visiting him as at all a heroic act of friendship. Those who are familiar with recent outbreaks of infectious disease, – especially of Small Pox – will be amazed at the easy way where he had several close neighbours, but no kind of isolation or disinfection was resorted to, and I feel certain that not a man left the College nor even shifted his rooms, in consequence. I was not the only [57] friend who used to go and see Chalmers; but beyond the fact that most men avoided entering his room, there was no great difference shown between this case of small pox and a bad cold, so far as practice or regulations went. In due time I caught the disease myself. I well remember the night when it came on (December 12, 1854). It was at the very end of the term, – in fact I had been thinking of going down that same day. Somewhere near my staircase was a noisy supper party at the rooms of a man* who was celebrating his 21st birth-day.

// * John Ward, elder brother of Sir A.W. Ward, Master of Peterhouse.[100] //

All night, as I lay tossing about in a feverous state, I heard the songs and cheers and smashing of glass. When I came down in the morning I had a most frightful headache and felt sure what was the matter with me. My friend Ponton coming in I got him to send a telegram to my father to say I was not well and should not return that day. Telegrams were not common in those days, and I remember my friend seemed a little surprized that I should resort to such an extreme course. The same morning Dr Paget came to see me. Knowing something about College accommodation, he immediately enquired where I slept, and finding out where my bedroom was he at once ordered a bed to be put up in the sitting room. Later on in the day my father arrived, and stayed with me for a day or two. It was not a severe attack, but quite bad enough to be extremely unpleasant. As with Chalmers, no precautions whatever were taken, and no nurse engaged. If as happened once or twice, a friend looked in I told him what was [58] [1855] the matter and he departed promptly. The bedmaker used to come in early in the morning, at her usual hour, and then go on to her other lodgers. Before I was well the men had all left for the vacation: except Chalmers and I, who stopped on in College over Christmas Day. We were the only men up at that time, and we had our dinner together in my rooms.

I have often thought of my experiences in this matter, and wondered how many leading articles would have been written, to denounce such an outrage of sanitary

---

98 [Frederick William Marsh Chalmers (1836–1898), entered Caius in 1854 but did not graduate.]

99 [The 'Mutiny' refers to the Indian Rebellion (also known variously as the Great Rebellion, the Indian Insurrection or the First War of Independence) of 1857, a major but unsuccessful uprising in India against British rule in 1857–58.]

100 [John Ward (1835–1874), entered Caius in 1853 but did not graduate. Adolphus William Ward (1837–1924), Peterhouse 1855; B.A. (12th Classic) 1859; Master of Peterhouse, 1900–1924; Vice-Chancellor, 1901–1902; knighted in 1913.]

science, if a similar case occurred at the present day. But though Paget attended me, and Dr Drosier[101] was resident in College, no one seemed to think that anything was amiss in the arrangements carried out.

Shortly after Christmas, when danger of infection was supposed to be over, I went for a week or two to Hastings with my sister, and then returned home. During Christmas and most of January the weather was mild, but about the beginning of February commenced the long and bitterly severe season [*del*: long] //well// known as the "Crimean winter", from the terrible sufferings of our troops before Sebastopol.[102] I felt the full severity of it in my exposed garret. For a month or more, I should say, the jug was frozen from top to bottom and my sponge resembled a brick. The only possibility of washing was offered by having a bason in the sitting room. Times must have been altogether rougher in those days than now, for quilts or any kind of extra covering were quite unknown, but it never [59] occurred to me or any of my friends that the bedroom was in any way objectionable. The Cam of course was frozen over for several weeks, and every one who could get along upon a pair of skates made the journey to Ely. It was a strange scene all the way: a path swept or kept clear by the skaters, lines of straw laid down from the ice to the doors of a few public houses on the touring path, and men meeting and passing each other the whole stretch from Cambridge to Ely.

At Easter came the "Little Go,"[103] which was then compulsory upon every one at the same time, viz., at the middle point of their career, at the end of their fifth term. The failures in my year were rather numerous, probably owing in part to the frost and the unusual opportunities for skating afforded during the term.

Nothing particular happened, so far as I remember during the May term of 1855. I worked hard at Mathematics, and again came out first, though Garrick was something gaining on me. I was elected to a better scholarship, – the foundations were then very various, and all were known, as they should be, by the names of their founders –, and as this did not grant me a right to rooms I changed my quarters. My new rooms were on the second floor of the old red-brick house which then occupied the site of the present Tower, and commanded a fine view down King's Parade.[104] The bedroom here was adjoined to the sitting room, being, in fact, a portion of this separated off by a partition. [60] In June I again secured a short walking trip in Wales. This time I was more fortunate, as I had the advantage of a charming companion in J.E. Whiting; and, as we took a knapsack between us, we moved about with perfect freedom. It was my first experience of that delightful and independent

---

[101] [William Henry Drosier (c. 1810–1889), Caius 1835; B.A. (23rd Wrangler); M.D., 1847, Fellow of Caius, 1845–89. Drosier was for many years the Medical Lecturer at Caius.]

[102] [Around the end of November 1854, a winter storm ruined the Allies' camps and supply lines, starving large numbers of men and horses. On the "Great Storm" see Guy Arnold, *Historical Dictionary of the Crimean War* (Lanham, MD: Scarecrow, 2002), p. 73.]

[103] [Officially called the Previous Examination, the "Little Go" was held in the second year of residence in college, one year before the degree examination.]

[104] ["Old red-brick house": the old building at the south end of Caius, next to Senate House Passage.]

mode of travelling, now so generally neglected. We began again at the Jolly Herring, Penmaenmawr, where we had several days of very bad weather. This did not however prevent us from wandering over the hills to the south of our lodgings, as I remember we succeeded in hitting the little lake a few miles off, whose water drains out at Llanfair Fechan. We also paid visits to the Holyhead Breakwater, then in process of construction, and to the Menai Suspension Bridge. Our second week was spent in rambling from place to place. We went to Llanberis, and up Snowdon; to Beddgelert; to Festiniog and Dolgelly etc.; and thence returned, on one magnificent day, the delights of which are still present to my mind, by the Conway valley to Penmaenmaur.

The Long Vacation was spent as usual at Cambridge. By this time I had made another change, and was now a pupil of the well known Isaac Todhunter. He was an excellent and learned mathematician, and a very industrious and systematic teacher; but I still feel convinced that my studies, so far as concerns the intellectual and educational profit which is to be made out of mathematics, would have been far more advan-[61]tageous with a teacher of another stamp. He always had the Tripos prominently in view, and seemed to have no sympathy with those who wished to turn aside to study some detached point which interested them or to speculate about the logical and philosophical problems that arose in the way. I suppose he considered it his duty to prepare us, to the utmost, for our impending examination, for, as I afterwards came to know, he really possessed wide knowledge in other directions, and some speculative taste.

In August of this year my father came down for two or three days on the occasion of my 21st birthday, and took me with him to Bury. As far as Newmarket, I remember, we drove, which gave me my first sight of the Devil's Ditch, and of the great racing place. The main inducement for choosing Bury was the fact that his cousins, Edmund and Sarah Harvey,[105] were living there. The former I had not seen since childhood. He was a very eccentric man, passionately devoted to his violin: in fact all the family had inherited eccentricity, or worse, from their father. We spent the Sunday at Bury, and saw the ruins of the formerly great Abbey, and then I returned to Cambridge.

In the course of this Long Vacation my great friend, Garrick, broke down, owing to a violent attack of inflammation of the eyes, and had to leave Cambridge for a year. I not only lost his companionship but the help of his rivalry, and as there was no one else in the College near to me in mathematics [62] I got rather more slack than I had previously been.

In September my father took me with him on a trip to Switzerland. It was my first visit to that country, and indeed, except for Boulogne and one or two coast towns in France, my first foreign tour. My father thoroughly understood the art of foreign travel and made a point of seeing everything that was to be seen. We went by Calais to Brussels and Cologne, thence up the Rhine to Strassburg and Freiburg in Baden. There we took the diligence through the Black Forest to Schaffhausen. Then to

[105] [No further information available.]

Zurich and Lucerne and Geneva. From Geneva we took the diligence to Lyons [sic], as there was then no railway. The route lies across the Jura range, and must have been very beautiful, but the best part was passed at night. The journey occupied some 19 hours, and next day I was quite stiff and sore with sitting and bumping for so long. My father on the other hand, who had been early broken in to long coach journies [sic] seemed as fresh as usual the next morning. From Lyons my father had intended going to see Arles and Avignon, but the cholera was raging so badly there, – it was severe at Zurich, and some other places in Switzerland –, that he changed his plan, and we came home via Paris. At Lyons we met my cousin, Mrs Russell Gurney,[106] with her husband, whom she had married three years before, and returned together. We saw a fair amount of the sights of Paris during the few days we were there. It was just when the great re-building of the City under [63] Napoleon III[107] was beginning to be conspicuous, and as Paris was the first large town that I had seen which presented any pretence to street-architecture, it made a great impression on me.

I ought to have enjoyed the tour in Switzerland more than I did, for my father was an active and experienced sight-seer, keenly enjoyed what he saw, and had historical knowledge sufficient to estimate the various ancient towns at which we stopped. The fault was mine[1], or rather that of my age, so far as there was any fault. I was in the stage when Switzerland is mainly regarded as a climbing or scrambling place, and when the most splendid views cannot be enjoyed from a carriage or a hotel window but most be climbed for. I had only one mountain walk. When we were at Zurich, my father went on direct to Lucerne by himself, and sent me over the Righi with a party of friends whom he had met on the way, one of whom was a Cambridge man of my standing. We slept at the little Inn, as it then was, at the summit, and came down on the Lucerne side. He brought <u>my hat</u> with him. On this point it will seem to the present generation an almost incredible piece of etiquette when I say that I took my tall silk hat with me from England, and wore it everywhere until I came to the foot of the mountain, where my father took charge of it for me. I do not think that I was [*del*: at all] //very// singular in so doing.

[64] On this tour I made some progress in conversational German, and took every opportunity of speaking to such people as I met. But I soon became convinced that I had no real aptitude for languages. With far less labour then I bestowed upon the task, many people, – my friend C.H. Monro,[108] for instance –would have made far more progress. My memory for words was perhaps about the average, but my ear

[1] Letter from H. Venn to his brother John. "Oct. 13, 1855. Yesterday I arrived at home…Dear John seemed to enjoy everything in his quiet way; – but he did not quite wake up as I had hoped."

---

[106] [**Emelia Russell Gurney** (1823–1896).]

[107] [Napoleon III was the first president of France, from 1848 to 1851; he became the Emperor of the French (1852–1870) upon seizing power by force.]

[108] [**Charles Henry Monro** (1835–1908).]

was not good in catching a sound or in reproducing it. On a subsequent occasion, in 1857, I acquired a fair conversational level for ordinary matters of travelling interest, but I never reached the point of being able freely to discuss subjects of general interest with the natives whom I met.

At the usual time in October I returned to Cambridge. About now I consider I really began to get some of the good of College life. I still belonged nominally to the Evangelical party, – I shall say something presently as to the extent to which this adherence was merely nominal and traditional –, but I had made other friends outside their ranks, with whom my time was mostly spent. General reading was also a little more resorted to. As regards prose works, it is true, I can't recall anyone, either up to this time or for a year afterwards, as having been read in a way to be remembered, except Roger's "Eclipse of Faith".[109] This was recommended by J.E. Whiting, and I thought a good deal of it at the time. The easy confidence with which the author disposed of the infidels one after [65] another, – infidels, mostly, of whom I had never heard but through him –, impressed me, and seemed at the time conclusive. But it was in poetry that I made most advance. Here I was very backward for my age. My father's sympathies were rather with the formal and classical school of the 18th century, though he had a great admiration for Walter Scott (his poems, that is, not his novels) and to the Lays[110] of this latter he introduced me early. But I now came under the influence of Byron, Shelley, and Tennyson. My introduction to Byron I owed to Garrick, as he was passionately fond of Childe Harold[111] and others of his works, and knew a good deal of them by heart. For several terms [*del*: he] // Byron// completely dominated me. I bought some volumes of his works, read them constantly, and learnt much of them by heart: – I could still, with a little effort, recall hundreds of his lines. Then followed Tennyson, whose rule was more complete and lasted for very much longer. My introduction to him I owe to C.H. Monro.
I remember dropping into his room in the Caius Court one evening, and his reading 'Fatima'[112] to me. The concentrated passion there displayed, and the setting of the desert scenery in which it is conveyed, powerfully affected me. I bought his Poems,[113] and soon, and for a long time he //in turn again// completely dominated me. In this I was not singular, in fact the only wonder is that it was so long before I

[109] [Henry Rogers' *The Eclipse of Faith: Or a Visit to a Religious Sceptic*, first published anonymously ('F.B.') in 1852. Rogers (1806–1877) was an English man of letters and Christian apologist. *The Eclipse of Faith*, his best-known work, staged a dialogue in which 'Harrington', the sceptic, attacks various forms of rationalism defended by his friend, named Fellowes, showing that all objections to revealed religion are futile.]

[110] [Walter Scott's *The Lay of the Last Minstrel*, a narrative poem in six cantos portraying the 16th-century society at the Scottish Borders, first published in 1805.]

[111] [Byron's *Childe Harold's Pilgrimage*, a long narrative poem in four cantos, published between 1812 and 1818. The poem, describing the melancholic foreign travels of a young man, contributed to the cult of the wandering Byronic hero, whose example was not lost on Venn.]

[112] [*Fatima*, first published in 1832, is a cross-dressing poem, speaking from the woman's point of view about desire.]

[113] [*Poems* was a highly-esteemed two-volume collection, published in 1842, containing new poems and reworked older ones. It went through almost twenty editions in the 1840s–60s.]

came under his influence, for Tennyson ruled the [66] young men of my generation to an extent which probably no other poet has ever attained.[114] My introduction to Shelley I owe to an accident. He was not really popular in my generation. The first time I ever heard his name, so far as I can recall, was in a notice at the Union, on which some young orator announced his intention of asserting Shelley's superiority to every other British poet. So ignorant was I of English literature that at first I thought there must be some mistake. I literally did not know the name of any such poet. However I soon made his acquaintance, bought a volume of his poems, and for some terms was deeply affected by him.

Those who are interested in the growth of opinions are probably familiar with the way in which [*del*: a mind] //men// may be profoundly influenced, emotionally and aesthetically by some author who has yet no logical or intellectual hold whatever upon them. I read Byron and Shelley in my room, and 'spouted' them when alone, but they either conveyed no teaching to me, or this teaching ran in a channel entirely distinct. The scoffs of Byron, and the occasional defiant Atheism of Shelley, scarcely shocked me, and certainly never in the least disturbed my religious opinions.

It was to C.H. Monro that I owe most in the way of speculation. He was a brilliant classical scholar from Harrow, skeptical in mind but extremely, in fact, morbidly [67] conscientious, and prone to Socratic disputation with his friends. If one once got into argument with him one was generally in for it till far on at night. Though anything but a mathematician himself, he was fond of probing our mathematical assumptions and results, and he often gave one much to think over before any satisfactory answer could be found.

I have said that I had none of those mental or moral crises, or periods of sudden development which many young men experience. As a speculative question of personal interest I have sometimes tried to ascertain what I really did believe at that time; but it will be best to try to answer that question when I come to the commencement of my clerical life a year or two later. Broadly speaking, I had been started on my career with a stock of evangelical convictions, – or rather strong sympathies with these, owing to home training and profound reverence for my father's character and abilities, – and these I never consciously lost, or even doubted, till several years afterwards. I imagine that these convictions were losing their power somewhat all the time, but there was no other set of influences to take their place. The only preacher in Cambridge then with any spark of originality or power was Harvey Goodwin (afterwards Dean of Ely and Bishop of Carlisle.) His sermons, as then preached, would now be [68] thought moderate but commonsense High Church; but in those days there was

[114] [Venn belonged to the mid-Victorian generation for whom Alfred, Lord Tennyson (1809–1892) was the poet who "sings for all" (see letter **134**). One of Tennyson's main themes, which appealed to Venn on both a personal and academic level, was the wide range of conjunctions between science and religion. Tennyson himself was a Cambridge man: he had been educated at Trinity College, also joining the Cambridge Apostles. On Tennyson at Cambridge see, for instance, John Beer, "Tennyson, Coleridge and the Cambridge Apostles," in Philip Collins, ed. *Tennyson: Seven Essays* (London: The Macmillan Press, 1992), pp. 1–35.]

supposed to be a flavor of Heterodoxy or Broad Churchism about them,[115] and most of those like-minded with myself rather shunned his church. I remember going occasionally, and rather preferring him; but conscience (as I understood it) told me that Clayton was the right man, and so I mostly went to hear him. He was a simple hearted kindly man whose platitudes were largely redeemed by his extreme earnestness, sincerity, and absence of any display or push. Religious enquiry and consequent skepticism must already have begun to spread amongst the more thoughtful students, but I never heard of any one at Caius who was influenced by any new lights of this kind, except indeed one or two who had been under Maurice[116] at King's College in London. But, as it happened, these were not at all influential men in the College, nor in any degree personal friends of mine. Certainly when one compares the calm of this period (1854–7) in the more sheltered colleges, with the storms which swept their way everywhere [*del*: three] //a few// years later after the publication of 'Essays and Reviews'[117] in 1860, the contrast is startling enough. Nor did the other modern influences, – Reviews and Novels, – do the work of religious teaching or un-teaching, so far as we were concerned. The only distinctly unorthodox review of any fame was, I suppose, the Westminster;[118] and of that neither I nor any of my friends [69] had heard a word until the appearance in it of F. Harrison's famous notice of 'Essays and Reviews' in 1860.[119] The Saturday Review[120] came into existence about this time (i.e. 1856); I remember a friend and cousin at Trinity (Alf. Elliott)*

// * Charles Alfred Elliott, son of Henry Venn Elliott. Afterwards a very distinguished Indian official //

showing me one of the earliest numbers, with the strange introduction that it was to be a rival of the 'Times'[121] newspaper, and that one ought therefore to wish it well. In its early days it marked a somewhat new power in journalism, in respect of the number of its clever and speculative articles on social subjects; but there was little if anything in it which would not now rank as entirely orthodox. As to Novels, we

[115] [Heterodoxy was the position associated with liberal, anti-dogmatic Broad Churchmen, such as those involved in *Essays and Reviews* (1860), who strove for a National Church which should be inclusive rather than exclusive.]

[116] [Frederick Denison Maurice (1805–1872), generally known as **F.D. Maurice**.]

[117] [*Essays and Reviews*, a collection of seven articles by six clergymen and one layman of the Church of England, appeared in the autumn of 1860, a year after Charles Darwin's *On the Origin of Species*. Their advocacy of 'free inquiry', applying rational and disinterested textual analysis to Christian theology and its sources, became the trigger and target of a nation-wide controversy. See Joseph L. Altholz, *Anatomy of a Controversy: The Debate over* Essays and Reviews *1860–1864* (Aldershot: Scolar Press, 1994).]

[118] [The *Westminster Review* was a quarterly publication established in 1823 by Jeremy Bentham as the official organ of the Philosophical Radicals. It appeared from 1824 to 1914.]

[119] [Frederic Harrison, "Neo-Christianity," *Westminster Review* 18 (October 1860): 293–332. For Harrison's role in the controversy see, for instance, Martha S. Vogeler, "More Light on *Essays and Reviews*: The Role of Frederic Harrison," *Victorian Periodicals Review* 12 (1979): pp. 105–116.]

[120] [The *Saturday Review of Politics, Literature, Science, and Art* was a London weekly newspaper of a liberal Conservative stamp. The first issue appeared in November 1855.]

[121] [The *Times* had begun in 1785; its political allegiance being neither overtly pro-Tory or Whig.]

did not, I think, read them much in those days. The best proof of this is that I cannot now recall the name of any novel whatever as having come out during my undergraduate years, though Dickens, Thackeray, Kingsley, and others, must all have published some of their works during that period.

It was, I suspect, a time of considerable mental sluggishness in the University. No one of the resident fellows or clergy attempted to see anything of the undergraduates, if we except good Mr. Clayton's well-meant attempts in this way. These consisted of breakfast parties, – very ill-assorted, for he knew nothing whatever of the character and affinities of the men in his charge at College, – and coffee or tea parties after hall. The nucleus about which these parties was [70] gathered was in every case some 'deputation' which had come to Cambridge to preach or speak for some Evangelical Society. They were not cheerful or profitable occasions, and the evening ones in particular were apt to be avoided if possible, as it was about as easy to escape the public meeting which followed on the tea as to desert after taking a recruiting sergeant's shil-ling. The bulk of us were, as I have said, sincerely and unsuspectingly orthodox. There was, I remember, a great turmoil at the Union Society about the question whether that place should be open on Sunday. Most of the strongly religious men probably did not belong to that society, but nevertheless the proposal was utterly rejected. It was not, indeed, till a good many years later that it was opened at all on Sunday.

1856

I do not remember anything of note during this academical year. In February 1856 we had the experience, – much more common in former times than at present –, of a fire in College. About 9 o'clock at night; as I was sitting and reading Shelley, before setting to work on my mathematics, the bedmaker came running in to say that the Hall was on fire. I rushed out and found two or three men already on the spot. Tradition and common sense prompted the immediate use of our slop-pails, of which of course there was one in every set of rooms. The [71] pumps were set to work, – there was no water then laid on –, we formed a double line up the staircase to pass the buckets, and after an hour or two of hard work the fire was suppressed.

Between the May term and the Long Vacation I again went for a short walking trip to Wales. This time my companion was Gould, – whom I first went to visit at his home in Sheffield –, with who I passed a pleasant ten days or so among the mountains. We went to scarcely any but the already familiar places, but I could never see too much of them.

In those old days readers for a tripos enjoyed the advantage of the Long Vacation and the subsequent October term, for revision of what they had read.[122] We were kept pretty steadily at work, and I remember nothing particular occurring during this time. I used to play chess occasionally with two or three friends, – a rather fool-

[122] [The Tripos examination, open to students who were in their eighth term at least, took place in January.]

ish pursuit at the time, as that 'game' demands quite as severe application as any mathematical subject, and therefore instead of being a relaxation was rather in the way of further study, but study which led to no result. Amongst other things, eight of us took part in a sort of tournament. We subscribed 5/0 each, and the winner was to have a set of Staunton chessmen.[123] I had all the struggle, being left in to the last. The winner was G.F. //S.?// Mathews[124] afterwards an Estate agent, and a prominent citizen of Birmingham.

During this summer my father, with my brother and sister, had paid a visit to Switzerland. In the beginning of September [72] they had returned to Paris, where I went out to join them. After a few days there, my brother and I went off to Belgium, where we spent a week or so. It was my first visit to Ghent, Bruges, and other towns in the Low Countries.

The last term of ten spent almost entirely in preparation for one examination, can never be a very pleasant one. It was devoted by every student to gathering up the remains left unfinished, doing papers systematically, and making sure of the weak and uncertain points. I remember one incident, towards the end of the time, which will show that my hereditary religious sentiments, though mainly retained by sympathy and authority, were still predominant. I was having tea with a friend, – E.A. Hillyard,[125] afterwards known as an extreme Ritualist[126] –, and talking about our prospects, when he asked me if I had seen a sort of prophecy or estimate of the chances of the leading men. As soon as I left his rooms I went to the Union to see the Paper. I found the notice there, -- one in the style of the "tips" published before a horse-race, – wherein my name was mentioned, after ten or twelve others, with a rather contemptuous remark that it was only referred to in consequence of my claims being pressed by the Evangelicals.[127] I was much disturbed at this confident forecast, and returning to my rooms, I registered at once a vow that if I secured a good place I would devote myself to the life of a Missionary: – for, with my home [73] associations, no other suitable offering seemed naturally to occur to me. I deferred the step from time to time, but it was several years before I ceased to regard the resolution as really binding upon me.

---

[123] [The chess set, named after the English chess master Howard Staunton, used for competitions.]

[124] [George Spencer Mathews (1836–1904), Caius 1855; B.A. (7th Wrangler); land-agent who became prominent in municipal affairs in Birmingham.]

[125] [See note **75**.]

[126] ['Ritualism' is often used to describe those ceremonial developments in the Church of England that were in the Victorian era believed to be making its services approximate more closely to that of the Roman Catholic Church. The concern among many clergy and laity was that those Anglicans who supported this move also supported Roman Catholic theological beliefs that were inconsistent with those of the Church of England. See Nigel Yates, *Anglican Ritualism in Victorian Britain, 1830–1890* (Oxford: Oxford University Press, 1999), chapter 1.]

[127] [The Evangelicals probably betted that Venn would achieve high rank in the Tripos.]

1857

Our degree examination was then carried on in January, at the Senate House. My elders were somewhat anxious as to what should be done to keep young men warm whilst they were at work at their papers, as in the days of my father and uncle the Senate House was not heated, and there was often no little suffering endured in the depth of a severe winter. I came out 6th Wrangler, bracketed with one Skelton of Queens'.[128] I think this was about my right place, as things were. Probably if I had begun with Routh,[129] in my first year, I should have been some places higher. And if I had read with Frost, I should certainly have had a more intelligent and philosophical grasp of mathematics. The former was no loss, the latter, I think, was.

The evil side of the system on which the whole time and attention, during three and a half years of study, is thus dominated by one great examination, is displayed in the re-action and disgust which sometimes ensues when that examination is over.[130] I felt this myself very strongly, and [74] whereas, before coming to Cambridge and for my first year or two, I had a real enjoyment in the subject, I never made any study of it afterwards: and in fact felt almost a repulsion towards it. I sold nearly all my mathematical books almost directly it was no longer necessary to study them. I think this was a great mistake. I do not believe in the value of any branch of education as a mere mental discipline. What has been painfully gained ought not to be let drop. Even if its study is not carried on to any further point, it should at least be retained, and utilized in other directions. When I came afterwards to write on Probability and on Symbolic Logic I felt very strongly the loss of time and of power in having to recover, some years later, what ought never to have been let slip. The same applies to Political Economy, and of course to the whole range of Physical Science.

As I have said, I think my place fairly indicated my relative attainments at the time. Very often, in the list of wranglers, there is one man or perhaps more than one, of great mathematical power, and who afterwards distinguishes himself in science. In my year this was not the case: Perhaps Gorst,[131] of St. John's, – afterwards well known in Parliament //as Sir John Gorst// –, had the best ability of any of the first

[128] [Thomas Skelton (1834–1915), Queens' College, 1853; B.A. (6th Wrangler) 1857; Fellow, 1858–9; Senior Professor at Bishop's College, Calcutta, 1863–7.]

[129] [Edward Routh (1831–1907), Senior Wrangler in 1854, belonged to the same new generation of private tutors to which also belonged Frost, **Todhunter** and several others.]

[130] [Neither mental nor physical exhaustion was uncommon among students reading for the Tripos. See, in this regard, Andrew Warwick, *Masters of Theory. Cambridge and the Rise of Mathematical Physics* (Chicago & London: The University of Chicago Press, 2003), chapter 4.]

[131] [John Eldon Gorst (1835–1916), St John's College, 1853; B.A. (3rd Wrangler); Fellow, 1857–60; Honorary Fellow, 1890. Gorst afterwards was M.P. for Cambridge Borough (1866–68), Chatham (1875–92) and the University (1892–1906), Under-Secretary of State for India (1886–91), Deputy Chairman of Committees, House of Commons (1888–91) and Financial Secretary to the Treasury (1891–92.]

eight or ten wranglers.[132] He did other things besides read mathe-[75]matics, and came out third. As regards the rest of us, I can't but think that more intense application, or more suitable coaching, might have brought out any of the first ten at the head of the list.

Summing up the results of a University education, from personal experience, I should say that I learnt something, though not much, from lectures; a great deal from my own study supplemented by private coaching; but most of all, – so far as the clarifying of ideas, their mutual relation and dependence, and the vital and rational interest felt in what was being studied –, to the long discussions held with friends, [*del*: and] //in// tea-hours and walking-hours. Cambridge probably has never been a place where personal influence, on the part of the teachers has been very powerful. For myself, I must say that I never came into close contact with any one of my seniors who personal influence upon me, intellectual or moral, was at all appreciable.

During the following two terms I resided in College. I was duly elected to a fellowship[133] about Easter, and moved forthwith from the bachelors' table to that of the Fellows. I cannot say that the society of the High Table was of a very profitable or stimulating kind. The old days of coarseness and drinking had gone by, and, with it, the formerly not unfrequent eccentricity. The resident body were intelligent, and fairly well educated; but there was [76] no one amongst them of such mark as to take the lead in conversation, and no one whose special study, or interest in any special subject, would serve from time to time to direct the talk. Ferrers[134] was probably the ablest of the body, and the man whose reading, – or at least whose memory of what he read –, was most extensive. But I can't remember, on any occasion, a discussion of such interest as to have left the slightest recollection in my mind. On religious matters, the only strong feeling felt, or at any rate expressed, was that of aversion to Evangelical opinions rather than of attachment to any other opinions. Ferrers especially was a leader in this direction, and never lost the opportunity of a gibe at the obnoxious opinions, especially if Clayton was present. J. Lamb[135] was a good-natured chaffing fellow, who always kept the table lively, and, in later days he took a strong interest in College business and in College antiquities. A.H. Green[136] afterwards obtained distinction as a Geologist, but this, I think, was due to his leaving College almost immediately, and having to work professionally on the Geological Survey.

---

[132] [The Senior Wrangler in 1857 was Gerard Brown Finch of Queens' College; Thomas Savage of Pembroke was 2^{d} Wrangler. Neither of the two afterwards excelled as a mathematician.]

[133] [Venn was elected to a Junior Fellowship.]

[134] [**Norman Macleod Ferrers** (1829–1903).]

[135] [John Lamb (1825–1880), Caius 1844; B.A. (11th Wrangler) 1848; Fellow, 1849–77; Dean, 1852–60; Hulsean Lecturer, 1860. Lamb was vicar of St Edward's, Cambridge, 1858–66 and Rector of Blofield, Norfolk, 1876–80.]

[136] [Alexander Henry Green (1832–1896), Caius 1851; B.A. 6th Wrangler) 1855; Fellow, 1855–68; Honorary Fellow, 1892. Green worked on the Ordnance (Geological) Survey (1861–75) before becoming Professor of Geology at Leeds College (1875–88) and at Oxford (1888–96). He became President of the Geological Section of the British Association in 1890.]

Those were the days of Reform, and our resident body, as it happened, was almost entirely composed of reformers.[137] I sided, on the whole, with them, almost from the first. But it must be admitted that the reform was wholly a [77] derivative of the Political movement of those days and the generation preceding. Some of the leaders in the University, – such as Whewell, and Peacock (then Deal of Ely) –, were men of wide thought and research in educational matters.[138] But in our College, and perhaps in most others, advocacy of reform simply arose from dissatisfaction, (sometimes due to personal causes) with details of management. Not a single man, I believe, had the slightest knowledge, or the slightest wish to know, what was done at other Universities or what was the past history of our own University. Free competition was supposed to be the one panacea for every evil, and therefore, as examinations were the main means for carrying out this competition, the multiplication and improvement of examinations assumed enormous relative importance.

It was about this period that I first began anything which could be called literary study. I had one or two mathematical pupils, but not enough //of them// fortunately, to occupy more than a very little time in the day. I took lessons in German, which I could already read fairly well –, and in Italian, in prospect of a tour on the Continent in the summer. I also made my first acquaintance with the principles of Art in //the works of// that most stimulating writer, John Ruskin, whose "Seven Lamps of Architecture" I carefully read and made notes of.[139] I also began the Stones of Venice and the Modern Painters.[140] Having two friends in College whose bad [78] eye-sight prevented them from reading anything but their books for necessary study, I used to spent some time in the day in reading to them. These were Garrick and Knight.[141] I remember, in this way, going through my uncle James Stephen's Lectures on French

[137] [Cambridge reform was a long, complex and delicate process, starting in the 1830s but only really taking off in the 1850s, with the appointment of the first Royal Commission (1850–52). The so-called Graham Commission resulted in the 1856 Act, which produced the first major revision of University and College statues since the reign of Elizabeth I. See Searby, *A History of the University of Cambridge*, chapters 12–14.]

[138] [William Whewell (1794–1866) and George Peacock (1791–1858), both Fellows of Trinity College, Cambridge. Venn's reference is to the fact that Whewell and Peacock were both not only actively involved in university reform at Cambridge, but also wrote educational treatises, reflecting on the principles of English university education. Whewell and Peacock were both on the side of reform, but differed over what aspects of change they were prepared to support – Whewell being on the conservative side and Peacock tending more towards the progressive camp, at least in the context of mathematics teaching. See, for instance, Harvey Becher, "Radicals, Whigs and conservatives: the middle and lower classes in the analytical revolution at Cambridge in the age of aristocracy," *The British Journal for the History of Science* 28:4 (1995): pp. 405–426.]

[139] [John Ruskin (1819–1900) was the leading art critic of the Victorian era. He also wrote on many other subjects, ranging from geology and architecture to ornithology and education. Ruskin's *Seven Lamps of Architecture* was an essay, first published in 1849, on the principles of architecture, which played a considerable role in the architectural movement known as Victorian Gothic.]

[140] [*The Stones of Venice* (1851–53), a three-volume treatise on Venetian art and architecture; *Modern Painters* (1843–1860), a five-volume work in which Ruskin defended the idea that art should accurately document nature.]

[141] [Probably Charles Knight (1836–1923), Caius 1855; B.A. 1859; subsequently held curacies in France, spending the last years of his long life in Canterbury.]

History,[142] and making my first acquaintance with Charles Kingsley in Alton Locke.[143] With a view to knowing something about the principal scenes of the great European war I went through Alison's History,[144] reading carefully the accounts of the fighting in Switzerland. The passage of the Alps by Napoleon and by Marshal Macdonald, and the guerilla warfare under Hofer in the Tyrol I studied with especial interest, as I was looking forward to a tour I those districts in the following summer. Whatever complaints may be made against [*del*: that writer] //Alison as a historian//, he has, to my thinking, one great merit, viz., that he so groups the various campaigns that one can at least remember the main outlines.[145]

I may add that about this time I began, almost for the first time, to make a few friends in other Colleges: – having no school-friends, and belonging to no University club or societies, my acquaintance had been hitherto almost entirely confined to my own College. Amongst these new acquisitions were A. Blunt, of Pembroke, – afterwards a very close friend –, J. Barton, of Christs, T.L.N. Causton, of St. Johns, and a few others.[146] It was a great gain, and I only wish that such friends had been more numerous. [79]

Far my most enjoyable experience about this time was my first real sight of foreign scenery. As already stated, I had been to Switzerland, for a couple of weeks or so, two years before. But in 1857 I had better opportunities. My friend A. Ransome[147] was then a medical student in Paris. I agreed to join him there, and then go on with him to Switzerland. My father, with his accustomed liberality, gave me £50 for my expenses on the tour: – a junior fellowship in those days was not sufficient for support, for it began at a value of about £60, without rooms or commons. I ought to have enjoyed my previous tour, with my father, more than I did. I cannot account for the fact, but it does seem as if one's emotions in the presence of new and impressive scenery cannot find adequate scope when one is in charge of those who are looked up to as superiors; any more than a boy can fully enjoy a game of any kind with

---

[142] [James Stephen, *Lectures on the History of France*, 2 vols. (London, 1851).]

[143] [Charles Kingsley's novel *Alton Locke* appeared in 1850. Written in sympathy with Chartism, the working-class male suffrage movement for political reform in Britain, it introduced Thomas Carlyle as one of the personages.]

[144] [Archibald Alison's *History of Europe, from the commencement of the French Revolution*, first published in 1835.]

[145] [Venn's implicit point might be that Alison wrote his *History of Europe* largely as a warning about the dangers of democracy. On Alison see Michael Michie, *An Enlightenment Tory in Victorian Scotland. The Career of Sir Archibald Alison* (Montreal & Kingston: McGill-Queen's University Press, 1997.]

[146] [Arthur Blunt (1835–1871), Pembroke 1854; B.A. 1858; afterwards Lecturer at St Andrew's, Holborn, 1861–63, and curate of Bishop's Stortford. John Barton (1836–1908), Christ's 1855; B.A. 1859; founder of the Cambridge University Church Missionary Society; afterwards vicar of Holy Trinity, Cambridge, 1877–93, where he introduced changes in the services on the ground that refusal to modernize would alienate the younger generation from the church. Thomas Lilford Neill (or Neil) Causton (1836–1899), St John's 1855; B.A. 1859; afterwards curate of Christ Church, Croydon, 1860–66, and vicar of St Matthew's, Croydon, 1866–99.]

[147] [**Arthur Ransome** (1834–1922).]

grown up people. Whatever the cause, the fact was that I felt a degree and kind of enjoyment and excitement in wandering about the first French town at which I stopped that I can still vividly recall. The town was certainly a most favourable one for the purpose. It was Rouen; and Rouen unmodernized and unrestored. I wandered about the streets for hours in the sultry evening not so much admiring the special sights in the way of the great churches, as imbibing the myriad small impressions of picturesque novelty, conveyed in every sight and sound and smell. [80]

I joined Ransome next day at Paris, and spent some ten days with him, before starting for Switzerland. This is the only time that I have seen anything of Paris, for one other occasions I have at most spent a day or two in the city, when on the way to Switzerland or the Riviera. As a medical student [*del*: he] //Ransome// naturally lived in the "Quartier //Latin//",[148] occupying one of the better of the Hotels for students, (Hotel Fleurus) in the neighbourhood of the Luxembourg. We saw nothing of the loose or rowdy life of the time and place and people, but otherwise we conformed to the customs of such a house. They provided us with our caffée and rolls in the morning, but for all other meals we went to the Restaurants of the district. We saw a good deal of the place in our rambles by day, – chiefly the standard old sights –, but made no acquaintance with the theatres, and still less with the nightly amusements of the students. I do not think indeed that Ransome had any friends amongst what novelists regard as the typical students of that day. My principal recollection is of the time we spent in the large bathing establishments in the Seine. The weather was intensely hot, and the baths were treated then, – as they may be now –, as an afternoon lounge, where one took coffee, – furnished by neat young women –, in the intervals between our swims. It was not, of course, much of the really old Paris that could then be seen: the destruction of that had been already in progress for some five or six years: but there was still a [81] vast amount of antiquity preserves, and I am glad to have been able to carry away some recollections of a past which has now almost entirely vanished.[149]

On June [*blank*] we started for Switzerland, taking the night mail for Strasburg. There was nothing of any particularity about our tour, except indeed the fact that Ransome was completely <u>lame</u> the whole time. He had sprained his ancle just before my arrival; and, confidently expecting that it would soon heal, he would not defer or abandon our tour. The result was that after the first few days, till he returned home some two months later, he could never put his foot to the ground. We got a village carpenter somewhere in Switzerland to make a crutch, – my German [*del*: just] sufficed to explain what was wanted –, and hobbling on this in the towns and riding on a horse amongst the mountains, he contrived to see about as much as any one ever did or could under the circumstances. We commenced with a fortnight or three weeks in the Oberland, where I made my first mountain experience by the ascent of

[148] [The Quartier Latin, on the then unfashionable side of the Seine, was long the hive of almost all students of law and medicine in Paris.]

[149] [The years between 1853 and 1870 were the years of Georges-Eugène Haussmann's renovation of Paris, commissioned by Emperor Napoléon III. It included, among other things, the demolition of medieval neighbourhoods and the building of wide avenues and large parks.]

the Schilthorn. This was really something of a job in those days, for the present village of Mürren then consisted of only a few huts. So I had the whole pull up, from Lauterbrunnen to the summit, some 7000 feet of ascent //i.e. more of actual ascent than is commonly made in one day, in climbing Mt Blanc//. We then crossed the St. Gothard, in a return voiture, from Fluellyn to Bellinzona, sleeping at Andermatt. The latter part of the drive gave me my first glimpse of Italian scenery. I shall never forget the [82] glories of colour and form which were displayed, as one exquisite valley after another opened out to right and left. We recrossed the Alps by the Bernardino and Via Mala to Choire, I on foot and Ransome in the diligence,[150] and back again by the Maloya to Chiavenna. We spent a quiet Sunday at Silva Plana, at the Post house where the diligence had put us down. Here, where one is now within a short step of so many huge and fashionable hotels, we thought we had reached the world's end; we came home however with the report that there was yet another place further up amidst the mountains, called [*del*: Pontresina] //San Moritz//, to which Germans resorted as a Bath. We spent a week at Milan and Venice, then swarming with white-coated Austrian troops. From Italy we went to the Tyrol, by the Lake of Garda, stopping at Trent, Rotzen, Sterzing, Meran, and Insspruck [*sic*]. At Munich we made a halt of a week, to see the various sights. We went, amongst other things, to the Opera one evening. Strange to say this was actually my first sight of the inside of a theatre. At Munich Ransome left me, having come to the sensible conclusion that the only hope of curing his ancle was to return straight home and there give it a thorough rest. So we parted, as I had still a fair portion of the £50 in my pocket, and wanted to spend it in seeing more both of the mountains and the towns of South Germany.

I still have my letters, which were written home during [83] this tour, and need not therefore give details of the tour.[151] It was a time of almost uninterrupted enjoyment of the keenest kind, and has given me a very strong conviction that to take young people abroad before they are able thoroughly to appreciate foreign travel may very likely mar their future pleasure in it. Ransome was an excellent companion, and never flagged in his enjoyment, in spite of the drawbacks of his compulsory lameness. We had not the slightest accident or contretemps of any sort, – unless it be a sharp attack of diarrhea which kept me for a few days in our Hotel at Zurich: and we had splendid weather for the greater part of the time.

We travelled in the then popular way, taking nothing with us but our knapsacks. We had no books with us: being out in the open air from morning to night there was not a moment left for more than reading the local newspapers, and writing home, and not much for that. Three months of outdoor work may seem to modern youth a long time for one suit of clothes to last (these not being new at the start), and we must have presented a rather ragged spectacle at the end of the time. I remember that Ransome went to visit Liebig,[152] the great chemist at Munich, who was a personal

[150] [A type of stage coach.]

[151] [Venn's 'Letters of a Foreign Tour' are held at the Church Missionary Society Archive at the University of Birmingham.]

[152] [Justus von Liebig (1803–1873), German scientist who made important contributions to agricultural and biological chemistry and who became known (or notorious) in Britain in the early 1860s

friend of his father, and spent the evening with him. After scrutiny of our respective coats it was concluded that mine was slightly less dilapidated than his own, so he borrowed it for the occasion. We may have carried out travelling principles too far; but it is certain [85] that we enjoyed ourselves intensely, saw an immense deal, and saw it in a way to stamp it for life on the mind. Moreover we saw it very cheaply.

Almost every day, with its various incidents, is still vividly impressed upon my mind, and I could even now, after more than 45 years (March, 1903) recall our successive journies stage after stage. Perhaps the following three or four scenes stand out most prominently from the rest. One of these was the walk from Bellinzona to Tusis, over the Bardoino [sic] pass. It was all along the high road, and I was alone, as Ransome had arranged to follow by the diligence. Perhaps the effect was due to the steady change of scene from the sultry Lombard plains which I left at 5 in the morning, to the bleak summit of the pass which I reached towards sunset, displaying itself through a never ending variety of splendid scenery. I felt no trace of fatigue through the thirty or more miles of distance, and the exercise of walking on an excellent road seemed to me to leave the mind free to absorb the beauty of the scenery. Another of these recollections, or rather a group of them, centres in Venice. We were there in the beginning of August and the heat was tremendous. We did little else than lounge among some of the palaces in the morning, rest (and sleep) during the middle of the day, and spend the evening and half the night in gondolas or on the great Piazza. Doubtless there were foul odours and mosquitoes in abundance, but we did not smell the one or feel the other. The superb Austrian bands, – [86] it was the time of foreign domination,[153] – filled the air with their melody, and we simply rambled, and listened, and looked at the water and the sky and the artificial beauties of the place. Another and very different but charming experience was in the old towns of the Southern or Italian Tyrol. We came slowly back by the diligence from Riva to Trent and thence to Botzen. The diligence always stopped at the old fashioned inns, and the joy of travelling, so far as it is due to the novelty of scenery and distinctiveness of national characteristics, was here at its height. We spent altogether some three weeks in the Tyrol, and during that time, I think, I met but one Englishman and that only for a few minutes. We never //otherwise// even met a person who spoke English. One detail which much impressed me was an evening service in a large church, – Trent Cathedral if I remember right. It was a litany of some kind in which the entire congregation earnestly joined. The whole scene seemed to put one back three or four centuries: the national costume, the crowd of earnest worshippers, my feeling of being apparently the one mere spectator there; all this in that vast church, as the evening darkened into night, gave me an indescribable feeling of

as a harsh critic of Francis Bacon. Liebig had been in Munich since 1852, when he accepted an appointment from the King of Bavaria to the Ludwig Maximilian University. **Ransome**'s father was a prominent surgeon with his own practice in central Manchester. The origin of his connection to Liebig is unknown.]

[153] [This was the time of the Austrian Empire (1804–1867). Venice was then part of the Kingdom of Lombardy-Venetia, a constituent land of the Austrian Empire dissolved in 1866, when it was incorporated into the Kingdom of Italy (1861–1946).]

sympathy and awe. It must be difficult nowadays for a traveler in most part of Europe to obtain similar sensations, so fresh and so strong.

[87] At Munich, as I have said, Ransome parted from me and went straight back to England. I then went southwards for a few days, in order to see something of the Bavarian Mountains. I stopped at Benedictbeuern, not far from the district since rendered familiar by the Ober Ammergau Play.[154] There are some beautiful lakes close by, and I much enjoyed the scenery. What I remember best is the view from the summit of the Benedictwand, a mountain about [*blank*] feet high. The prospect consists of two totally distinct and contrasted halves: to the north an uninterrupted view of the vast plains of Bavaria: to the south nothing but a sea of mountains, range beyond range. From thence I returned slowly home by Nuremburg, Frankfort and the Rhine; ending with a week in Holland, by way of contrast with the mountains. The total cost of the three months' trip had been a few [*del*: shillings] //pounds// within the £50 allowed me.

I spent about one year more in Cambridge but did not find the society very congenial so far as the fellows' table was concerned. The old habits of [*del*: free] //loose// speech and strong drinking were of course quite passed by: but the new habits of speculative activity or of professional industry in tutorial work had not yet penetrated into our College. The residents were as a body, men of decent lives and moderate industry, fairly interested in the ordinary college routine; but there was no one of marked originality, intellectual or moral, (except [88] Ferrers, of whom more hereafter) and none of special geniality or personal influence of any kind.

I can't think that the exclusive study of mathematics for three whole years, – which I, following the routine, had practised, – was a good thing in itself. But I feel sure that the wisest course for one who had already gone so far would have been to try to preserve the best of it by confirming or commending the study of some other subject to which this might be a good introduction: e.g., Physical science, Inductive Logic, Probability, &c. Unfortunately for myself I knew nothing of such subjects as these, till some time later, and I had committed the blunder, in my reaction after the Tripos, of selling off at once, a few days after the examination, almost my whole stock of mathematical books. This was a serious mistake. I have always felt since that whatever has been carefully studied for several years should be retained as far as possible, and be adapted directly to further use, instead of being preserved only unconsciously as a habit or an improved faculty.

The two authors with whom I made acquaintance this year who had most effect upon my mind were Ruskin and Buckle,[155] – Mill seized me a year later. Coming fresh from Switzerland and Italy I read Ruskin with intense interest, and thus acquired the first glimmering of such slight acquaintance or love of art as I possessed. I do not suppose I had ever been [89] to an exhibition at the Academy and

---

[154] [The Oberammergau Passion Play is a passion play, staging Jesus' passion, played by the inhabitants of the village of Oberammergau Bavaria, since 1634.]

[155] [Henry Thomas Buckle (1821–1862), author of the unfinished, two-volume *History of Civilization in England*, published separately in 1857 and 1861. Sometimes called the father of scientific history, Buckle, a self-trained historian, was a major figure in the positivist movement in historical scholarship, applying the methods of natural science to history with an eye to discovering general laws of development.]

had only looked at a few foreign picture gallaries [*sic*] as a matter of travelling custom. But Ruskin's main influence, so far as I was concerned, was not in teaching one to rightly admire nature at second hand in pictures, but to provoke one to criticise and account for its beauty at first hand in itself. I am speaking of his two great works, the Stones of Venice and the Modern Painters: with his later productions, in order directions, I have never had more than very slight sympathy.

To Buckle I owe an immense debt. He was undoubtedly the author who first gave me an impulse in the direction in which my later studies have lain, and which, but for him, they might never have assumed. His faults, as well as his merits, made him highly suggestive to one in my then state of mind. His parade of almost universal learning was very imposing, and the paradoxical character (or at least paradoxical statement) of some of his views forced the attention and interest of every reader. The particular doctrine of his which first directed me towards Inductive Logic was that of the necessity of human actions as indicated by the statistical regularity of apparently the most casual and capricious acts.[156]

1858

The summer of 1858 I spent at Penmaenmaur with my father brother and sister, my aunt Jane and cousin Caroline Stephen. [90] I there renewed my acquaintance with that splendid stretch of scenery which lies to the south of Penmaenmaur, towards Snowdon. I then first realized the extreme loneliness of that district as soon as one gets off a few of the more beaten tracks. I remember starting early one morning and walking over Foel Fras and by the edge of the precipices over Llyn Dulyn till I got to Carnedd Llewellyn and Carnedd Davydd; from which I scrambled down (a very venturesome and even dangerous scramble it proved) into the road a little below the foot of Lake Ogwen and thence walked round home by the road through Bethesda. There is no other walk of any length in Wales to compare with this for loneliness and grandeur, especially whilst one is on the ridge which connects Carnedd Llewellyn and Carnedd Davydd. I did not see a human being from the time of leaving the village of Penmaenmaur till I got down into the high road: and scarcely a living thing besides the few sheep and a raven which accompanied me for a long while croaking overhead. The impressive solitude was added to in those days by there generally being a buzzard or two high up in the air, wheeling round with plaintive screams.

It was about this time, viz., the early autumn of 1858 that my 'preparation' for taking orders, – such as it was, – may be said to have begun. I suppose I was about as fit, neither more nor less, [*del*: than] //as// most of my contemporaries. The bishop's examination,[157] at any rate for fellows of colleges, who were treated separately in the Ely Diocese, was

[156] [This doctrine was defended in chapter 1 of the first volume of the *History of English Civilization*, which provided 'proofs of the regularity of human action'.]

[157] [At that time, formal requirements for clerical education were minimal. As a College fellow, Venn could be ordained on the title of his fellowship, which made the bishop's examination for him the 'merest farce'.]

the merest [91] farce. The only test deserving the name was the so-called 'Voluntary Theological' which was in fact not voluntary at all.[158] I went back to Cambridge[1] for the month of September to work up for this, and really I think, this month represented almost the only preparation for my profession which I underwent.[159] I had never attempted a sermon or address of any kind, written or spoken, to any class of persons: of course I knew not a word of Hebrew: knew nothing of the tenets or constitution of any of the dissenting bodies in contact with whom every Church of England clergyman has to work: and beyond the minimum of experience gained from having taught in the Jesus Lane Sunday School[160] during one long vacation I had experienced no sort of relation with the poor from a religious or any other point of view. But in all these respects, nevertheless, I was not, I think, behind my contemporaries. Where a whole system is in fault, one [92] must not too strongly blame individuals. But it is impossible not to feel that something more might have been expected on the part of Clayton. He was the sole tutor, with the good income, and the influence which that post conferred: he was himself a parish clergyman: and he was the current representative of Charles Simeon in //College// position and //religious// opinions. But, to the best of my memory, not one hint or suggestion of help did he ever give to any pupil of his who was expecting to take orders.

I was ordained deacon at Ely Cathedral in November 1858, and went at once to a curacy under Mr James Faithfull at Cheshunt, Herts. The post was an honorary one, so far as payment went. My father thought this would be best, – I suppose from

[1]I and one friend, H.W. Southey, had the College to ourselves, the whole place being then even more empty than it now is in September. I dined in solitary state at the Fellows' table, and Southey at the Bachelors'. In those days the game on the College estates was regularly shot by several of the fellows, but it was etiquette for them to send a portion at least of what they killed to the College table. As a result I feasted on partridges every day by myself; the head-waiter (afterwards well known as Mr Palmer the college butler) standing with superb dignity behind me. It was rather oppressive to a novice like myself. Southey and I worked hard all day, and went for a stroll in the evening. The grand comet[161] of that year was just then in its glory. I well remember the sight of it and even its reflection in the river by King's College.

[158] [After the bishop's examination, all students hoping to be ordained had to pass the Cambridge Voluntary Theological Examination, taken in the Michaelmas term of the fourth year and consisting of eight papers. Since the Bishops required all candidates for orders to pass this examination it was, therefore, a compulsory one.]

[159] [Venn's experience underlines the general impression that theological training given at Cambridge in the 1850s was not particularly intensive or challenging. Venn's judgment, however, was a slight exaggeration, as candidates for ordination also had to show that they had attended lectures delivered during one term by at least two of the three Theology professors; also, colleges offered a number of informal opportunities for religious education. At the same time, it is true that formal requirements for clerical education was minimal and that practical and pastoral training for young curates was limited.]

[160] [Founded in 1827 by students from Queens' College, the school, which in the 1850s was located in King Street, relied on volunteer workers like Venn to teach children through reading Christian works.]

[161] [Donati's Comet, a long-period comet first observed on 2 June 1858 by the Italian astronomer Giovanni Battista Donati.]

the belief that I should have more leisure on these terms, – and as it required him to help my slender fellowship stipend, I did not like to object. Here I staid just one year.

I have often tried since then, especially after abandoning the clerical status, to understand definitely what my actual state of belief was at the time. In a few words I should say my position was this. I had naturally a decidedly speculative and sceptical turn of mind; but I was timid, emotional, and profoundly influenced by the opinions and practice of those whom I much reverenced: above all by those [93] of my father. I was in fact like a body launched with great force in a direction against the stream, whose original impulse is powerful enough to keep it in the primitive direction for a considerable time. For four or five years, at any rate, I should have unhesitatingly called myself an 'Evangelical', and my preaching was distinctly of that character. My general belief in the truth and necessity of Revelation was of course entirely unshaken, – I held the customary conviction of the religious world at the time, that "Infidelity" was almost confined to the French and other continental nations, and to the wicked amongst our own countrymen, – but I had never seen religion presented in any other form either at first or second hand (i.e. amongst persons whom I had known, or in biography or history) except in the Evangelical guise. Accordingly it never entered into my head to question the truth of this particular creed.

My want of clerical preparation was vividly brought home to me as soon as I got to Cheshunt. I was to preach the next Sunday, – in the afternoon, as being the least public occasion, – but when the Saturday came I found I simply could not get a sermon ready I tried to fill a number of sheets of paper with remarks which could be spun out for 20 minutes if possible: but no platitudes which I could venture to put down into writing would come into my head. My only resource was to get [94] the other curate, – a man considerably my senior, – to change places with me and let me take the address to a set of rustics in a hamlet 2 miles off, Gough's Oak, which was a part of the parish. Being less afraid of these poor people I did manage to hold out for the required 15 minutes or so, but if there was any one of these sheep who had been accustomed to frequent dissenting chapels, or had, himself conducted a class or meeting of any kind, it must have crossed his mind that he and his shepherd had better have changed places. By the next Sunday I did manage to get ready, by beginning early on Monday and compiling something each day. But my father, with his customary devotion came down on the Saturday evening, through the cold and the densest of fogs, to assist and encourage me.

My year at Cheshunt was on the whole a very happy and prosperous one; and I got through more work, in reading, then in any other year of my life probably. There was no dining out, at least for the curate, there; and croquet or other clerical dissipations were unknown. I had also no friends there except the Faithfulls, but I never felt the life the least dull. Long walks [*del*: by day] //in the afternoon//, the parish was very extensive so that much of this was in the way of duty, – and steady reading in the //morning &// evening, filled my time. The vicar's family consisted of Mr and Mrs Faithfull and their three young children. The two former were charming persons, and the only [95] complaint I have to make against them is that they were too kindly and sensitively courteous to give me the constant hints and advice which I sorely needed.

Owing to the size of the parish, and the number of scattered hamlets in it there were a large number of cottage lectures, classes, etc.; at least during the winter months. In fact I think I had something of this kind almost every evening in the week except Saturday. It was a very good thing for me, for the practice of talking to a dozen or so of poor people when one is sitting on a chair amongst them, renders it almost impossible to talk quite such irrelevant platitudes as can be vented from the elevation of a pulpit. In the summer the work became much lighter: for all the parish, [*del*: except those] //with the exception of the small tradesmen & labourers// who lived in the strip along the high road, was purely agricultural in employment and character.

It was a time of considerable progress mentally, but scarcely spiritually, to me. I mean that I did not at all change my religious convictions here, but simply gained more theological knowledge. I worked hard at Hebrew, at the Greek Testament in the commentaries of Alford and Ellicott and others, and also at a few other theological works.[162] That admirable old Commentary, Bengel's Gnomon, (I wonder if the present day clergy ever appeal to it) was steadily employed.[163]

Intellectually however the great event in my life was [96] that I became acquainted at this time with the works of J.S. Mill. I then first heard, not of his works merely, but even of his name. I remember the occasion vividly. My friend C.H. Monro had come down to spend a night with me a few weeks after my arrival at Cheshunt, and had brought with him the then recently published volume on Liberty.[164] We began talking about the subject, and my only association with the name was that there had been a theological professor or preacher called Mill[165] at Cambridge some years before, and I wondered if it was the same man. Monro left the book behind him for me to read, and I found myself from the beginning to end in an entirely new world of thought and feeling. I was not shocked or in any way religiously disturbed, but simply interested and almost bewildered. Other books which I read at the same time were the following: – Mansel's Bampton Lectures,[166] first introduced to me by W. Berkley, who had heard some of them at Oxford. I was carried away by the rather elaborate and pompous rhetoric of the work, and almost overwhelmed by the display of learning, and the

---

[162] [Henry Alford (1810–1871), Dean of Canterbury, 1857–71, and theological and textual critic whose chief fame rested on his monumental edition of the *New Testament in Greek* (8 vols.), produced between 1841 and 1861. Charles Ellicott (1819–1905), English bishop, theologian and Professor of Divinity at King's College London.

[163] [Johann Albrecht Bengel's *Gnomon of the New Testament*, first published in 1742 as *Gnomon Novi Testamenti*.]

[164] [**John Stuart Mill**'s *On Liberty* (1859), the famous plea for the protection of individual liberty.]

[165] [William Hodge Mill (1792–1853), first principal of Bishop's College, Calcutta and later Regius Professor of Hebrew at Cambridge (1848–53).]

[166] [Henry Longueville Mansel (1820–1871), Oxford philosopher working in the tradition of Aristotle, Kant and Thomas Reid. His controversial Bampton Lectures for 1858, *The Limits of Religious Thought*, sought to protect divine revelation from human reason by arguing for the unknowability of God. In doing so, Mansel rejected the German tradition of absolute idealism that had emerged in Kant's wake.]

//summary// way in which he knocked over one heterodox German after another. It was the first glimpse I obtained of the Hamiltonian[167] doctrine that the human mind cannot grasp the Infinite, which Mansel seemed to push to the conclusion that we simply did not know what we were speaking of when we said that God was just or merciful. Mansel however made me acquainted with other books, of which one of the most useful [97] was M'Cosh's Method of the Divine Government.[168] M'Cosh was a follower of the old-fashioned philosophy of his countrymen, Reid and Stewart,[169] acute, sober, and well-read. His work was in some ways an amplification of Butler's analogy.[170] I read it again and again and am still surprised that it was so little known or studied by clergymen of the day. Butler himself was of course an old friend of mine. He was the only good author, in the Divinity way, put into my hands at school. I still have a small copy which I bought in order to carry in my pocket when travelling. It was to this time also that I date my first acquaintance with that remarkable book Boole's Laws of Thought.[171] This introduction was due to Todhunter, my old Cambridge tutor, who had a profound admiration for Boole. I read the book with care, and could follow the mere mechanism of his processes. But to rationalize them was a far other business, and I did not see my way to do this till after the third or fourth study of the book, in 1878–9.[172] In fact no one attempted to do this at the time. Jevons,[173] a professed logician, and whose characteristic methods were really a simplification of Boole, did not attempt this, and regarded Boole's formulae as the mathematicians used to regard "impossible quantities"; i.e. as absolutely uninterpretable processes which led to correct results. I think my Symbolic Logic was the first attempt at such an interpretation.[174]

---

[167] [Mansel had adopted William Hamilton's Kantian doctrine concerning the limitation of human knowledge, which said that man cannot know the absolute or infinite.]

[168] [James McCosh's *The Method of Divine Government: Physical and Moral*, which appeared in 1850, enjoyed wide-spread fame in religious circles as an update of William Paley's apologetic science in light of the kind of evolutionary thinking found in Robert Chambers's *Vestiges of Creation*.]

[169] [The Scottish Common Sense philosophers Thomas Reid (1710–1796) and Dugald Stewart (1753–1828).]

[170] [Venn's reference is to Joseph Butler (1692–1752), another representative of the tradition of natural theology and apologetic science, and his famous *Analogy of Religion* (1736).]

[171] [George Boole (1815–1864), self-taught mathematician, philosopher and logician who spent most of his short career as the first Professor of Mathematics at Queen's College, Cork in Ireland. He is mostly remembered today as the founder of the tradition of algebraic logic, through his book *The Laws of Thought* (1854).]

[172] [It was actually in 1876 that Venn published his 'Boole's logical system' in the newly-established journal *Mind*. See John Venn, "Boole's Logical System," *Mind* 1:4 (1876): pp.479–491.]

[173] [**William Stanley Jevons** (1835–1882).]

[174] [In brief, whereas Jevons sought to divest Boole's system of its mathematical dress, insisting on the primacy of logic over mathematics, Venn defended the view that logic, though not a branch of mathematics, as Boole had suggested, could be generalized on the basis of the analogy of its symbols with those of mathematics. The main goal of Venn's *Symbolic Logic* (1st ed., 1881; 2nd ed., 1894) was to show that all symbols and operations of Boole's calculus could be given a fully logical interpretation.]

It was a queer state of things altogether, and I am [98] inclined to doubt if there was another curate at the time, in the whole of England, who was exactly in my position: – preaching the Evangelical creed every Sunday, without the slightest insincerity or suspicion, and studying the works of Mill for hours every [added: week] day. Having begun with the Liberty I naturally went on to his other volumes. Here again I was indebted to Monro[175]. On a subsequent visit, – he walked over from Barnet where he lived – he brought the first volume of the Logic[176] with him. I turned over the pages, and at once lighted on subjects which deeply interested me. I borrowed the volume, but quickly bought the work for myself, and proceeded to study it, making (as I frequently did in those days) a careful abstract of it from beginning to end. I then went on to the Political Economy,[177] which I read through and analysed with equal care.

1859

In the early summer of this year I went for a short holiday in the English Lakes. It was not actually my first visit to that delightful district, for I had been there with my father in 1865, but it was the first occasion on which I was able to explore the wilder parts on foot. At that time I had a kind and hospitable aunt there, Ann,[178] second wife of my uncle Joseph Sykes –, who was living in the house at Grasmere called Allan Bank*: a beautiful house, and in an excellent position for exploring the country.

// * A fairly old house, standing in its own large grounds, on a hill to the N.W. of Grasmere Lake. My aunt rented it. At one time S.T. Coleridge had occupied it. //

The rest of the summer and autumn I spent at my curate's [99] post. I had moved from my winter lodgings near the church, and taken quarters in a cottage at Goff's Oak, two miles off, in a high part of the parish which looked over the Lee marshes and the hills towards Epping Forest. That part of Hertfordshire was then remarkably retired and beautiful, and I had time to explore it fairly well, for the country people were all at work out of doors; there were no lectures in summer; and my fellow-curate took most of the visiting in the town district along the high road. Two or three of my friends paid me an occasional visit. Berkley was one of these. I remember he brought down with him Tennyson's Idylls of the King,[179] then just published, which we discussed at night. The next day we spent in a long ramble of 16 miles or so to St Albans, and thence back again. The whole way was along beautiful lanes, by the side of woods and parks. I had not

---

[175] [**Charles Henry Monro** (1835–1908).]

[176] [**John Stuart Mill**'s *A System of Logic* was first published in 1843, and went through many editions in the following decades. A groundwork for Mill's own work in economic, ethical, social and political theory, it became a staple of English university education. See Antis Loizides, ed., *Mill's* A System of Logic. *Critical Appraisals* (London: Routledge, 2014).]

[177] [**Mill**'s *Principles of Political Economy*, first published in 1848, with many later editions.]

[178] [Ann Sykes (née Butler, of Ackworth) (1841–?).]

[179] [Tennyson's *Idylls of the King*, published between 1859 and 1885, is a cycle of twelve narrative poems which retells the legend of King Arthur.]

realized how pretty a county Hertfordshire is, or was. In my enforced leisure, – for the residents then took little or no notice of a curate –, I got through a large share of reading. I continued my Hebrew and other theological work; my study of Mill; made acquaintance (through Berkley) with Coleridge's Aids to Reflection,[180] which I read again and again, and indexed carefully. The collected editions of his works in 10 or 12 volumes was then being issued, and I was so struck by the first of these which fell into my hands that I went on through the whole set.[181] Besides the purity of his composition there was [100] much in his personality and experience which attracted me: – his residence in the Lake District, his vagrant pedestrian habits interspersed with nights of study, and the singular mixture of subtlety and humour which pervaded his various Essays.

About the beginning of November I quitted Cheshunt, spent a few days at Cambridge, was ordained priest at Ely, and then went to Hastings for the winter. Our household in Highbury was just then somewhat broken up. My brother was at College, in his second year; my sister had developed some serious weakness in the chest, and was spending the winter in lodgings at Hastings, with our cousin Mary (Sykes) Hope.[182] My father was mostly at Highbury, but generally joined us for the Saturday and Sunday.

I led the same life here as at Cheshunt. I acted as a temporary curate for a*

// * Some years later he assumed the title of a baronet, for which it was not supposed that he had the slightest claim. //

Mr Tilson Marsh[183] – a rather vain and foolish evangelical, who had the church under the cliff at St. Leonards (he was best known as the brother, – some people said the <u>sister</u> –, of the then well-known writer Miss Marsh, author of the Life of Hedlye Vicars and other such books.)[184] I divided my time between study and walks. My favourite ground was the fine mass of hill and cliff which lay about Fairlight, but I also made my first experience of Eastbourne, or rather of what is now the site of [101] Eastbourne.[185] I spent one long day in the early spring in walking from Hastings to Beachy Head. This could then be effected almost without passing a single inhabited house, except the Coast Guard Stations, and now and then a Martello tower occupied by an artilleryman. Such a deserted strip of coast is hard to

---

180 [Samuel Taylor Coleridge's (1772–1834) *Aids to Reflection*, published in 1825; an unsystematic collection of commentaries and aphorisms stressing the importance of religion as personal revelation.]

181 [Probably *The Complete Works of S.T. Coleridge*, 7 vols., edited by W.G.T. Shedd (New York: Harper & Brothers, 1853).]

182 [Probably Joseph Sykes's eldest daughter Mary Sykes, married to James Hope.]

183 [William Nathaniel Tilson Marsh (1815–?), of Oriel College, Oxford, assumed the title of baronet at some point in the 1860s, apparently without legal authority. Marsh was the son of William Marsh and Maria Chowne (born Tilson).]

184 [Catherine Marsh, *Sketch of the Life of Captain Hedley Vicars: Abridged from the Memoir*, published in 1855; a popular work chronicling the short life of a Christian soldier.]

185 [During the second part of the 1860s, Venn would sustain two households, one in Cambridge and one in Eastbourne, on the south coast, for his wife and son.]

find now along our southern and eastern coasts. Bexhill did not then exist, except as a little village perched on a hill a mile from the sea: seaside Pevensey comprised only some coast guard houses, and perhaps a cottage or two: and as to Eastbourne, anyone approaching it along the beach found only an old-fashioned hotel, a few lodging houses, and I think a little wooden jetty, by the present "Splash point". I strolled on to Beachy Head, where I found a solitary coastguard, and returned to Hastings by train.

During the mornings and evenings I was mostly at work indoors. Pearson on the Creed[186] represented Theology, and Locke's Essay,[187] and Bacon's Works[188] found me in Philosophy, all being carefully abstracted. The habit of thus analysing the books I read was an indirect good result of my mathematical studies, and I continued it occasionally for a number of years. The good of such a plan does not consist, I think, in the result, but in the process. The analysis itself may as well be soon thrown away, or at most revised on a second reading, but the attention required in making it is very useful.[102] 1860

A notoriously wet and cold year, and for me a rather idle or rather desultory one, much of it spent in the open air. Between March, when I left Hastings, and October, when I began at Mortlake I had no regular curacy, but occasionally helped a friend here or there. The spring was mostly spent at Highbury, with my father. The two books of this time which made the most effect, temporary or permanent, were probably Essays and Reviews, and the Origins of Species.[189] I read them both, but I cannot say that they produced the full effect that they should have produced. They certainly gave no shock to my religious convictions.

The first part of the summer, June, was mostly spent at the Lakes. I ascended Scawfell Pike, in the last week of June, with the unusual experience of finding a vast sheet of snow near the summit. I spent the inside of a week at William Ribson's, Wastdale Head. He and his brother, who had the only two Inns there, – if one may use the word Inn –, provided between them some 8 or 10 beds: a rather scanty provision sometimes when a fine day brought two or three more parties than usual over the mountains. The board and lodging were rough and scanty. Eggs and bacon were the provision for meal after meal. On the occasion in question, after some days of this diet I walked over Sty Head pass to Rossthwaite, to enjoy the unusual luxury of a mutton chop: on the strength of [103] which I walked back to Wastdale.

About the beginning of July I went on from Cumberland into Scotland, stopping for a few days in Edinburgh, and then making my way to Braemar, where I was to meet my father and sister, and the Stephens. From Perth I went on foot, by Dunkeld,

---

186 [John Pearson's influential *An Exposition of the Creed*, first published in 1659.]

187 [John Locke's *An Essay Concerning Human Understanding* (1690), inaugurated the tradition of British empiricism.]

188 [Either Basil Montagu's *The Works of Francis Bacon* (1825–44) or the Spedding-Ellis-Heath edition of *The Works of Francis Bacon* (1857–1859).]

189 [*Essays and Reviews* (London: John W. Parker and Son, 1860); Charles Darwin, *On The Origin of Species by Means of Natural Selection, or the Preservation of Favoured Races in the Struggle for Life* (London: John Murray, 1859).]

Blair Athol, and Glen Tilt. I was at about my best as far as walking was concerned, at this time, and made some expeditions which if nothing extraordinary were quite respectable. One was from Braemar to the top of Ben Mac Dhui, thence down the other side to the so-called Shelter-stone, and back again to Braemar. Another was on leaving Braemar, when I walked past the Wells of the Dee, just to the west of Ben Mac Dhui, and thence on the little Inn at Aviemore. Of course there was no Railway then, so I had to wait at the Inn till the coach came by which brought me into Inverness about 11 at night. They call the distance 40 miles, and I don't suppose it can be much less. What most struck me was the savage loneliness of this great district devoted to deer-stalking. I was alone, and from the time of quitting the highroad near Braemar till I reached the banks of the Spey at Aviemore I do not think I saw a human being. I suppose that we and those behind us must have been less anxious than our successors of the present day. I know if any one near to me had started by himself over such a wild and lonely district I should have wished for a speedy telegram to know of his safe arrival.

[104] The rest of the summer I spent at my uncle's at Hereford. He and his sister were away for some time after I arrived, so I had the house to myself. But his friends near by were as kind as ever, and I certainly had no dull time. My direct business there was to take his place at St Peter's, during his absence, in so far as such a substitution was not utterly ludicrous. What defects his congregation found in my doctrine I do not know, but they must have noticed a terrible contrast so far as voice and delivery were concerned.

In the autumn of this year I commenced my second spell of curateship, viz., that of Mortlake. I feel the same difficulty in describing, and even in myself understanding, what exactly was my mental position at this time. The fabric of evangelical belief, which I had, so to say, taken bodily over from my father, or at least adopted under his influence, was still standing apparently untouched. No specific doctrine in it was rejected; if it had been, I should soon have heard of it from Manley. And yet it was being actively attacked on every side by nearly all the books I was reading, and implicitly contradicted by much of my conversation with my friends. I feel certain that there was no insincerity on my part, but any of my friends who heard me preach, – J.R. Seeley[190] soon became one of these friends –, must have been somewhat puzzled to reconcile what they heard me say in the pulpit [*del*: and] //with// what they heard me say when walking in Richmond Park. As I have [105] remarked, there was no insincerity, in the sense of my saying in public what I did not believe, but I must admit that there was reserve, in the sense of not saying a great deal which I did believe.

Different mental constitutions must greatly vary in this respect. Some persons seem to be able to specify doctrine after doctrine which they once held, and to assign the exact date at which each was doubted or rejected. When I read Newman's Apologia[191] he

[190] [**John Robert Seeley** (1834–1895). Best known as the author of *Ecce Homo: A Survey in the Life and Work of Jesus Christ* (1865), controversial for its focus on Jesus's moral and political life, and *The Expansion of England* (1883).]

[191] [John Henry Newman's *Apologia Pro Vita Sua* (1864) described his departure from the Anglican church and conversion to Catholicism.]

seemed to me to afford a remarkable instance of a mind of this constitution. I suppose the fact is that I was in no sense a theologian. My creed was not founded on intellectual convictions. It had been accepted as a whole, and was only held together by my emotions, of which emotions one of the chief was reverence for my father.

I must say here a few words about my vicar, Mr. Manley.[192] He was not actually a hypocrite: in fact during my long experience of evangelicals at home I never met with any one who really deserved that name. They are common enough in fiction, and I have no doubt that they are not infrequent in actual life, but they did not come within my circle: very probably they fought shy of my father and my uncle. But I may say with truth that [*del*: he] Manley was a thorough humbug. He was a little over thirty when I first encountered him. He was a very presentable man in his appearance, //extremely neat in his dress// oily and affectionate in his manner, of very fair ability though without any pretence to [106] learning, could compose a decidedly good sermon when he chose to take the trouble, and had as good a delivery as I ever heard in a pulpit. He was incredibly indolent, as I soon found out, and was, – as I got to know some time afterwards –, decidedly addicted to secret drinking.*

// * He afterwards had a church at Tonbridge. My brother assures me, from information supplied by those who were well acquainted with him there, that his failing was not 'drink' in the current sense of the term. But that he was a victim to the habit of imbibing drugs of some kind or other seems certain. //

This last fact doubtless accounted for the shyness and frequent lassitude which his friends regretted. The full measure of his indolence I did not discover till he went away for his summer holiday. He then gave me a list of sick persons to be visited in his district, – the parish was divided into two parts, between us, for visiting purposes, neither of us interfering with the work of the other. The list [*added in pencil*: he gave me] was carefully drawn up, with [*added in pencil*: special] marks against the names of those who required visiting at frequent intervals. I went to them of course at once, and found to my surprize that most of them had not seen him for months.

Our mutual relations were certainly peculiar, for after the first few days I was never once asked to step inside his house during all the time, nearly twenty-two months, of my stay with him. We met on Sunday morning, when he preached and I read prayers. The afternoon service I took, whilst he stayed at home and these parts were reversed in the evening. A few words of oily speech, and an affectionate shake of the hand in the vestry, once a week, constituted almost our whole [107] intercourse from month to month. There were no parish meetings whatever, at least none to which I was asked. But for the fact that my father had taken a house in the parish, mainly in order to be near me, I should have left the place [*del*: before long] //very soon//. But I had no direct cause of quarrel with my vicar; after the first week or so he never made a remark on anything I said or did; and the fact of thus being left entirely to myself rather suited me than otherwise. My father, who very soon took

[192] [John Thomas Manley (1830–1901); Trinity 1848; B.A. 1852; curate of Clapham, Surrey, 1852–55; perpetual curate of Mortlake, Surrey, 1855–64; vicar of Tonbridge, 1864–1890.]

Manley's measure, probably held the same view, for he never recommended me to leave. The people also in the village were for the most part kindly and hospitable.

In most respects I went on here much as at Cheshunt. I read for hours every morning and evening, and walked for hours most afternoons. The visiting in the half-parish allotted to me did not amount to much, though I think I did my duty pretty fairly there; and the total absence of any kind of supplementary service, or of any of the innumerable parish meetings and committees which occupy so much time nowadays, left me abundant leisure. The actual state of things there was of course an abuse, but I doubt if it was more so in one direction than the present state of things, in some parishes, is in the opposite direction. I perhaps took too much time for study, but I have small doubt that the current curate takes far too little.

[108] I soon get to know the beautiful country which lies in every direction in Surrey. Richmond Park was an inexhaustible recreation ground for those who were fond of walking, and Kew and Hampton Court were also near by. As the summer came on, and friends sometimes came to stop at my lodgings for a night or two, I extended my range. Increased wealth, and suburban railways, had not yet dotted the country over with new hamlets and houses, and the waste area of furzy common and pine-covered hill, seemed to stretch endlessly to the south and west. Three or four of these excursions will never fade from my mind. One was with Ferrers, afterwards Master of the College, and at that time a keen University reformer and an interesting person generally. We pounded along through Kingston, Guildford, and along the Hogs Back, to Farnham, whence we returned by train. Ferrers had plenty of remarks, or at least answers to questions, but he had nothing of what could be called talk, and that long walk is not stamped at spots here and there, in my memory, with recollections of the details of our conversation. It is far otherwise with a ramble which I took with a very dear friend Arthur Blunt, – he died young, of consumption. We started through Richmond Park, crossed Epsom Downs, descended Mickleham Hill, – where we were simply startled by the strangeness and beauty of the juniper woods –, thence over Box Hill, – I never before saw box trees except singly or in small clumps –, and home by [109] train from Reigate. As I recall that long and happy summer day one detail after another of our discussions starts to my memory, and I could localize the spots almost to a few yards: how, near Merton, we got on to the doctrine of Atonement, and whether it was just that one should suffer in place of another, which I supported and he rejected: how, as we wound round the descent from the down, by Mickleham, church, we got onto Tennyson's Love and Duty,[193] – we were both devoted admirers of Tennyson: and how again, along a little lane behind Surbiton which made a short cut into the Leatherhead Road, I happened to quote Carlyle's grim metaphor of our generation 'being like a group of apes grinning and chattering over a dying fire in the forest, which they know not how to light

[193] [Tennyson's *Love and Duty*, which appeared for the first time in the 1842 edition of *Poems*. A blank verse poem, it tells the story of two lovers, parted by a sense of duty, contemplating a lonely future with the hope of their love's fulfillment in the future.]

again',[194] and how profoundly this struck him. On another occasion a foolish fancy seized my brother and myself to walk out and see the sun rise on Leith Hill; well known as the loftiest summit in the South east of England. It was on the 1st. or 2nd. of May, 1862. I rather thinking that what prompted me to this exploit was an account that De Quincey had written of the beauty and calm of a long night's walk that he had once taken after an interview with Coleridge.[195] We started about 10, went through Richmond Park and Leatherhead, //where we// ate some food, which we had brought with us, on a door step, [*del*: in Dorking] and finally reached the summit a little before the sun was timed to rise, viz., 4.34. Till early morning nothing could be more delicious. [110] The night was wonderfully calm and mild, and the very hedge rows the whole way seemed almost vibrating with the incessant song of the nightingales. The actual sunrise, as visitors to mountain summits generally find, was a failure. Nothing could be seen but the clammy mist which wrapped the hill top. So we returned to Dorking, and got some breakfast at the first public house which opened its shutters.

// *Twilight was appearing as we passed through Dorking. We feel asleep in the heathen at the top of Leith Hill, & woke to find the sun has risen. //

The society at Mortlake was exceedingly respectable, but rather dull; or at least seemed so to me. The only eminent resident//s// [*del*: was] //were// Professor Owen,[196] the great naturalist, who lived in a small cottage in Richmond Park, which was granted to him by the Queen //& Sir Henry Taylor,[197] author of Philip Van Arteveld, &c.// But the resident of real importance, so far as I was concerned, was J.R. Seeley. He was then just leaving Cambridge, to take a mastership at the City of London School, and I got to know him slightly at once, for his family lived in the parish. I took walks with him, and was much struck by his wide knowledge and his brilliance of intellect, but it was not till some years later that I began to be intimate with him. It was hardly likely indeed that one whose mind was then teeming with the ideas which took the world by storm in 1865, in the pages of <u>Ecce Homo</u>, should have been really intimate with the curate of Mr. Manley, and all that he stood for in the way of religion. [111]

---

[194] [This passage is found in Book III, Chapter 3 ('Gospel of Dilettantism') of Thomas Carlyle's *Past and Present* (1843): "The men of the Dead Sea, when we next went to visit them, were all 'changed into Apes;' sitting on the trees there, grinning now in the most *un*affected manner; gibbering and chattering very genuine nonsense; finding the whole Universe now a most indisputable Humbug! The Universe has *become* a Humbug to these Apes who thought it one."]

[195] [Venn's reference is probably to Thomas de Quincey's recollection of his conversation with Coleridge in his *Recollections of the Lakes and the Lake Poets* (1839).]

[196] [**Richard Owen** (1804–1892). For some thirty years, Owen lived in the cottage in Richmond Park, near the Sheen gate.]

[197] [Henry Taylor (1800–1886), dramatist, poet and men of letters, whose circle included James Stephen, John Stuart Mill and Thomas Carlyle. Among his plays were *Philip van Artevelde* (1834), which brought him fame and recognition.]

1861.

This year was rendered memorable in the Church by the clamour against the Essays and Reviews. They had appeared a year or two before, but it took a little time for them to become known generally among the clergy, – in fact I think it was a famous review by F. Harrison in the Westminster //Review//, which first made both that volume and his own name generally known. The prosecution before the ecclesiastical courts of two of the writers in the heretical book also aroused much attention.[198] My cousin, James Stephen, was counsel for one of the authors attacked.[199]

The popular outcry about this volume ought to have done more than it did to bring home to me the real inconsistency of my position. If I had had any even occasional intercourse with Manley there would have been an outbreak between us, for the mere abstention from denunciation of the hated views would have been enough, in highly orthodox circles, to attract attention. As it was I freely expressed my approval, of much in the general spirit of the volume, and even of some of the details, amongst my own private friends. Amongst other forms of protest adopted at the time was a declaration of belief in verbal inspiration (it practically came to this) which was signed by a large majority of the clergy.[200] How it was that Manley did not send it round to me with a request to sign it [112] or depart, I do not know, for much pressure was put on curates and other dependent clergy in many places. I belonged to a local clerical society, of which Coplestone,[201] rector of Barnes, was the head, – Manley would have nothing to do with it –, which was fairly liberal for such a gathering. The declaration was produced, and I presume signed by every one but myself, but no pressure whatever was put on me to add my name.

Two of the books which had perhaps the most effect upon me about this time, – an effect largely due to the fact that the subject matter of them was entirely new to me –, were Maine's Ancient Law and John Austin's Jurisprudence.[202] My

[198] [These two writers were Rowland Williams and Henry Bristow Wilson. For context, see Victor Shea and William Whitla, *Essays and Reviews: The 1860 Text and Its Reading* (Charlottesville & London: University Press of Virginia, 2000), chapter 4.]

[199] [In 1861, Venn's cousin **James Fitzjames Stephen** (1829–1894) was counsel for Williams and Wilson in their heresy trial.]

[200] [It was in 1862 that the London Evangelical clergyman, Alexander McGaul, had pointed to the Bible as being, not merely containing, the Word of God as the crux of the matter. This phraseology (rather than 'Verbal Inspiration') was inserted as one of the two central points of the Declaration signed by a majority of the clergy in 1864. See J.L. Altholz, "The mind of Victorian orthodoxy: Anglican responses to 'Essays and Reviews', 1860–1864,' in Gerald Parsons, ed., *Religion in Victorian Britain. IV: Interpretations* (Manchester & New York: Manchester University Press, 1988), pp. 28–40.]

[201] [Reginald Edward Coplestone (1811–1878) had become rector of Barnes in 1839.]

[202] [Henry James Sumner Maine's *Ancient Law* (1861) aimed 'to indicate some of the earliest ideas of mankind, as they are reflected in ancient law, and to point out the relation of those ideas to modern thought'. It went through twelve editions in Maine's lifetime. John Austin's *The Province of Jurisprudence Determined* was first published in 1832 but only became influential after his death, when a second edition was published, in 1861. The book contributed much to the develop-

introduction to the first I owed to Monro (the occasion being a walk across the common which then lay to the south of Surbiton on the way to Leatherhead.) I felt at first rather bewildered by the breadth and novelty (novelty, so far as I was concerned) of the speculations set before me: e.g., that our familiar convictions as to the Family and the Nation, Individualism, rights of property and bequest, &c., &c., were just those only of the modern Western world: all this being driven home by a style which was impressive by its dignity and simplicity. Austin came to me through Albert Dicey,[203] it being of course a professional subject to him. Here again was a new conception to me: that of an abstract theory of Rules [113] of Action of which Ethics and Law should be sub-divisions. It opened a door to me for a good deal of future speculation: the germs which I thus picked up developed into one of the later chapters of my Empirical Logic many years afterwards.[204]

I also made acquaintance with one or two theological books of a kind which is familiar enough to modern readers, but which was utterly new to one with my bringing up; books, that is, which discussed Bible History and religious doctrines from the point of view of the cultivated layman. Two of these were Strachey's Hebrew Polity,[205] and a series of pamphlets, by T. Hughes and his friends and associates, called Tracts for Priests and People.[206] They advised us, – to use a phrase[1] once so denounced by opponents of the Essays and Reviews –, to "read the Bible as we should read any other book".[207]

As to lighter literature, Tennyson was the one poet who continued to dominate me. I read and re-read him until I knew hundreds, or rather thousands, of lines by

[1] Almost the first reference I heard to Essays and Reviews was the enquiry by a lady of the congregation if I knew of the shocking book just out which advised people to read the Bible like any other book. The phrase occurred as the summing up of Jowett's mildly suggestive contribution to that volume, and was at once seized on, as a convenient weapon, by opponents.)

---

ment of legal positivism, arguing that legal systems should be studied analytically and not, as in the case of natural law approaches, with reference to morality.]

[203] [**Albert Venn Dicey** (1835–1922).]

[204] [The last chapter of Venn's *Principles of Empirical or Inductive Logic* (1889) dealt with 'Speculation and Action; or the Logical and Scientific view of the world as modified by our practical tendencies'.]

[205] [Edward Strachey's *Hebrew Politics in the Times of Sargon and Sennacherib; an Inquiry into the Meaning of the Prophecies of Isaiah* (1853). Sir Edward Strachey, 3rd Baronet (1812–1901) was a man of letters, landlord of the Somerset estates of his uncle, Sir Henry Strachey, 2nd Baronet, and a disciple of F.D. Maurice.]

[206] *Tracts for Priests and People* (8 vols., 1861) contained contributions from various writers, including F.D. Maurice. 'T. Hughes and his friends and associates' refers to Thomas Hughes (1822–1896), a member of the Christian socialists led by F.D. Maurice, and the Working Men's College in London of which he was one of the founders.]

[207] [Jowett's famous phrase, summarizing the method of recent German biblical criticism. Jowett and all other Essayists set out to read the Bible like any other book, but in the confident belief that in doing so they would find that it was *not* like any other book – i.e. that its spiritual value would not be weakened but strengthened.]

heart: we used sometimes in those days to beguile the way, on long walks, by reciting our favourite pieces. Such a taste as this [114] was not likely to be dulled by contact with Seeley, who when at Cambridge was popularly supposed to have been able to restore by memory the whole of In Memoriam, had that poem been lost.[1] In respect of novel reading I was still very slack, though some of the masterpieces of the most popular writers were coming out in the early sixties. Of Charlotte Brontë I had not heard, till I happened by accident on Mrs. Gaskell's life of her,[208] when I procured and read Jane Eyre, Shirley and Villette. About this time Albert Dicey forced one or two of George Elliott's novels, then just published, upon my attention. But I doubt if I had then read any of either Dickens' or Thackeray's novels.

This avoidance of novel-reading did not spring from sheer lack of taste in fiction: there was somewhat of the survival of the Puritan notion that such reading was frivolous and unnutritive. I felt as if it was eating cake every day for breakfast. When in after years I came to find that not a few of the very ablest and most learned men I knew were almost omniverous [*sic*] in their capacity for swallowing and digesting current novels, I recognized how foolish my earliest sentiments had been.

[115] I the summer of this year, 1861, I had a pleasant ramble with my brother in Brittany. It was an introduction to still rather primitive district, for the railway did not then extend further than Rennes; beyond that we had to go by diligence, or trust to our own legs. We took as usual nothing but our knapsacks with us. We entered the province by Mont S. Michel, thence followed the north coast to Brest and returned by Quimper and Rennes to St. Malo. We had a delightful ramble amongst such old-world, and then little-visited, towns as Morlaix, Goingamp, Quimper, &c. Nothing that I have seen in Devonshire was so much like a rabbit-warren. We tried making our way across country, but soon had to give it up and keep to the high-roads, for the lanes were intricate beyond description; mere footpaths, or trenches, zig-zagging about the woods and moors and orchards. And hardly any peasant whom we met could speak a word of French. I got there to understand the significance of the struggle in La Vendee between the Republican soldiers and the royalist peasants.[209] Hugo's Quatre Vingt Treize, and still more Balzac's Chouans,[210] cannot be fully understood by those who have not seen the country. From Brest I walked out to Pointe St. Mathieu, one of the most westerly headlands of France, – some 32 miles there and back, – but the day was too calm for me to see the big blue rollers from

[1] This learning of favourite pieces of poetry was an early habit of mine, and as my taste varied my collection of pieces became rather miscellaneous. At one time Milton had been among my favourites. I remember when at Meran, one wet and weary Sunday, telling my travelling companion that I would send him to sleep: whereupon I inflicted upon him the whole of the first book of Paradise Lost.

[208] [Elizabeth Cleghorn Gaskell, *The Life of Charlotte Brontë*, 2 vols. (London, 1857).]

[209] [The War in the Vendée (1793) was an uprising, headed by the Catholic and Royal Army, against the Jacobin government in Paris taking place during the French Revolution (1789–1799).]

[210] [Victor Hugo's *Quatrevingt-Treize* (1874) and Honoré de Balzac's *Les Chouans* (1829) are both set in the region of Brittany, with the former focusing on counter-revolutionary revolts and the latter on post-war uprisings.]

the Atlantic for which I had hoped. We crossed the Channel, on our return, via St. Malo and Newhaven; [116] the route taken by a little screw steamer, devoted mainly to eggs, but which found room for four or five passengers.

In the course of this year I made my first attempt at anything in the way of speculative writing. Beyond letters and sermons I had never hitherto ventured upon anything in the way of an essay, either private or for publication. The point I now tried to work out was one that had struck me at an early date when reading Mill's discussion, in the latter part of his Logic, of the abstract possibility of a science of Sociology, and in particular of predicting the course of human actions.[211] I tried to show that a vital factor had been left out of consideration here, viz., the influence of the forecast itself on the action of the agents concerned. My production was fairly acute and original, though terribly rough in diction and construction, and my cousin Fitzjames Stephen, to whom I submitted the MS, was well pleased with it, and by a letter to Mr. J.A. Froude, the then editor of Fraser's Magazine, procured its admission into that Journal.[212] I wish I had appealed on other occasions to my cousin, for though his manner was outwardly brusque and even rough, he was kind hearted, and would I am sure have given me a helping hand. He was keenly interested in similar questions himself, and was thoroughly familiar with most of the editors and writers concerned with the best periodical literature.[213] [117]

1862

Early in this year I came to the conclusion that I could no longer remain with Manley. I was gradually coming to see what a humbug the man was, and moreover my own gradual alienation from his creed was being forced home upon me, partly by my own reading and reflection, but probably still more by the indirect influence of several of my friends. I could not talk for hours together with such men as J.R. Seeley, W. Berkley, A. Blunt, and my cousin Albert Dicey, without recognizing that what I said to them was not the same as what I said in the pulpit. I do not mean that there was a contradiction here, at least not a direct one. But it is one thing to utter conventional platitudes to a congregation,[1] most of whom if they

[1] I hardly ever preached except to the afternoon congregation, which in that wealthy suburban district consisted almost entirely of schoolboys and schoolgirls, children, and women servants. I used to reckon that if they had been given their choice 9 out of 10 of those present would have walked straight out of the Church.

---

[211] [Book VI of **John Stuart Mill**'s *System of Logic*, entitled 'On the Logic of the Moral Sciences', was devoted to the establishment of a "science of mankind". It contained chapters dealing with the relation between human actions and causality (Chapter 2: 'Of Liberty and Necessity') and with the possibility of a science of social phenomena (Chapter 6: 'General Considerations on the Social Science'). Mill admitted that predicting human behavior is nearly impossible – e.g. there are too many variables involved – but he still believed that approximate generalizations made certain predictions possible.]

[212] [Venn's paper, entitled 'Science of History', was published in *Fraser's Magazine* in October 1862. See letter **15**.]

[213] [Like Venn, **Fitzjames Stephen** was profoundly influenced by **Mill**'s *System of Logic*, especially its last book, where Mill postulated the possibility of a science of human nature analogous to the physical sciences. He published several papers on the topic. See James A. Colaiaco, *James Fitzjames Stephen and the Crisis of Victorian Thought* (London: Palgrave Macmillan, 1983), chapter 4.]

attend at all, take everything for granted, and another thing to be pulled up for any loose word or argument. I thus came gradually to see that much of what I said in public was only held by tradition or from respect for those I reverenced; not from rational conviction. An opening occurred about this time which gave me a convenient opportunity of leaving Mortlake. At the previous general College meeting, Dec. 1861, I had proposed a scheme for [118] making the office of Catechist more permanent. The office, as I have explained in the College History, was an ancient one, but for a good many years the duties had been extremely light.[214] There was a prize awarded annually for 'Moral Philosophy', and the catechist gave in the course of the May term, some half-dozen lectures in supposed preparation for it. This showed an improvement over the 18th century practice, when almost every lectureship was an absolute sinecure; but, as the catechist was changed every year, and confined his lectures to such elementary books as the little compendiums of D. Stewart and Abercrombie there was still much to be desired. The stipend though small, (about £50) was sufficient to justify a good deal more work than was actually performed, so it was proposed to make the office permanent and to assign to the catechist the duty of teaching Logic and Political Economy to such of the Indian Civil Service[215] candidates as might be in residence: – the Moral Science Tripos was scarcely taken into account at the time. The proposal was mainly mine, but was supported by Ferrers, – no one else took the slightest interest in the subject –, but at the time I had not the least intention of returning to Cambridge myself. At the Easter meeting the seniors accepted the scheme, and the post was offered to me. I could not come at once, so a substitute was appointed on the old arrangement and I undertook to begin work in the following October. I intimated to Mr. Manley my proposal to leave him [119] in three months, a proposal which he received with expressions of regret, uttered with his usual oily civility, though I suspect he must have been almost as glad to hear it as I was to

[214] [At Caius, the office of Catechist has been filled up since 1608, the tenure being annual, and the holder constantly changed. During the eighteenth century, as Venn explained, 'this office, like most others in college, seems to have been filled, – so far as the appointment and stipend were concerned, – by one fellow after another, as a mere matter of routine. The first sign of reawakening interest is in 1838. In a college order of that year [...] it is decreed, "that the Catechist give at least eight lectures in Moral Philosophy to the junior sophs, in the Easter Term, and examine them in the subject of his lectures in the May Examination." In 1840 still further encouragement was given to the subject in order that "one Wortley exhibition (£20 for one year) should be given to the best answerer in the Examination in the Moral Philosophy." Things continued on this footing, a fresh appointment being made every year, until 1862; when the recent establishment of the Moral Sciences Tripos, and the arrival of students who were preparing for the Indian Civil Service, called for some alteration. The tenure was in consequence made permanent, and it was decided that lectures should be given on Logic and Political Economy; but the title and the stipend remained unchanged'. John Venn, *Biographical History of Gonville and Caius College. Vol III* (Cambridge: Cambridge University Press, 1901), on pp. 238–249.]

[215] [The Indian Civil Service (ICS) was the elite higher civil service of the British Empire in British India, in the period between 1858 and 1947. After the competitive examination for entry to the ICS, candidates studied one or two years in the United Kingdom, either at Oxford, London, Dublin or Cambridge.]

make it. I began at once somewhat more steady work on the subjects and books which I should have to lecture on. I now first hit upon the diagrammatical device[1] of representing propositions by inclusive and exclusive circles. Of course this device was not new then, but it was so obviously representative of the way in which any one, who approached the subject from the mathematical side, would attempt to visualize propositions, that it was forced upon me almost at once.[216]

In July I resigned the curacy, and so ended finally what could be called regular clerical work; for though I did help Lamb at St. Edward's, during a few months some years later, there was but little of the professional clerical life involved: I took my departure quietly; delivered no farewell sermon, and received no testimonial. Such gifts were not so frequent then as now, and most of the people interested in church affairs were too much in sympathy with the vicar to pay attention to his unsatisfactory curate. Moreover I had made no attempt to make myself popular, and certainly not to ingratiate with any opposing party in the parish.

[120] Of course my whole connection with Manley and Mortlake was a mistake both on my father's part and my own. I can only say that if I had felt that a matter of conscience was at stake, I should never have gone there, or should have soon quitted the place when I found what it involved. But I have explained my peculiar position at the outset of my career: my rather hazy acceptance of the evangelical creed, – the only creed that had ever been set before me –, buttressed up as it was by a profound reverence for my father and for other great and good men – and every eminently good person whose intimate life I had been familiar with, held the same doctrines. I continued to call myself an evangelical, and to suppose that I really was one, and it took a long time to part company with old traditions. First went some of the accretions to the creed, like Sabbatarianism: then some of the outworks like Verbal

[1] The special, and almost entirely new device, of shading the segments of the circles in order to represent the possibilities excluded by the propositions, did not occur to me till much later.

[216] [Venn would first introduce his diagrams in an 1880 paper: John Venn, "On the Diagrammatic and Mechanical Representation of Propositions and Reasonings," *The London, Edinburgh, and Dublin Philosophical Magazine and Journal of Science* 10:59 (1880): pp. 1–18. He described and used them more extensively in his *Symbolic Logic* of 1881. There he explains that it was his dissatisfaction with Euler's diagrammatic scheme that led him to invent his own diagrams. Since Venn uses circles, as Euler did, Venn diagrams have been said to be a misnomer. But Venn made very different use of them; instead of representing directly the actual relation of the classes, Venn first draws a primary diagram representing all possible sub-classes obtained by intersecting the terms involved in the argument. He then adds distinctive marks to represent propositions according to the emptiness or occupation of the sub-classes. This two-step representation sets Venn's scheme apart from that of Euler. For a discussion see, for instance, Amirouche Moktefi and Sun-Joo Shin, "A history of logic diagrams," in Dov. Gabbay, Francis Jeffry Pelletier, and John Woods, eds., *Handbook of the History of Logic. Volume 11. Logic: A History of Its Central Concepts* (Amsterdam: North-Holland, 2012), pp. 611–682; and Lukas M. Verburgt, *John Venn: A Life in Logic* (forthcoming at The University of Chicago Press), chapter 10.]

Inspiration[1] and the truth of the Mosaic Narrative.[217] The more central doctrines, like the total depravity of man, were for some time nominally accepted, but I gradually came to see that my studies of Mill and Austin and Bentham were evacuating these doctrines of all significance.[218]

What would have been the result in my case if I had been thrown under the influence, as curate, of some modest Broad-[121]church vicar whose character and abilities I could have really reverenced, I cannot say. I was not naturally more skeptical than hundreds, or perhaps thousands, who find no difficulty in adhering conscientiously to the Church. I was skeptical rather than inquisitive in purely theological [*del*: matters] //doctrines//, about which indeed I felt no very keen interest, though I think I may call myself very reverent and even religious in temperament. I should probably have turned aside from speculative doctrines, and given whatever emotions I possessed full play in the direction of those important moral social and practical questions which can find abundant scope for more active and powerful natures than mine. This was the lot of some of my friends, of Arthur Blunt, [*del*: with] // curate to// his cousin; the rector of St. Andrew's Holborn, and William Berkley [*del*: with] //curate to// Mr. Rogers at Bishopsgate. Free expression of one's opinions, is, I feel sure, one of the best ways of forming and stiffening them. This free expression I never had; partly owing to the potential coercive restraint of my vicar, and partly owing to the actual restraint which my reverence for my father imposed upon me.

Between leaving Mortlake and commencing work at Cambridge I had a pleasant tour in Switzerland, – with knapsack, and on foot, – in company with T. Dalton,[219] afterwards [*del*: my brother-in-law] //Mathematical Master of Eton//, and Arthur Blunt. On my return I joined my father and party at Llanberis, where they were staying with the Stephens. I secured several good rambles with Leslie Stephen about Snowdon, which we managed to ascend from the Crib Coch [122] side, – not much of a feat for him who had come back to England after the first successful ascent of the Schreckhorn. We also tried to ascend the Tryfaen from the Glyder Vawr, but he

[1] I remember an eloquent passage in one of Manley's sermons in which he enumerated several of what he held to be fundamental doctrines, concluding thus, – "...and the full inspiration of God's Blessed Word; on all which subjects to err is fatal, and to doubt is to die".

[217] [Sabbatarianism; the doctrine which says that the Sabbath should be observed in accordance with the Fourth Commandment, which forbids work on the Sabbath because it is a holy day. Verbal Inspiration: the doctrine which says that every word of the Bible is divinely inspired and was thus the authority of faith. Mosaic narrative; the narrative of the creation, including the notion that the earth is only about six thousand years old.]

[218] [All these thinkers belonged to what was then often referred to as the Utilitarian Positivists, who attacked Christian morality as an ethics of self-denial and passive obedience rather than individual self-fulfillment and activity.]

[219] [Thomas Dalton (1838–1894), Caius 1857; B.A. (21st Wrangler) 1861; Assistant Master at Eton, and subsequently Senior Mathematical Master and House Master, 1861– 94; author of several mathematical textbooks.]

was fairly beaten here, along the route we tried: which was rather amusing considering his recent mountain triumphs.*

// *He granted his disgust at being thus baffled, with the remark, 'I wish I had Melchior here' – Melchior Anderegg was a very famous Swiss guide. //

I returned to Cambridge rather early in October, as it was the year of the British Association meeting there. There was nothing very exciting at the meeting that I remember. It was in our Hall that Huxley made his violent attack on Owen whom he plainly charged with unfairness if not with direct tampering with the evidence before them.[220] But the celebrity of the encounter the year before at Oxford when Huxley had let himself loose on the Bishop, made such an encounter as this seem by comparison flat.

From this time onward I need not mention the books which were formative of my mind. My study in one direction became more continuous, and the books I had to read were too numerous for any one in particular to have an appreciable effect. In fact the new influencing elements were rather to be found in personal intercourse than in the study of print.

I soon got to work at lecturing, and in place of the half dozen or so of hours during one term in the year, which had been the traditional allowance given by the catechist, I lectured three times a week every term. I soon found that the Indian Civil Service students, for whose wants the lectureship had been reconstituted, were far too few to occupy [123] me, and that the only prospect of becoming useful was to throw myself into the Moral Science Tripos work.[1]

After a few terms I had a very fair class. Two or three of the men were of marked ability, but the bulk of them belonged to a class since unfortunately lost to the Tripos. They were generally men of fair attainments and industry, but whose previous training put it out of the question that they could do much in Classics or Mathematics, and who had no taste for Natural Science; these being then the only other portals to an honour degree. Most of them contemplated taking orders, and often showed a keen interest in ethical questions. I have very agreeable recollections

[1] It had existed as a 'Tripos' nominally, i.e., as an examination, for 10 or 12 years, but it was not until 1860 that it was put on a serious footing as a means of obtaining a degree.[221]

[220] [The clash between the advocate of Darwinism Thomas Henry Huxley (1825–95) and **Richard Owen** and his supporters, known as the hippocampus controversy, took place at the annual British Association meetings of 1861 and 1862. Owen had identified several anatomical features of the human brain, and in particular the hippocampus minor, as 'peculiar' to humanity's special mental powers, allowing them to rise above animality, representing the image of God. Huxley denied that there existed an anatomical chasm between animals and humans; arguing, on the contrary, that there is an animal-human proximity, supporting the theory that mankind and the higher apes are related by descent. See Nicolaas A. Rupke, *Richard Owen: Biology Without Darwin* (Chicago & London: The University of Chicago Press, 2009 [1994]), chapters 5–6.]

[221] [The Moral Sciences Tripos has been established in 1851, in addition to the Mathematical and Classical Triposes already in existence. For the history of the Moral Sciences Tripos see David Palfrey, "The Moral Sciences Tripos at Cambridge University, 1848–1860," unpublished PhD-Dissertation, University of Cambridge, 2002.]

of the work of teaching such students. It need hardly be said that successive 'improvements' of the Tripos have changed this, and turned the curriculum into a far more professional one; advanced, absorbing, and highly specialized.[222] It now possesses all the Cambridge merits (and defects) of such a course. Another cause of (to my thinking) serious loss to the Tripos, lies in the institution of several other such degree examinations; which so to say, sunk deeper wells into the same stratum and drew off most of our own supply. The Theological Tripos at once diverted some of our most agreeable and industrious students, [124] and the Historical Tripos still further thinned their numbers.[223] It may seem paradoxical if I add to this that the teachers were themselves somewhat of amateurs and diligent learners, for they had of course graduated in the old course of Classics or Mathematics. I am certain that we did not know anything like so much as the average lectures now does, but we had a naïf interest in subjects which were opening out before us. I cannot but suspect that something of what our pupils lost by our lack of learning they must have found replaced by our sumpathetic [sic] interest in their own struggles. At the time of my appointment there was only one other Moral Science lecturer in Cambridge, Joseph Mayor,[224] who had just been similarly appointed. A few years afterwards H. Sidgwick, and A. Marshall were added to the number.[225]

I soon found two or three men in the College prepared to read for the Tripos, and arranged my teaching for their wants. For some years the numbers increased, until at one time (about 1867) I was able to say that there were more men in the College reading for the Moral Science, than for the Classical Tripos. From 1867, for two or three years, my lectures were thrown open to the University, and my classes in Logic and Political Economy became rather large, up to eighteen or so if I remember right.[226] Not a few of these were from Trinity, and of course they ceased to come as soon as Sidgwick was appointed lecturer in his own College.[227] [125] In addition to those who attended lectures I generally used to have in those days a few private pupils. Some of these were men of great ability. It is with a feeling approaching to

[222] [For instance, one of the first innovations was to separate the subjects for examination into two groups, the one embracing Moral Philosophy, Mental Philosophy and Logic, the other History and Political Philosophy, Political Economy and Jurisprudence.]

[223] [The Theological Tripos was established in 1871, the History Tripos in 1873. During the decades that followed, the number of Triposes offered expanded further. For instance, 1903 saw the creation of the Economics and Political Sciences Tripos.]

[224] [**Joseph Bickersteth Mayor** (1828–1916).]

[225] [**Henry Sidgwick** (1838–1900). Alfred Marshall (1842–1924), St John's, 1861; B.A. (2nd Wrangler) 1865; Fellow, 1865–77 and 1885–1908; College Lecturer in Moral Science, 1868–77; Professor of Political Economy at Cambridge, 1884–1908; President of the Economic Section of the British Association, 1890, and one of the founders of the Royal Economic Society.]

[226] [Venn prided himself on the claim of having been the first to introduce the inter-collegiate system into Cambridge in his own department of the Moral Sciences around this time. See John Venn, "College life and ways sixty years ago," in John Venn, ed., *Early Collegiate Life* (Cambridge: W. Heffer and Sons Ltd., 1913), pp. 253–282, on p. 256.]

[227] [**Henry Sidgwick** was appointed Lecturer in Moral Sciences at Trinity College, Cambridge in 1868 and Praelector of Moral and Political Philosophy in 1875.]

shame, as if the relations even then might almost have been inverted, that I recall the fact that A.J. Balfour[228] the present prime minister, was one of these. He struck me at once as a man of great acuteness, but I certainly did not sufficiently realize his extraordinary philosophical ability. What he could do was somewhat marred by his languid and almost indolent manner: but his penetrating remarks prevented one from ignoring that one was dealing with no common man in matters of speculation. Other men of mark whom I had the privilege of teaching or trying to teach, were the present Sir Charles Dilke, Lord E. Fitzmaurice, A.T. Lyttelton (Bishop of Southampton), Professor Maitland, Dr. W. Cunningham, and a number of others.[229] Whatever may have been the sentiments of the students in the Moral Science Tripos of those days, I can answer for it that the teachers enjoyed a time that was both pleasurable and highly profitable. The numbers were of course small,[1] but the proportion of men who rose to distinction afterwards in Church or State, or as professors and teachers, was remarkably large. I may add that when, in after years, [126] the teachers became more numerous, they formed a very homogeneous body, in constant and friendly intercourse with each other. One excellent consequence resulted from this. Of the 'tyranny of examinations' we had no experience whatever. We had the entire management of the 'Faculty', if one may call it so.

[1] It will however startle some Cambridge men of later date to record that I was once asked by Sir George (then Dr.) Humphry, how it was that we succeeded in attracting more men to the Moral, than they did to the Natural Science Tripos. This was about 1866, before the advent of Michael Foster and the rise of the Physiological School.[230]

---

[228] [Arthur James Balfour (1848–1930), Trinity, 1866; B.A., 1870; Prime Minister of the United Kingdom, 1902–1905; Chancellor of the University, 1919–1930; President of the British Academy, 1924–29. Both conservative statesman and philosopher, Balfour published numerous papers in *Mind* and delivered the Gifford Lectures at the University of Glasgow, which formed the basis for his book *Theism and Humanism* (1915).]

[229] [Charles Wentworth Dilke (1843–1911), Trinity Hall 1862; B.A. (Law Tripos, 1st Class) 1865; President of the Union Society, 1864 and 1866; succeeded his father (Sir Charles Wentworth, Trinity Hall 1827) as 2nd baronet, 1869. A Liberal and Radical politician who supported labour and feminist movements, Dilke represented the radical wing of the Liberal Party. His aspirations to high political office ended in 1885 after a notorious divorce case. Arthur Temple Lyttelton (1852–1903), Trinity, 1870; B.A. (Moral Sciences Tripos, 1st Class) 1873; Hulsean Lecturer, 1891; Archdeacon of Winchester, 1900–1903. Frederick William Maitland (1850–1906), Trinity 1868; Moral Sciences Tripos, 1st Class, 1872; B.A. 1873; Downing Professor of the Laws of England at Cambridge, 1888–1906. William Cunningham (1849–1919), Caius, 1869, migrated to Trinity, 1872; Moral Sciences Tripos, 1st Class, 1872; B.A. 1873; Fellow of Trinity, 1891; Hon. Fellow of Caius, 1895; Chaplain at Trinity, 1880–91; University Lecturer in History, 1883–91; Hulsean Lecturer, 1885; Professor of Economic Science at King's College, London, 1891–97; Lecturer in Economic History at Harvard, U.S.A., 1899; Archdeacon of Ely, 1907.]

[230] [Michael Foster (1836–1907), educated at University College, London; Hon. M.A., 1871; Praelector in Physiology at Trinity, 1870; Fellow, 1871; Professor of Physiology, 1883–1903; Professor of Practical Physiology at University College, London, 1869; Fullerian Professor at the Royal Institution, 1869; President of the British Association, 1899; M.P. for London University, 1900–1906. It was chiefly owing to Foster that the Physiological Society was founded in 1875. On Forster's work and influence at Cambridge see Gerald L. Geison, *Michael Foster and the Cambridge School of Physiology. The Scientific Enterprise in Late Victorian Society* (Princeton: Princeton University Press, 1978).]

We could practically arrange the curriculum as we pleased; and we were the examiners, or we appointed them. Of course such a state of things is impossible were the students are trained in different institutions, – as in the great state examination Board called the London University, or where the teaching is strictly collegiate, as in Classics and Mathematics in Cambridge. But when, as in the case of the Moral Sciences at the time in question, every candidate for an examination had been taught by the same group of lecturers, examinations sank to their right level, being simply regarded as a means of testing the students in what they had learned; the test being practically applied by those lecturers themselves. Having since been during four years myself an examiner at London University, I feel the difference keenly. Of course the schedule of subjects, and to a certain extent the list of books, was assigned, but I can say with confidence that I was never in the slightest way constrained by any thought whether what I delivered would ultimately assume the form of a question in a paper, nor did any pupil ever show that this consideration affected his interest in [127] what he heard. I sincerely believe that the Tripos of those days was, as a literary and speculative course, an admirable subject of study for those concerned with it.

One of the experiences, to which I look back with most interest, in those days, was a small gathering of a few friends at Professor I. [*sic*] Grote's,[231] at Trumpington. It has been called the 'Grote Club', but this is far too ambitious a name for such a small and friendly meeting.[232] It was simply an Essay club, composed of a few Cambridge men who were interested in philosophical speculations, and had not then been long in existence. Small clubs and other societies for special purposes, have enormously multiplied of late,[233] but they were rather scarce then, and I personally felt the benefit of this opportunity the more, because as it happened I had never had the advantage of joining any such gathering before. J. Mayor and H. Sidgwick were the more prominent members and soon after my arrival in Cambridge they asked me to join them. Others who generally, or occasionally appeared were J.R. Mozley, Aldis Wright, and R.B. Somerset.[234] We used to dine at Grote's vicarage house, for this purpose, once or twice in the term, and afterwards an essay was read by one of the party and discussed by the others. They were

[231] [**John Grote** (1813–1866).]

[232] [See letters **40** and **119**. On Grote and his influence at Cambridge see John R. Gibbins, *John Grote, Cambridge University and the Development of Victorian Thought* (Exeter: Imprint Academic, 2007).]

[233] [On clubs, societies and coteries in Victorian Britain see William C. Lubenow, *'Only Connect': Learned Societies in Nineteenth-Century Britain* (Woodbridge: The Boydell Press, 2015).]

[234] [John Rickards Mozley (1840–1931), King's, 1858; B.A. (12th Wrangler and 5th Classic) 1862; Fellow, 1861–69; Professor of Pure Mathematics at Owens College, Manchester, 1865–85. William Aldis Wright (1831–1914), Trinity, 1849; 18th Wrangler, 1854; B.A.1858; Fellow, 1878; Librarian, 1888–1912; Senior Bursar, 1870–95; Vice-Master, 1888–1912; as a nonconformist, Wright was ineligible for election to a Fellowship before 1878; Shakespearian and Biblical scholar. Ralph Benjamin Somerset (c. 1834–1891), Trinity 1852; B.A. (39th Wrangler) 1857; Fellow, 1859; Junior Dean, 1866–69; Junior Proctor, 1868–69; vicar of St Michael's, Cambridge, 1868–75.]

delightful and profitable evenings, but certainly the most edifying part of the proceedings was the criticism and summing up of Grote himself. As a moderator, in the old scholastic sense, he was simply admirable. His knowledge was wide and accurate, and his judgment [128] quick and keen. No sophism could escape him, and if there had been the slightest attempt on the part of any of us towards fine writing, – which I don't think there was –, he would quietly but promptly have checked it. I do not think his lectures were very successful, and he never published anything during his life, but his influence for good in the way just described, if narrow was by no means feeble. Personally, I owe much to these gatherings which however did not last long as Grote was already in bad health and died in the summer of 1866.

I saw a great deal of Sidgwick in those days, and owe much to his influence. But even then his dominant characteristic was the critical. Philosophy was then an occasional relaxation with him: his absorbing study when I first knew him was Theology, or more accurately, linguistic foundations for Theology.[235] He was working hard at Arabic, which he had been studying in the summer at a German University, with the view to proceeding to Hebrew and thence making a thorough critical study of the Old Testament History and Religion. He was also devoting time to the study of Comte's works,[236] an author about whom I had of course heard a good deal indirectly through Mill, though Sidgwick was the first person whom I met who had any direct knowledge of his philosophy. He was remarkably advanced for this age, – he was only twenty-four in 1862, when I first made his acquaintance –, both in extent of knowledge and [129] still more in maturity and sobriety of judgment. Whether in still earlier youth he had ever shown the common excess of zeal which overshoots the mark in praise and blame, I cannot say, but I never knew him without philosophic calm.

About Christmas of this year I took a pleasant trip with Seeley in the South of England. It was the first occasion on which I came to know him as a travelling companion, and to realize the full charm of his character and manner. We went to Romsey, Winchester, Salisbury, and Christchurch, at all which places his already good historic knowledge was at my service. From Salisbury we paid a visit to Stonehenge; reaching it as it should be reached, on foot, and walking back to our Inn in the evening. We were almost overwhelmed with the impressiveness of the scene, as we stood by the giant ruin in the approaching dusk of a winter's afternoon, and looked at the waste of downs around: not a single inhabited dwelling, nor a single human being, in sight. We got some tea in a little inn or farm house, hidden in a hollow a mile or so off, and then walked backed to Salisbury.

---

[235] [On **Sidgwick**'s early intellectual occupations see Bart Schultz, *Henry Sidgwick. Eye of the Universe. An Intellectual Biography* (Cambridge: Cambridge University Press, 2004), chapter 3.]

[236] [Auguste Comte (1798–1857) the French positivist whose work had a decisive influence on Mill. The precise nature of Comte's influence on **Mill** is a vexed question. See, for instance, Angèle Kremer-Marietti, "Introduction: Comte and Mill – The Philosophical Encounter," in *The Correspondence of John Stuart Mill and Auguste Comte*. Translated from the French and edited by Oscar A. Haac (New Brunswick & London: Transaction Publishers, 1995), pp. 1–32.]

It was in 1863 that I first began to put together rough notes of what grew into my Logic of Chance.[237] New facts continue to accrue all through life, but I suppose most people will agree that new ideas are very scarce after a comparatively early age, and that the germs at least of nearly all that they finally produce can be dated back to a period before the [130] mind had lost its early receptivity and spontaneity. The root idea of my Logic of Chance, so far as it is anything more than an exposition of the familiar mathematical rules of Probability, was taken in as I read, on a summers day at Penmaenmaur in 1858, the often quoted statement of Buckle about the impossibility of checking the statistical regularity of human actions.[238] The passage puzzled me at the time, and soon began to raise many doubts and much vague suspicion. But I had then read nothing whatever of Logic, and no more of Probability than was to be found in Colenso's Algebra.[239] Soon after I took the curacy at Cheshunt I wrote to my old coach, Todhunter, for advice as to what books should be read. Amongst others he recommended De Morgan's short treatise,[240] which I read repeatedly and found most helpful. When a few months later I happened on Mill's Logic, it was no wonder the book produced such a deep impression, for the ground had been fully prepared. Every word that Mill had to say about Laws of Nature, especially as applied to human actions,[241] either forced into clearness my own vague reflections, or supplied the links to bring them into connection or a firmer basis on which to rest them. From 1862 I began from time to time trying to clear my thoughts by writing them down on paper: and at last worked out the small volume which was published in 1866.

[237] [John Venn, *The Logic of Chance. An Essay on the Foundations and Province of the Theory of Probability, with Especial Reference to Its Application to Moral and Social Science* (London & Cambridge: Macmillan and Co., 1866); second edition in 1876, third edition in 1888.]

[238] ['In a given state of society, a certain number of persons must put an end to their own life. This is the general law; and the special question as to who shall commit the crime depends of course upon special laws; which, however in their total action, must obey the large social law to which they are all subordinate. And the power of the larger law is so irresistible, that neither the love of life nor the fear of another world can avail any thing towards even checking [i.e. stopping] its operation'. Henry Thomas Buckle, *History of English Civilization. Volume 1* (London: John W. Parker and Son, 1857), on pp. 25-26.]

[239] [John William Colenso's textbook *Elements of Algebra*, first published in 1848, contained a chapter on the 'Doctrine of Chances'.]

[240] [Augustus De Morgan, *An Essay on Probabilities, and Their Application to Life Contingencies and Insurance Offices* (London, 1838). De Morgan (1806–1871), a former student of Whewell at Trinity (4th Wrangler, 1827) working at London University, contributed much to the introduction into Britain of Continental probability theory. See, for instance, Adrian Rice, "'Everybody Makes Errors': The Intersection of De Morgan's Logic and Probability, 1837–41," *History and Philosophy of Logic* 24:4 (2003): pp. 289–305.]

[241] [Mill, *System of Logic*, Book VI ('On the Logic of the Moral Sciences'), Chapter 2 ('Of Liberty and Necessity'). In brief, **Mill** argued that the law of causality applied in the same sense to human actions as to other phenomena.]

[131] 1863

The summer of 1863 was mostly spent with Ferrers in the Lake District, and it seems a convenient opportunity to say something about him. Those who only knew him during the time of his mastership,[242] especially during the latter part of his life, will have formed a very inadequate conception of what he was in earlier days. His remarkable appearance and character forced him upon the notice of every one whom he met, and some of his characteristics were of the kind most apt to be impressive upon young University men. A tall upright man with a somewhat shambling gait; a decidedly self-reliant manner; a shrill, almost cracked voice which emphasized the force of every epigram he uttered; all this attracted attention on the first occasion of meeting him. He possesses an astonishing verbal memory; far the best that I have ever known. At a dinner party in his rooms, one night, a dispute arose about the exact words of // "the military officer", as reported by Mark Tapley // in Martin Chuzzlewit.[243]*

// * "What more, says I? 'Snakes more', says he, 'rattlesnakes'. 'You're right to a certain extent, stranger. There air some catawampons chawers in the small way too, … cut him down, for he means wenom." //

Ferrers undertook to write it down, which L. Stephen betted he could not do accurately. On reference to the original it was found correct to a letter and a stop. He could, I believe, have rolled off the list of wranglers in every year recorded in the Calendar. As to Shakespear, his verbal knowledge was astonishing: it was practically impossible to make a quotation from any play without his at once supplying the context. With [132] nucleus which is clear and firm, with a large area about it which is hazy or which requires time for verification. With Ferrers there seemed no haziness whatever. If he remembered a thing at all, he seemed to retain it to the minutest detail.

And this amazing memory seemed almost wasted. His special subject, pure mathematics, was just that in which it was least helpful. Of course it saved him time and trouble to have formulae and previous results ready at hand, which other men had to work out afresh; and every fellow-examiner was familiar with the reception of some ingenious invention of [*del*: their] //his// own, – "Yes: very pretty: but it was set 18 years ago by the junior moderator in his problem paper". Many men with such a gift would have taken up subjects where it would have been all in all, history of languages. Of the former of these he probably retained all he had read, but he made no study of it; in the latter he was decidedly weak; as far as I could ever ascertain he knew no German whatever. Certainly during all my intimacy with him I never knew him put his memory to any but what might be called secondary purposes.

---

[242] [**Ferrers** was Master of Caius between 1880 and 1903. He was succeeded by Ernest Stewart Robers (1847–1912).]

[243] [This passage appears in the September 2, 1843 instalment of *The Life and Adventures of Martin Chuzzlewit*, a novel by Charles Dickens, originally serialised between 1842 and 1844.]

In one direction he was really admirable, namely, as a mathematical lecturer; certainly in his earlier years. I have nothing to complain of my two previous lecturers, Croker and Mackenzie,[244] but when Ferrers began his work about the middle of my career it was almost a revelation to me of what was possible in the way of clear arrangement and lucid exposition. [133] If he had only a trifle of enthusiasm, in the shape of a readiness to help us out of the formal lecture hour, or to encourage us to think of something beyond a Tripos, his help would have been invaluable.

In his performance of duties of a specific and routine character he was extraordinarily particular, and prided himself on never failing. I remember his once telling me that, for (I think) ten years, he had not missed a single morning chapel; "What an excellent gyp you must have had" was the not very fair retort. On one occasion on a Sunday afternoon's walk we took a wrong turn, and found ourselves about 10 miles from Cambridge with less than 2$^{1/2}$ hours before chapel. He was then dean, but most men in that position would have taken it quietly, knowing that there would be certainly other clergymen present to take his place. But he started off at once, in cap and gown, with his strange shuffling gait, to run home and succeeded in returning in time.

This characteristic made him a keen reformer, – up to a certain point. I have already said something about the shortcomings of the University reformers of that day, and those remarks apply pre-eminently to Ferrers. The College examinations had been hitherto conducted, – as I should now say they ought to be –, as a business of no very serious import. All the fellows in residence were expected to help in them, and they were regarded merely as a means of testing the men's [134] knowledge. They were now reformed after the likeness of the Tripos: regular examiners were appointed and paid: they met ever and ever again, revising their papers and testing their problems, until an amount of time and money was spent which to my present thinking was quite disproportionate to the end.

One great deficiency in him was a total want of all that can be called spontaneity. This showed itself very prominently during the time that he was tutor and master. In all those many years I do not think that there was a single improvement, or change, great or small, that was due to his initiation. He did not oppose, in fact he sometimes supported, what was advocated by others, – without his help there could probably never have been a place found for a Moral Science Lecturer[245] –, but he never made a suggestion himself. I do not even know that the little reform at the outset, in the matter of examinations, was his doing, though I have taken it for granted. This char-

[244] [**Ferrers** was Mathematical Lecturer at Cambridge between 1865 and 1880. The post was offered to him by the then Master, Edwin Guest. Ferrers' predecessors were the following: Joseph Morrison Croker (c. 1825–1891), Caius 1836; B.A. (8th Wrangler) 1840; Fellow, 1841–56; Dean, 1849–61; 'Mathematical Lecturer for many years'. Charles Frederick Mackenzie (1825–1862), St John's 1844, migrated to Caius 1845; B.A. (2nd Wrangler) 1848; Fellow of Caius, 1848–62. **Todhunter** also worked as a Mathematical Lecturer at Caius around the time that Mackenzie migrated to Caius.]

[245] [Venn's office of Catechist was transformed into a Moral Sciences lectureship in 1867.]

acter was displayed in conversation, or rather by the lack of what deserves to be called conversation. Nothing could be quicker or more to the point than his reply to a question or his correction of an error, but there he stopped. There was nothing in him of what was so conspicuous in Seeley or Sidgwick; – the answer that prompts another question or gives a broader aspect to what had been advanced. His reply reminded one of an echo from a wall; an instantaneous response and then a complete silence. I remember one almost ludicrous [135] experience. We were walking over Grisedale Pass from Grasmere to Patterdale when –, conversation flagging, as it was apt to do under the conditions just described –, I hazarded some rather feeble remark about the angular shape of a mountain in the distance: – "Exactly the remark you made when we walked over here in 1861" was the reply. This did not encourage a flow of talk.

Six weeks in the Lakes, with the customary weather of the country, spent by two men together, ought, one would think, to bring them closer together; if it did not cause a quarrel and separation. But I literally knew no more about his opinions or about his private life, at the end of the time than I did at the beginning. We spent almost the whole day walking amongst the mountains, and we were alone together nearly every evening. I found, rather to my surprise, that he had nothing in him of the nature of the genuine student, and that his early rising and diligent lecturing, conscientious and exact as they were, were only what might be called an "office" habit. During our holiday he came down late, went to bed always at ten, had hardly any books with him, and really seemed to me to read little more than the Times. I used to think, with a certain regret, what sort of use some men would have made of such a magnificent implement as his memory.

[136] I seldom met a man with less of sympathy, reverence or enthusiasm: – I speak of course from the outside, for I know nothing of his domestic life. I do not think I ever heard him utter a word of hearty admiration of any cause, or of any person: – unless perhaps of the mathematical power of Professor Cayley.[246] In particular he seemed as destitute of any interest or respect for the past as the most vulgar New York storekeeper. In his early days as fellow and lecturer the University was in the throes of that reform which was marked by the first Commission.[247] Ferrers was a typical representative of what I should now call the narrow and common-place aspect of that reform: one of those who, provided scholarships and endowments were thrown open to free competition, cared not a snap of the finger if every link which retained the names and wishes of the departed donors was snapped. And by the irony of fate this man became the holder of a mastership with five and a half centuries of unbroken history and tradition behind it!

[246] [Arthur Cayley (1821–1895), Trinity 1838; B.A. (Senior Wrangler) 1842; Fellow, 1842–52; Tutor, 1843–46; Hon. Fellow, 1872–5; re-elected Fellow, 1875; Sadlerian Professor of Mathematics, 1863–95; Fellow, 1852, and President, 1872–4, of the Royal Astronomical Society; President of the British Association, 1883; contributor to almost every field of pure mathematics.]

[247] [The first Royal Commission, the Graham Commission, was set up by Lord John Russell, Prime Minister and Bertrand Russell's grandfather, in 1848.]

In consequence of this his influence was of anything but an inspiring kind in the combination room. The conversation largely consisted of a string of jokes sneers and repartees, especially directed against those who were academically above him. When I first knew him he was a junior fellow, and the stream of his sarcasm used to play on the "senility", as he termed the seniority or Governing body. Once there, [137] his object of attack was the Master, – the learner and courteous, but old-fashioned conservative Dr. Guest[248] –, for whom he invented abusive nick-names, and on whose ability and attainments, and even personal cleanliness, he showed his epigrams. It was the language and attitude towards authority on which a fifth form boy generally begins to look back with some shame. I regret to say that Ferrers's dominant character and dominant seniority, infected his juniors more or less; and some of us only got to know in after years what a learned man and courteous scholar then adorned the College. I fell the more readily under [*del*: his] //Ferrers'// influence at the beginning (i.e. in the person of the Master), owing to the fact that he was especially kind to me, showed a certain interest in the subjects on which I was occupied, and was indeed the only man in College who was at all disposed to encourage the study of the Moral Sciences.

My life during 1864 was one of steady work in my own department. I was appointed an examiner[249] in the Moral Sciences Tripos that year, which gave me the pleasure and advantage of working with three such colleagues as Grote, Todhunter, and Westlake[250]; – and, I may add, the pleasure and profit of examining such men as F. Myers and [*blank*] Jardine.[251] The candidates that year were not numerous, but were of remarkable ability and attainments.

In the winter of this year, during the Christmas Vacation, I paid a visit to Rome, my first, and in fact only visit to that city. Of course I had not the requisite knowledge, either of ancient or of mediaeval history, fully to understand and [138] appreciate the place. But there was much to be learnt there by those who chose to work, and the Rome of those days had an indescribable charm about it. They were the last days of the city as our fathers and grandfathers saw it; as Gibbon and Byron saw and

248 [Edwin Guest (1802–1880), Caius, 1819; B.A. (11th Wrangler) 1824; Fellow, 1824–52; Master, 1852–80; Vice-Chancellor, 1854–55; an intimate friend of Goethe.]

249 [The task of the Examiner was to set the papers for the Tripos examination.]

250 [John Westlake (1828–1913), Trinity 1845; B.A. (6th Wrangler and 6th Classic) 1850; Fellow, 1851–60; Whewell Professor of International Law, 1888–1908; one of the founders of the Working Men's College, Great Ormond Street, 1854.]

251 [Frederic William Henry Myers (1843–1901), Trinity 1860; B.A. (2nd Classic and Moral Sciences Tripos, 1st Class) 1864; Fellow, 1864; Assistant Tutor, 1865–68; College Lecturer in Classics, 1865–69; poet and spiritualist; one of the founders, alongside his friends Sidgwick, Edmund Gurney and several others, of the Society for Psychical Research, 1882. Probably William Jardine (1841–1873), Christ's 1861; B.A. (22nd Wrangler, Moral Sciences Tripos, 1st Class) 1864; for some time Judge of the High Court at Allahabad, until his death.]

described it.[252] The French occupation[253] had made no perceptible change beyond the presence of a red-legged sentry here and there. It seemed as if nothing had been built and nothing destroyed for centuries, and within the walls were wide open spaces with huge almost deserted palaces scattered within them. We, – Thomas Dalton was my companion –, worked hard during the month we were there: gave the whole of every morning to the churches, palaces and museums, the afternoon to long rambles about the city or excursions in the Campagna, and wrote our diaries and studied Murray[254] and other handbooks for some hours in the evening. When Dalton left me to return to Eton I went on [*added in pencil*: southwards] and had a week at Naples. The weather was bad, but I got three out-door excursions; – ascent of Vesuvius, Pompeii, and Lake Avernus and district. I returned home by steam boat to Genoa, and thence over the Mont Cenis. I am glad to have seen something of the old-world made of crossing the Alps in the winter. We then still had to go by diligence, and the sight of the vast uniform wilderness of snow, during the long calm starlit January night was very impressive

As I looked forward to a final clerical career, as rector of a College living, I undertook again a spell of curateship. [139] All that was distinctive of evangelicalism had now fallen off me. I consider myself to have been, and should very possibly, if asked, have then described myself, as a very moderate Broadchurchman. Lamb, of our College, was at that time vicar of St. Edwards, a small parish with almost no poor. I stayed with him for some six months or so, //as his curate//.

1865.

In the summer of 1865 I had a pleasant tour //with my brother// in Switzerland, the last of my old-fashioned pedestrian expeditions with knapsack on back. We did not attempt any very long excursions or big mountain climbs, – the ascent of the Titlis, and the pass over the Tschingle glacier between Lanterbrunnen and Kandersteg, represented our utmost efforts –, but we enjoyed ourselves greatly and really saw a good deal of the country. Readers of Alpine will connect this summer with the famous first ascent of the Matterhorn, and the terrible accident which ensued.[255] At the time we were at Zermatt this expedition was being planned, and we naturally heard much of it. With the best known of the victims, – Lord Francis Douglas, – we picked up a slight acquaintance. It was a Sunday morning and we were at a late table d'hôte breakfast. One of those athletic visitors who were to be

[252] [Venn's allusion is to the eighteenth-century historian Edward Gibbon's *The History of the Decline and Fall of the Roman Empire*, seven vols. (1776–1789) and to Byron who in his *Childe Harold* made many references to his stay in Rome in 1819.]

[253] [It was in September 1864 that the Kingdom of Italy and the French Empire signed the September Convention, which ruled that all French troops would be withdrawn from Rome within two years.]

[254] [John Murray, *A Handbook of Rome and Its Environs* (London, 1853).]

[255] [After sharing in what was the first ever ascent of the Matterhorn, Lord Francis Douglas (1847–65) died on a fall on the way down. For background see Henry Day, "A Matterhorn postscript. John and Henry Venn's visit to Zermatt in 1865," available from alpinejournal.org.uk.]

found at most mountain resorts, – skinned by exposure to the sun and almost in rage –, [140] enquired what day it was. He was sitting opposite us. We got into conversation with him, and it was obvious at once what a superior man he was, – he had just before come out at or near the top of the Army entrance examination. He told us that he proposed taking part in the forthcoming expedition, having just, by the narrow shave, escaped a fatal fall on another mountain.

The next day we went up to the Riffel Hotel, and engaged guides to take us over the Weiss Thor to Macugnaga. However the weather turned bad, and not wishing to delay we returned to Zermatt and walked round instead by the valley to Saas. Just after leaving Zermatt we met the whole party who were dashed to pieces not many hours afterwards, – Hudson, Hadow, and the famous guide M. Croz.[256] Our own guide, Taugwald,[257] joined their party, and was one of the few survivors. Going into another part of Switzerland, we heard nothing of the events of the next day or two till more than a week afterwards.

1866.

Most of 1866 was spent in passing my first book, the Logic of Chance, through the press. I made the experience, which probably has been shared by most of those wo have not had the discipline of writing for the periodicals, of the difference [141] between a MS which looks in fair order and the same MS when set up in type. I had however the great help of my friend, and former tutor, Todhunter, who scrutinized most carefully every proof sheet for me with most beneficial results. It is a rare capacity, that of giving really efficient help in this way. Any one, of course, can point out where he differs from you, ad can set his finger on actual errors and slips. But it takes a very skilful [*sic*] critic to suggest just those slight alterations which can be admitted when a book is already in type and which may yet make all the difference as regards the signification of a paragraph. I have had cause to appreciate this skilled help twice; once, in this first edition, by Todhunter, and again in the second, by C.J. Monro[1],[258] brother of my Caius contemporary.

[1] A man of rare ability and subtlety of judgment, whom natural reserve, and still more, the crippling fate of wretched health, prevented from reaching celebrity. He was an excellent classical scholar, and one, I think, of the two or three of his time at College who had a touch of real genius for mathematics. Others besides myself, – the late Dean Farrar for one –, could have told of the generous way in which his critical powers were always at the disposal of his friends.

[256] [Rev. Charles Hudson (1828–1865), Douglas Robert Hadow (1846–1865)) and Michael Croz (1830–1865).]

[257] [Peter Taugwalder (1820–1888), a Swiss mountaineer and guide. Along with his son, Taugwalder was one of the group of seven that made the first ascent of the Matterhorn.]

[258] [**Cecil James Monro** (1833–1882).]

In the summer of this year, 1866, I took a couple of Moral Science pupils to North Wales. They were pleasant fellows, and keen for all out-door excursions, and one of them (J.R. Mowatt)[259] of more than average maturity for his age. We stayed at first for a short time at Tal y Bont a small village about mid-way between Conway and Llanrwst. The accommodation was very rude and primitive, but it came up to our real wants, [141] and the situation, – the bottom of a narrow valley which ran up to Lakes Dulyn and Melyn Llyn, and from which access was easy to Carnedd Llewellyn –, gave openings to many and delightful excursions. The latter part of the time we spent under more civilized conditions, in lodgings at Llanrwst, where I had the opportunity of exploring the beautiful group of lakes which lie a few miles to the west of that town. Mowatt was an excellent walker, and ready for anything. One of our excursions was to the top of Snowdon. We began by ascending Crib Coch, and then crawled along the extraordinary sharp ridge which connects this peak with Snowdon. The modern rock-climbers, who have recently exploited our mountains, have doubtless discovered plenty of small pinnacles, and gullies, of far greater difficulty of access, but as a continuous "knife-edge", with magnificent precipices on each side, it stands, I must think, almost alone in our country.

The great charm however of this way in Wales, and that which has left its deepest imprint on my memory, was the fact that Seeley was with us. As I have said, I had known him for some five or six years, and known him under the stimulating and intimate conditions afforded by short walking expeditions. But now we were together for a month, and the consequent profit and pleasure to me were past all reckoning. I summed up some of my reminiscences of this summer in a short biographical notice of him for the Caian, from which the following is quoted.[260]

He was an ideal companion on a walking tour; for, although he left arrangement entirely to me, he never indulged, as similarly complacent persons are rather apt to do, in subsequent complaints or belated suggestions. Nothing ever put him out, amidst the occasionally trying conditions of mountaineering in Welsh weather.

Many a lake, valley, and mountain top in that district are indelibly associated in my mind with our conversations, and would, I do not doubt, recall the very details if I were to go there again. There was scarcely a subject on which he could not, and would not, talk, and talk well; and we ranged freely over all the topics on which young men are fond of exercising their minds. He was always ready with a theory were the facts belonged to me, and would supply both theories and facts over the wide range which was his own. Everything came up in turn: literature, history,

---

[259] [James Mowatt (1844–1931), Caius 1864; B.A. 1868; equity draftsman and conveyancer; rode on horse-back through the Balkans to Constantinople.]

[260] [John Venn, "Some Personal Reminiscences of J.R. Seeley," *The Caian* 9:3 (Lent Term, 1895): pp. 164–170.]

metaphysics, aesthetics, poetry, – his interest in Goethe was already very keen –, and even science: for, although he had no acquaintance with physical details, he took much interest in the principles, and would often quote, and argue about what he had heard from De Morgan and others of his colleagues at University College. One night, I remember, as he remarked next morning we "shook the roof" about midnight, with a discussion on Political Economy.

One of our excursions rises before me with special vividness. We left Llanrwst at mid-day, took the picturesque short [144] cut over the hills to Cape Curig, where we picked up John Mozley and his brother;[261] thence along the road to Lake Ogwen, scrambled up past Idwal tarn and the Twll Du, and so over into the foot of Llanberis Pass. Thence we turned back to the little inn at Pen-y-Gwryd, where had tea, and parted with the Mozleys. We started homewards at nine, and the church clock was striking midnight as we walked into Llanrwst. The roar of the Llugwy beneath us, deepening as the night advanced, and the voices of the owls wailing in the thick woods which clothe the sides of the valley, as our talk slackened and we resigned our thoughts to the mere suggestions of nature, have stamped an image on my mind which cannot fade. I shall never take that walk again but I know that if I did, I should feel with him who recalled the hours spent with a lost friend in the valley of Cauteretz

All along thy valley, down thy rocky bed,
Thy living voice to me was as the voice of the dead;
And all along the valley, by rock and cave and tree,
The voice of the dead was a living voice to me[262]

When we were about to leave Llanrwst, my brother joined us there, and he and I and Mowatt determined to make our way to Hereford by the Wye. We had a boat sent for us from Hereford to Hay, whence we rowed down the river to the city, stopping to lunch at Bredwardine. The whole of that river is beautiful, and when, – as in our case –, the river is full, owing to a wet season, the navigation over the shallows and [145] rapids is very easy. After staying a few days at Hereford with my uncle and aunt, my brother and I made up a party of friends there and continued our Wye excursion down to Chepstow. We stopped the night at Tintern. There can be no similar length of river in England into which is crowded so much of natural beauty and grandeur, and so many objects of interest: – Hereford, Ross, Goodrich Castle, Monmouth, Tintern Abbey, Windcliff, and Chepstow Castle.

---

261 [Either Herbert N. Mozley (c. 1842–1919), King's, 1860; B.A. (16th Wrangler, 7th Classic 1864; Fellow, 1863–87; Assistant Master at Eton, 1864–97; Dante scholar; or Herbert Newman Mozley (1838–1906), King's 1857; B.A. (19th Wrangler); law 'coach'.]

262 [Opening lines of Alfred Lord Tennyson's *In the Valley of Cauteretz*, published in 1864, which he wrote during a tour in the Pyrenees (August–September 1861).]

In October I got my book out. It met with very fair success, considering that it was not, and never could have been regarded as, a text-book for any popular examination. I had a curious illustration in later years of the indirect way in which a scientific book brings its return to the author, if indeed it ever brings any return. The direct pecuniary result was a loss of some £10 after the whole edition had been sold off in eleven years. One indirect result was a request in after years from the editor of an American periodical to write a long article on 'Chance' for him.[263] I was away from home, and had no books, but just sat down and wrote in a week or two an abstract of my work, couched in popular style. For this essay rechauffe I received £60.

About this time the Professorship of Moral Philosophy at Cambridge fell vacant by the death of Grote. It will show how very little organised the subject of Philosophy then was in the University that I stood for this post, and was considered [146] to have a very good chance. My special subject was of course Logic, treated not from the historical or Aristotelian point of view but rather from the scientific side. With any thing like the present systematic division and arrangement of the philosophical Faculty[264] this would have been recognized by myself and by everyone else, and I should no more have thought of offering myself then than I did seventeen years afterwards against Sidgwick.[265] But in those days this Professorship represented the sole University teaching post in any department of Mental or Moral Science, and as I had been for four years almost the only lecturer in any of those sciences, I thought I was entitled to offer myself. Wisely, the electors chose F.D. Maurice. This rather startled the residents, as Maurice never, so far as I know, having set foot in Cambridge since he quitted it forty years before, had never been mentioned as a candidate. Ten years earlier there would probably have been a protest raised against his heterodoxy; as it was he was cordially welcomed by almost every one. It was to Kingsley,[266] I believe, who was one of the electors, that the entire credit was due of thinking of him and securing his election. It need hardly be said that the good influence of Maurice was social rather than professional. Under such an examina-

[263] [John Venn, "The Foundations of Chance," *Princeton Review* (September 1878): pp. 471–510.]

[264] [The Faculty of Philosophy at Cambridge emerged in the early twentieth century from the Board of Moral Sciences Studies, established in the 1860s.]

[265] [The Knightbridge Chair of Moral Philosophy fell vacant again upon Maurice's death in 1872, when it was assumed by the ultra-conservative Evangelical theologian **Thomas Rawson Birks** (1810–1883). His appointment led to a controversy, as it was seen as a retrograde step by all those who had sympathized with **F.D. Maurice**. Birks died in 1883, the year in which Sidgwick was elected to the chair.]

[266] [Charles Kingsley (1819–1875), Trinity 1838; B.A. (Classical Tripos, 1st Class) 1842; chaplain to the Queen, 1859–75; Regius Professor of Modern History at Cambridge, 1860–69; author of *Alton Locke*; his controversy with Newman gave rise to Newman's *Apologia*.]

tion-driven system as that of our English Universities, there is little opening in most departments, for a professor, unless he condescends to adopt the familiar and almost individual teaching [147] of a College lecturer. It is true, as already remarked, that the cramping influences of examinations at that time were but little felt in the Moral Sciences, but they existed nevertheless.

Science and Common Thought (Lecture at Caius House, Battersea, Oct. 1889)

John Venn

[1] There are two notions, I apprehend, which are still more or less common about Science & its relation to common Life. One of these is the old-fashioned view that Science does not in any way concern working people, but belongs rather to those who can sit in chairs & read & write by the fire-side; such people, they admit, can be useful, but their discoveries, it is supposed, have to pass through several hands before they come down to practice & be a matter of interest or concern to workmen. Such a notion, I hope, will not find favour amongst any who are thoughtful & studious enough to care to attend lectures in this room. But short of this there is a view which I hold to be almost equally wrong. This is the view that Science is something distinct from common Life, that is something which any one who wants to understand it must go out to learn as he would a new language or a new trade. Just as the carpenter who wanted to learn engineering, or the engineer who wanted to learn French, would have to begin at something new to him, so they think that the workman who wants to know something about Science must take in hand [2] a subject with which he had never meddled before.

I am going to try to prove that these notions are both wrong.[267] I am going to try to show that Science – the scientific way of doing work & of looking at work, – is close about us all: that the good workman is already, whether he knows it or not, something of a "scientist', & that in proportion as his work improves it is only as he has a measure of speculative science in his head & of practical science at his fingers' ends.

I. Before seeing what Science is, we must take account of one thing which it is not. Many people seem to suppose that it is complete knowledge, that the scientific man, so to say, 'knows all about the subject'. Nothing could be more erroneous. In fact though it may seem paradoxical it is really true that the more you know about a subject the greater is the proposition of your ignorance about it. This is so if we mean to refer to definite points of ignorance that is, to inability to answer questions which we can understand the meaning of. There are many questions which I could pose & could not answer, but which you could not pose at all, because you could not know what the question meant. [3] And I am sure that there are an immense number

[267] [This theme was part of Venn's life-long search for a pragmatic criterion of belief, linking knowledge to action, that applied to religion, science an everyday life. Cf. John Venn, *On Some of the Characteristics of Belief, Scientific and Religious. Being the Hulsean Lectures for 1869* (London & Cambridge: Macmillan and Co., 1870), especially the Appendix; John Venn, *The Principles of Empirical or Inductive Logic* (London: Macmillan and Co., 1889, chapter 25. For the historical connection between Venn and pragmatism cf. Cheryl Misak, *Cambridge Pragmatism: From Peirce and James to Ramsey and Wittgenstein* (Oxford: Oxford University Press, 2016) and Lukas M. Verburgt, "Pragmatism at Cambridge, England before 1900," *British Journal for the History of Philosophy* 29:1 (2021): 84–105.]

of questions which I could not claim to be ignorant about, because I have not even reached the point of knowing what those questions mean, which some of the great leaders of science – Sir Thomson or Prof. Huxley,[268] say, – would frankly answer by saying 'I don't know in the least'.

There is a common metaphor here which seems to me misleading. People talk of the "circle of knowledge", & say that as this circle increases, so will the circumference, which marks the dividing line of ignorance, increase too. This seems to suggest that within that circle the knowledge is tolerably complete. I think a more appropriate metaphor would be that of a cobweb. You know how the spider begins. He sketches a few threads between remote points, & then proceeds to connect these by cross threads, & so to work the whole up into a consistent web. But however many of these connecting links he may make, it is still a web & not a surface; the gaps can never be really filled up so as to leave no blank spaces between.

The notion that scientific knowledge ought to be complete knowledge is, as I have said, absurd. How absurd it is can easily be shown by a few examples. When we come to look at such instances we find how eccentric, how apparently capricious is the boundary line between the facts we do & those which we do not know. Suppose, for instance, [4] you were looking a railway train, or a ship, some way off, which was coming "dead on" towards you, so that no motion was perceptible. You would think it very odd if any one were to say that he could not tell within 100 miles how far off it was, but that he could tell you pretty accurately how fast it was moving. Now that is just the position of the astronomer as to the stars. Look at Orion on a clear night. Does it not seem strange that the astronomer has not a guess, within millions of millions of miles how far off any one of those stars may be, but that he can say, 'This one is approaching us at a rate of about 30 miles in a second'; 'That one is receding at about half that rate', & so on. Like again as regards the nature of the star. He has no notion how big it is, for whether there are any planets revolving about it, but he can tell you that there is iron on it, sodium, hydrogen gas, & so forth.

The same contrast runs through all our science, & becomes more striking with every fresh step of advance. Here & there we find comparatively isolated facts, solidly & accurately established, & side by side with them whole regions of the blankest ignorance. Of course every advance of science means that more & more of these strange facts are discovered & established: all that it means is that the proportion they bear to those left undiscovered does not seem to diminish. [5]

II. The first characteristic, I take it of scientific, as contrasted with popular, ways of thinking & performing is accuracy, what is called 'quantitative' accuracy. I think we might almost say that the power to answer the question "How much is there of it?" precisely, is coextensive with science; that he who can do this is scientific, & he who can't, is not. Take a few instances. Common feeling tells us that it is hotter in summer than in winter; it also tells us, but less certainly, that one day is warmer than another. But if any one wants to know how much hotter one day, or one part of a day,

[268] [The physicist William Thomson, Lord Kelvin (1824–1907) and the biologist and comparative anatomist Thomas Henry Huxley (1825–1895).]

is than another, he must make use of an instrument, viz. of a thermometer. If we want to know whether one thing is heavier than another, up to a certain amount you can tell by simply lifting them both; but if you want to know precisely you must make use of a balance. So with measuring lengths, say of rods or bars; of areas, say of fields or building plots & so on.

So far is plain enough; but what you must attend to here is mainly two things.

1. That science here introduces nothing new, nothing different. It is exactly the same thing, only more full & accurately done, as that, which the common person does. The letter-sorter who decides in a moment, as he lifts a letter, that it is over an ounce in weight, is really doing exactly the same kind of thing as the chemist who decides whether there is a grain of arsenic in a suspected dish. But because it is necessary that the latter should be much more accurate & delicate than the other, an instrument is [6] appealed to, & so we call the one, & not the other, scientific. The carpenter who has to divide a book-case into a dozen equal compartments is doing just the same kind of thing as the scientific instrument maker who has to divide & circle into 360 equal parts or degrees. But we should think that the one did his work well if he got within an eighth of an inch; we should think that the latter does his work very badly if he did not get within the 1000th of an inch. In respect of scientific work it is not the kind, but the degree which is different.

2. That though the process is of essentially the same kind, yet the superior accuracy will often lead to results which are quite unattainable to common sense. Take an example. Every one here has read something about that wonderful "Forth Bridge",[269] & every engineer knows what an astonishing product of skill & workmanship it is. The best way perhaps of realizing its magnitude is by comparing it with the Eiffel Tower which our neighbours across the Channel have made such an outcry about. If you were to take two of these Eiffel Towers, & hold them out horizontally by their bases & if they did not break by their own weight (which they would pretty certainly do) these two would together only make a little more than a single arch of that bridge. This however is not the point I want to speak about. It is accuracy more than mere magnitude with which science is concerned. Now, as the arches of the Bridge were built up from opposite piers in order to meet in the middle, it [7] was of course necessary to make an exact fit in the middle. What 'common work' would do here would probably be to try to cut a piece for the centre by actually fitting it in there, & then drilling the rivet holes where it stood. What 'scientific work' does is to calculate exactly the length that is wanted & to drill the holes precisely where they will fit opposite each other. This was done, & here arose a curious point. When these central pieces came to be put into their places, it was found that the rivet holes on one side of the bridge did fit, but that those on the other did not; the difference was small (an inch or so, I think). How was this? Iron, as you know, expands under heat, & of course the two pieces had been measured & drilled under

[269] [A UNESCO World Heritage Site, Forth Bridge is a railway bridge in the east of Scotland designed by Sir John Fowler and Sir Benjamin Baker and built between 1882 and 1890. It has a total length of 8,094 feet. When it opened, it had the longest single cantilever bridge span in the world.]

the same temperature. But when placed on the bridge, one was on the sunny side & the other on the cold side, the former accordingly expanded the most & therefore ceased to fit. But the engineers knew, – & here came in the characteristic of scientific work & of its precision, – that if they waited a few hours until the two sides had become equally warm the pieces would fit up on the bridge as they had fitted down in the workshop. And so it proved. At any rate a certain slight persuasion from a Hydraulic Press soon put everything accurately into place.

This example is perhaps already better known to some of you than it is to me, but when I read the account a few weeks ago I thought it would be hard to find a better instance of the characteristic of accuracy as belonging to all scientific work.

[8] Accuracy then is one 'note' or 'mark' of scientific work; but important as it is, it is not I think the one we have most to discuss now. If I were asked for the really distinctive characteristic, I should seek it rather in the analytic nature of scientific work. Let me explain a little what I mean by this word.

There are some highly ingenious & skillful people in the world, of whom the Chinese are an example, who can copy with wonderful accuracy work which they have never seen before. I believe that if one of those men were shown a clock he would succeed in making another just like it; & what is more his clock would go. I do not mean that it would go like one made by a skillful workman here, but it is an astonishing thing to succeed at all. But what is wanting to their work? It is not accuracy; that they must secure if the hands are to make a single revolution. You will see what is meant if you put the problem in this way. Tell one of those merely imitative men to make a slight alteration in the size or shape of the clock & he would be utterly baffled. Now this distinction runs through all our work. Every master or foreman knows the difference between workmen who can only be treated to work precisely according to a patter in every detail & those who can be trusted to make some slight variation in one point, & – here comes the rub – the corresponding slight variation in all other points which may be entailed by the former.

This characteristic power is what is commonly called Analysis, & when we are dealing with [9] some natural phenomenon, instead of a piece of human workmanship, it is generally called "resolving a fact into the Laws of Nature upon which it depends".

The reason why it is necessary is easily to be seen. Nature is excessively complex, & never presents the same thing twice over. We talk loosely about the 'same event' happening again & again, but this is only very loose-talk: the moment we come to look closely into the matter we see that there are many & important distinctions between them. The weather one day is not really like that on another; either there is more rain, or more sun, or a different wind. One thunderstorm differs from another in innumerable details; & so on. It is only in human work that precise recurrence is ever to be found. And even here it is only in the case of small & comparatively insignificant objects. One screw or one nail can be made quite indistinguishable from another, but it is only with elaborate machinery that one watch can be made just like another. No two big railway bridges are like each other: they are certain to differ in length, height, weight to be bore, foundations, & so forth. And I have no doubt that any engine driver or stoker would tell me that no two locomotives, even of the same company, are precisely alike. There are sure to be little differences in their consumption of coal, their powers of running, & so forth. I am quite sure that any one who has to do with any kind of machine on a large scale, hill agree with me, & will say that though two such things may look exactly [10] alike to an outsider, yet the man who has to work them all day knows that there are an immense number of small differences which have to be taken into account.

Now this, I take it, is the great reason why a process of Analysis is necessary. We can't find the same thing twice over, but if we take the whole to pieces we shall find that these pieces are alike in many respects. Accordingly the man who understands how much each part contributes to the whole will not be puzzled by a change which would thoroughly baffle any one who looked at the thing as a whole.

Take an example to illustrate this. Every one knows something about the nature of a common suction pump, & I suppose also almost every body now-a-days knows that such a pimp will not suck up water more than about 30 feet. Any ordinary carpenter could make a pump like one which was shown him, & might feel sure that it would work successfully. But now suppose that the pump was meant to draw up <u>oil</u> instead of water. It might be one of those great wells of natural mineral oil which abound in certain parts of America. Suppose that the oil rose naturally to within 30 feet of the surface, would the pump draw it up? You might give a guess, & say that the oil is lighter than water & would therefore rise higher; but this would be rather risky work if the pump had to be sent 50 miles up country in accordance with a contract. [11] Now what does Science do here, in order to make sure? It proceeds to <u>analyze</u> the complex phenomenon into its parts (explain with diagram]

This analytical process is so important & so universal in Science that we will take two or three more examples. There is the more reason to do this, because as we shall see in a moment, there is such a tendency in most minds to slur over the difficult by the use of a familiar <u>word</u>: to make a word do the work of a careful examination & separation into parts of some complex whole.

Begin with an extremely simple example. Why is it difficult to walk on ice? "Because it is slippery" will be the ready & immediate answer of almost every one who had not been taught to be very careful not to make words take the place of real explanations. Now I am far from saying that such an answer is worthless. Very few answers that people seriously give to questions are quite worthless. When you call the ice "slippery" you do give some help; for you class it with muddy pavements, polished boards, smooth marble pavements; & so on. And any answer which puts a number of like things together, & thus makes us think of them in the same light is some help towards further comprehension. But there is no <u>Analysis</u> here & therefore there is no real Science. What then ought we to do here? What would be the scientific explanation? (Explain)

Again: why does a balloon rise in the air? The spontaneous answer of common sense, of course, is "Because it is lighter than the air". As before, this is some help, for it classes the balloon with a cork rising in the water, with smoke rising in the chimney, & so on. But in Science, we want something more. Accordingly we proceed as before to separate this apparently simple fact into the separate facts of which it may be built up. (Explain with diagram)

I think this will help to throw light upon a fact which sometimes perplexes & troubles intelligent persons who want to understand the processes which go on about them. They ask for an "explanation", & the scientific man gives them an answer which seems to them far more puzzling than the fact which they wanted explained.

Sometimes for instance the fact which you want to have explained is one which is rather new & strange. You ask "Why it happens?" hoping that though the fact itself is strange at any rate the reason for the fact will make it look more simple & familiar. If you throw a bit of camphor upon a vessel of clean water it begins to flit backwards & forwards in the strangest way, as if some [13] one with an invisible magnet were drawing it first one way & then another.[270] And this it seems to do entirely of its own accord. [Explain]

Another example. Any one who is accustomed to work in metal knows that cast iron & wrought iron when turned on a lathe, act in a very different way. The chips of cast iron come off cold; those of the wrought iron come off burning hot. Why is this? The scientific man can give a reason, but I am afraid that the reason will not, in popular parlance, be exactly what is called an 'explanation". It would be hard, I imagine, to give a reason without introducing the doctrine of Energy. [Explain]

One more example. Any one, who has walked on wet sand, say by the sea-shore, must have observed how every time he puts his foot down on the sand the ground seems to dry up at once all about his feet. He dries up a patch a foot or so across. Now how is this? The first & natural answer would be, Why of course he squeezes the water out of the sand by the pressure, just as you squeeze the moisture out of a sponge. Not a bit of it; what we really do at the time is just the opposite. So far from squeezing the water out of the sand we have really squeezed it in. As this will seem rather puzzling, I must ask you to look at a diagram [Explain]

[14] One might go on all night adding examples of this kind. But these will be abundantly sufficient for our purpose. What I wanted to show here was simply that whereas common sense is generally apt to look at a thing as a whole, just as it presents itself to the eye, science is always analytic. That is, it always endeavours to divide that whole up into parts. Very often these parts may be extremely unfamiliar to us. And then it follows that the explanation, as the scientific man understands it, does not at all become an explanation as the popular man understands it, because it does not make the things easy & simple at once to him. Nature is constructed on quite other principles than saving us trouble in understanding all about her operations with the least possible trouble to ourselves.

III. This brings us to a third & last characteristic of all science & scientific thought: at least to the last which I have time to talk about now. Science always tries to generalize; that is to state the fact in the widest form, & not merely in the narrow form in which it may happen to present itself to our notice at the time.

---

[270] [Example probably derived from the work of Charles Tomlinson, Lecturer on Physical Science at King's College School, London, who in the early 1860s made a number of experiments to investigate the phenomenon, which 'has long been one of the waifs of science'. Charles Tomlinson, *Experimental Essays* (London: Virtue Brothers & Co., 1863), on p. 7.]

This characteristic is closely connected with the last. Whenever you analyze a single [15] fact into its parts, you will find that each of those parts is more general than the concrete fact. That is, they are of wider or more general prevalence.

Go back through our last example to see this:

The Pump
Walking on Ice
The Balloon
The Piece of Camphor
The Iron Chips
The Sand Squeezed Dry

This generalization, then, is the stamp of the scientific way of thinking. Remember, as I said before, it is not something totally different from popular ways of thinking. It is only that way improved & carried out more fully. So with this generalization. I said that people often slur over a difficulty by the use of a name; that instead of explaining why ice is difficult to walk upon they just put in a word & call it "slippery": – that instead of explaining the action of the pump they just say that it "sucks up" the water. But I said also, if you remember, that the mere use of such a word was one step towards generalization, that when you call the ice "slippery" you do, whether you think of it or not, class the ice with other slippery things, with polished floors, muddy pavements, & so forth. Now what the scientific man tries to do is to carry out this process to the utmost. He is determined never to stop short until the quality in question is traced out all through nature as far as we know nature. [16] For instance, every one knows the feeling which we call weight. You lift a stone, a log of wood, another man, & so forth, & you feel the same kind of thing. So far popular thought goes. It helps us with the common words 'weight' & 'heavy', & these words group all these things together, & are really a generalization up to a certain point. But science won't stop with a word like that. It won't be content to be limited to this word. It was this tremendous step which has immortalized the name of Newton. He agreed thus: – what we call 'weight' is nothing more than the "pull" of the earth upon the stone or log.: well but does not the earth pull the moon as well as any pebble? Does not even one fixed star pull another? Does not, in fact, every morsel of matter throughout the whole universe pull every other morsel? And so he generalized what popular thought calls 'weight' into the grand doctrine of "Universal Attraction" or "Gravitation".

These then, I take it, are the three main characteristics of the scientific way of thinking.

1. That it is more accurate: that is, that it endeavours to measure accurately not only things which popular thought measure but roughly, but also many things which such thought does not attempt to measure at all.
2. That it is always Analytic: that is, that it does not merely deal with things or events in what we call a concrete way, or as wholes, [17] but that it analyses them up into their component parts, & studies these parts separately. And power comes

of this analysis, for it is only in this way that we can apply our knowledge to new cases distinct from those which we have already obtained.

3. And thirdly it is general rather than particular. We are never satisfied with examining a property or quality just up to the limits of the case now before us. We seek to generalize it to the utmost, & so to express the fact, as to cover every case of it which can possible occur. We lift a stone, & feel it heavy, & we go on to generalize this conception until we reach the notion of Universal Gravity. We pull an india rubber band & feel it spring back; we go out to generalize this till it covers the case of solids like ivory & glass, & fluids like air, & so we get the general conception of Elasticity & so on in every other direction.

I began by insisting that Science was not something out of the way of ordinary working people, but only the extension & improvement of characteristics which we are every day concerned with. I will come back to the same ground again with a few concluding examples. I shall purposely choose them extremely simple, & if they seem to you absurdly simple, well that will only show all the more that Scientific principles be about us on every hand whether we think of them or not. I won't invent imaginary examples but will just recall two or three that recur to [18] me as having presented themselves within my own experience.

1. Rails chalked
2. String over nail
3. Bracket on Windsor ledge

(Simple; yes, when pointed out!)

These are all the examples I have to give, but I could easily go on for an hour multiplying them. Science & the scientific way of thinking & working may be compared to a ladder the bottom of which any one can put his foot on, & the top of which (if indeed it could be said to have a top) has far up out of sight among the clouds. But it is one & the same ladder for us all. And we must remember that any workman who has to do such a trifling thing as hang a picture or fix a bracket, if he does the work thoughtfully has got his foot some rounds up on the ladder high up upon which are such men as Newton & Darwin.

Part II
Selected Correspondence

Note to the Reader

Part II contains a selection of 150 letters from Venn's correspondence. The selection is chronologically ordered and divided into four periods: I. Student: 1853–1857; II. Curate: 1858–1962; III. Moral Scientist: 1862–1899; and IV. Looking Back: 1900–1923. All letters are numbered for purposes of reference, using the style **123** in continuous text. The heading of each letter gives the relevant manuscript information (abbreviated name of the archive, collection and item number), the full name of both sender and recipient, and the actual or presumed date and place of writing, which are placed at the head of the document regardless of their location in the original.

All editorial annotations and interventions, whether in the text or in footnotes, are enclosed within square brackets. The original spelling, grammar, punctuation, capitalization are retained as found in the original letters, except for removing the apostrophe from the possessive 'it's' and using a full-stop '.' instead of a dash '–' to mark the end of a sentence; omissions, mutilations, and illegible words or letters have been rendered through the use of the following textual devices: (a) when a word or words in the original text must be omitted in the printed text due to omission, mutilation or illegibility, the omission is shown by a word or phrase placed in square brackets in italics, such as: [*torn*], [*illegible*] or [*remainder missing*]; (b) when there is no question as to a missing word, it is supplied within brackets in [roman type]; (c) when a missing word can only be conjectured, the omission supplied within square brackets is followed by a question mark; (d) the abridgement of a passage is shown by dots placed in square brackets […].

All references to Venn's *Annals: Autobiographical Sketch* (abbreviated as *Annals*), reproduced in Part I of the present volume, are to the original page numbers, which appear within square brackets in the text. All cross-references to notes are to notes in Part II.

L. M. Verburgt, *John Venn: Unpublished Writings and Selected Correspondence*, Studies in History and Philosophy of Science 56,
https://doi.org/10.1007/978-3-030-79829-1_2

The reader is referred to the 'Biographical List of Names', found at the end of the book, for short biographical sketches of senders and recipients of letters. When only one letter was exchanged, biographical information about the sender or recipient is provided in a footnote.

List of Letters

I. Student: 1853–1857

L. M. Verburgt, *John Venn: Unpublished Writings and Selected Correspondence*, Studies in History and Philosophy of Science 56,
https://doi.org/10.1007/978-3-030-79829-1_3

24. John Venn to Henry Venn of C.M.S., 24 April 1863
25. John Venn to Henry Venn of C.M.S., 8 May 1863 [= 1865]
26. Henry Venn of C.M.S. to John Venn, 31 July 1863
27. John Venn to Henry Venn of C.M.S., 17 October 1863
28. Henry Venn of C.M.S. to John Venn of Hereford, 10 February 1864
29. Henry Venn of C.M.S. to John Venn, 10 February 1864
30. Henry Venn of C.M.S. to John Venn, 6 July 1864
31. Henry Venn of C.M.S. to John Venn, 30 October 1864
32. John Venn to Henry Venn of C.M.S., 7 March 1865
33. Henry Venn of C.M.S. to John Venn, 11 March 1865
34. Henry Venn of C.M.S. to John Venn, 15 April 1866
35. Henry Venn of C.M.S. to John Venn, 2 May 1866
36. Henry Venn of C.M.S. to John Venn, 1 August 1866
37. Henry Venn of C.M.S. to John Venn, 4 August 1866
38. Henry Venn of C.M.S. to John Venn, 13 October 1866
39. Henry Venn of C.M.S. to John Venn, 27 October 1866
40. John Venn to J.B. Mayor, 4 November 1866
41. Henry Venn of C.M.S. to John Venn, 6 April 1867
42. John Stuart Mill to John Venn, 4 February 1868
43. John Stuart Mill to John Venn, 18 March 1868
44. Henry Venn of C.M.S. to John Venn, 29 April 1868
45. Henry Venn of C.M.S. to John Venn, 18 February 1869
46. Henry Venn of C.M.S. to John Venn, 29 March 1869
47. Henry Venn of C.M.S. to John Venn, 7 September 1869
48. Henry Venn of C.M.S. to John Venn, 9 September 1869
49. Henry Venn of C.M.S. to John Venn, 15 September 1869
50. Henry Venn of C.M.S. to John Venn, 17 January 1870
51. Henry Venn of C.M.S. to John Venn, 17 January 1870
52. Henry Venn of C.M.S. to John Venn, 12 August 1870
53. Henry Venn of C.M.S. to John Venn, 17 August 1870
54. Alexander Bain to John Venn, 18 March 1871
55. Henry Venn of C.M.S. to John Venn, 29 April 1871
56. Henry Venn of C.M.S. to John Venn, [August 1871]
57. Henry Venn of C.M.S. to John Venn, [7?] January 1872
58. Henry Venn of C.M.S. to John Venn, 1 April 1872
59. John Stuart Mill to John Venn, 14 April 1872
60. Henry Venn of C.M.S. to John Venn, 16 April 1872
61. Henry Venn of C.M.S. to John Venn, 22 April 1872
62. William Stanley Jevons to John Venn, 9 October
63. John Venn to William Stanley Jevons, 11 October 1874
64. John Venn to William Stanley Jevons, 23 March 1876
65. William Stanley Jevons to John Venn, 25 March 1876
66. John Venn to William Stanley Jevons, 28 March 1876

IV. Looking Back: 1900–1923

I. Student: 1853–1857

1.
CMS/ACC81 C34/12
Henry Venn of C.M.S. to John Venn
3 August 1854
Penmaenmawr, Wales

My dearest son,

Many happy returns of tomorrow on which day I think that you will receive our letters. It will be the first birthday which you have spent away from your father's home or society.[1] In some sense you have now a home of your own, a habitation in your own right – so rapidly are you pushing into manhood & preparing to take your stand in the ranks of society – my wish & my prayer for you continuously is that you may act a decided part – in the cause of holiness of religious truth - & of love to the savior. That all men may know what you are about. That each year may find you advanced in the ranks of society in the best sense – a more marked man in the respects I have noticed – & more honored of last in having talents & opportunities for his work committed to you. Though I feel the loss of you here at present I rejoice that my deprivation of your society is occasioned by your own choice of the best employment of your time - & my ability to give you the advantages wh. you enjoy. My stay here will be very short – next Tuesday or Wednesday at furthest – then I hope to spend the next Sunday (that is Sunday week) with you in Cambridge. But I cannot tell at present so you must not say anything about it.

I fully reckon also upon our taking an excursion together the last fortnight of Sep. either to Belgium or to the south of France to see the Roman remains of an Amphitheatre at Nimes a Roman Bath & Temple & the finest Aqueduct wh. remains. But when we meet we can lay out our plan.

Henrietta & Harry[2] will tell you of our jog-trot expedition, compared with your mountain ramble. But I made many enquiries of routes wh. I think you will like to take on your next Welsh excursion.

Farewell my dearest son. May God answer all your mother's & your father's prayers & your own for the sake of our great Intercessor.

Most aff^{y} yours,

H. Venn

[1] [After a trip to Wales, Venn had returned to Cambridge in July to use the Long Vacation for study.]

[2] [Venn's brother and sister, Henrietta Venn (1832–1902) and **Henry Venn of Walmer**.]

2.
CMS/ACC81 C34/28
Henry Venn of C.M.S.to John Venn
19 July 1856
11 Highbury Crescent

Dearest John,

I had hoped that I should have spent tomorrow with you – but work has so thickened upon me, that I cannot get away – & I have no chance of crossing the channel before the week after next. I promise you a whole day at Cambridge before I go, probably the 27th – I have been chiefly detained by an African project – of opening the Niger River to lawful commerce & civilization.[3] […]

I have not yet found a Tutor for Harry[4] though I have written to several – some have too many boys – others are too careless as Tutors – & leave their pupils too much to themselves I want one who will make Harry "work up" thoroughly – I care little whether it be a book of Virgil or Homer – or any thing else so that he may acquire a habit of perfect accuracy such as your Mathematics will give you.

All the influence I gain in committees &c arises from the habit I early formed of "getting up" a question thoroughly: applying the maxim to particular questions – which is <u>in</u>applicable to large subjects – nil actum reputans, si quid superesset agendum.[5] But as old Boston[6] used to say of pumping the cistern full at Highgate 'it requires one to stick at it!'.

It is a habit which once formed applies to every thing – religion as well as science – personal & ministerial religion; in our missions many missionaries have for years been in the habit of itineration – taking long & pleasant journies [*sic*] & preaching in the cool of the day to such as would attend – we have long felt that this was not the way to "evangelize" a land. At East two missionaries one in the south, one near Bombay struck out the new plan of undertaking a District – & going so systematically to work – village to village – 3 or 4 visits to each in the course of the year – a weekly visit wherever there was a spirit of enquiry – so that not one person c^d plead that he did not hear of the Missionary – or that he was gone again before he had satisfied the inquiries. This true & sound way of itinerating has been devised & carried out by only two Missionaries out of several hundred whom we have

[3] [The largely unsuccessful 1841 Niger expedition, initiated by the Society for the Extinction of the Slave Trade and for the Civilization of Africa and backed by the British government, had aimed to make treatises with the natives there and persuade them to stop participating in the slave trade. Another expedition was undertaken in 1857, advocated by the C.M.S., to which Henry Venn was then officially attached. See Wilbert R. Shenk, *Henry Venn: Missionary Statesman* (Eugene: Wipf & Stock, 2006), pp. 67–79.]

[4] [The reference here is to a College tutor, who were then primarily responsible for the tuition of pupils.]

[5] ['Thinking that nothing was done, if anything remained to be done': Lucan, *Bellum Civile* 2:658.]

[6] [James Boston, the Venns' gardener at Hornsey Lane, Highgate, where they had lived between 1838 and 1848.]

employed, & they were the two highest wranglers who have joined us – Ragland 2[d] wrangler - & Frost also very high.[7]

It is a most instructive illustration of the value of a habit of doing every thing "thoroughly". Quidquid agis age.[8]

May you my dearest son be enabled so to sanctify the habits which are ow forming as to make your talents available for the Lord's service into whatever departments He may call you.

Ever most aff[y] yours

H. Venn

3. CMS C34/29
Henry Venn of C.M.S. to John Venn
9 December 1856
11 Highbury Crescent

My very dear John,

In no term since you went to Cambridge have I found it so difficult to get away & spend a pleasant Sunday with you. For the last 4 weeks I have intended to do so till the last day. Besides my usual work I have had 2 or 3 special questions on my hands which have taken up much time. I send you 2 papers to amuse you – the 3d. & most troublesome is not yet forward enough to send it to print.

But I never forget you. Twice every day I commend you by name in prayer to God that he may give you luck. Success in the degree as he may see profitable for you – but above all that he may fit & lift your mind & body for such future service as may be most for his glory: & if it be consistent with his will that the service may be in the ministry of the gospel to which from the day of your birth your mother & I commended you – in submission to the Lord - & I to this hour.

If I can get away on Saturday I shall run down for the Sunday, taking up my abode at the Bull.[9] Ord[10] of your College has spoken to Clayton[11] about offering himself as a Missionary. There seems some doubt of his wife's health. I should like to see them both & have written to him to say I may come.

---

[7] [Thomas Gajetan Ragland (1815–1858), Corpus Christi 1837; B.A. (4th Wrangler) 1842; Secretary (Corresponding Committee) of the C.M.S., 1845–54; helped to found the N. Tinnevelly Mission, 1854. Andrew Hollingworth Frost (1820–1907), St John's 1838; B.A. (11th Wrangler) 1842; C.M.S. Missionary at Nasik, Bombay, 1853–67; Secretary of the C.M.S., 1869–70. Henry Venn's allusion is to the so-called Itinerating (i.e. travelling) Missions, started by Ragland, Frost and several others in 1854.]

[8] ["Do what you do well".]

[9] [Probably 'The Black Bull Inn' in what was then called Bishopsgate Street.]

[10] [Charles Renshaw Ord (1830–1919), Caius 1852; an extreme Calvinist, Ord was Venn's main companion in his first year at Caius. See Venn's *Annals*, p. 35.]

[11] [**Charles Clayton**, the Evangelical tutor of Caius and Simeon's successor as vicar of Holy Trinity, Cambridge.]

The Rev Joseph Gedge[12] (also a 2^{d} son of Mr G. of Birmingham) who took his degree at Trinity a few years ago has offered himself so that we have now a larger number than usual of graduate candidates. In all other respects also the work is most encouraging.

I shall rejoice to see Panton with you at X^{mas}. I have much wished to make his acquaintance for his Father's sake, as well as for his own.[13]

Thank you for sending me the exact days of the examination.[14] I reckon upon them day by day as you do. On the 7th & following days of Jan. I shall have all the Ass. Secr.[15] in town & on the 14th the Annual Clerical Meeting at the C.M. College.[16]

Ever my beloved son

Your most affec. Father
Henry Venn

4.
CMS C40/6
John Venn to Henry Venn of C.M.S.
[January 1857]
Caius College, Cambridge

My very dear Father,

Our 'three days'[17] have come to an end at last of which I am very glad though I have not done so well as I expected on the whole. On Tuesday morning we had Euclid & conic sections in which paper I got on tolerably well, but made a mess of the Algebra & Trigonometry in the afternoon. On Wednesday morning we did the Statics & Dynamics in which I got on moderately well, but again made a mess of the afternoon paper on Hydrostatics & Optics. This morning it was Newton & Astronomy in which I did moderately & the same in the general Problem paper this afternoon. The next part of the examination does not begin till Tuesday next. As the post is going out soon I must leave off. Give my love to all. Ev. y. most affect. son,

J. Venn

---

[12] [Probably Johnson Hall Gedge (1831–1875) who was born in Birmingham, entered Trinity in 1849 and graduated B.A. (18th Wrangler) in 1853. This Gedge was also Associate Secretary of the C.M.S. for Yorkshire, 1864–70.]

[13] [Venn's friend David Brooke Panton (1832–1891), Caius 1852; son of the Rev. Richard Panton (c. 1803–1860), Archdeacon of Surrey, Jamaica.

[14] [The Mathematical Tripos examination took place annually in January, in the Senate House.]

[15] [Association Secretaries of the C.M.S., which held an annual conference.]

[16] [The Church Missionary Society Training College in Islington, founded in 1820 to prepare missionaries of the C.M.S.]

[17] [The Mathematical Tripos examination consisted of three parts of three days each – with each day divided into two parts (9–12am and 1:30–4pm). The first three days (6–8 January) were confined to the more elementary questions and problems. In 1857, the Moderators were William Magan Campion and William Walton, and the Examiners Samuel George Phear and William Henry Besant. See *Cambridge Examination Papers: Being a Supplement to the University Calendar for the Year 1857* (Cambridge: Deighton, Bell and Co., 1857).]

II. Curate: 1858–1862

5.
CMS C34/35
Henry Venn of C.M.S. to John Venn
6 February 1858
11 Highbury Crescent

My dearest John,

Your remark in your last letter that you were reading the Romans[18] with Garrick,[19] & that you supposed that you shd not fully understand it till you had read it several times over filled my heart with joy. To master the first part of the Romans has always appeared to me to feed upon the marrow of divinity – & to lay the only sure foundation of future usefulness in the Church of Christ. It is impossible to understand that wonderful argument fully without a deeper impression than flesh & blood sanguine of the utter ruin of one's nature. This is the key. It is somewhat like the Newtonian theory of gravitation – it solves a whole set of problems.

I pray therefore for you & your companion in study & for dearest Harry that you may all make sure progress in the mastering of the Epistle to the Romans being taught of God what are the first principles of your faith.

You will be glad to hear that I have just received a cheque for £5000 from an old lady for one special India Fund; & I have got it with permission to employ any part in temporal relief connected with spiritual objects – so I shall be able to do a little in India of the cotton business in Africa.[20] Though I must try sugar rather than cotton.

I have to preach a funeral sermon for Bp. of Calcutta[21] on Sunday. I hope there is a good prospect of Pratt[22] being appointed successor: but nothing can be yet known.

Ever my dearest son

Your most aff[t] Father
Henry Venn

---

[18] [The Epistle to the Romans, the sixth book in the New Testament, composed by Paul the Apostle. A text to which Evangelicals like Henry Venn attached particular importance as it was seen as revealing the only method by which men can be saved, namely through the gospel of Jesus.]

[19] [Venn's College friend James Percy Garrick (1834–1919), Caius 1853; B.A. (9th Wrangler) 1858; Fellow, 1858–81; tutor, 1875–6; later rector of Blofield, 1880–1912, and Honorary Canon of Norwich, 1899–1919.]

[20] [One of **Henry Venn of C.M.S.**'s many projects at the C.M.S. was to develop trade in African products, with an eye to checking the slave trade. He did this mainly through the encouragement of natives in Africa to clean, press, and pack their cotton for the English market.]

[21] [**Daniel Wilson** (1778–1858), Lord Bishop of Calcutta, and formerly vicar of Islington and head of Venn's school in Highbury, had died in Calcutta on 2 January 1858.]

[22] [John Pratt (1809–1871), Queens', 1828; Caius, 1829; B.A. (3rd Wrangler) 1833; Fellow, 1836; a gifted mathematician, Pratt had been a pupil of William Hopkins and himself taught Harvey Goodwin as private tutor; he became Chaplain to David Wilson, Bishop of Calcutta, 1846, and was Archdeacon of Calcutta between 1850–71. Pratt was nominated for the position of Bishop upon Wilson's death, but George Cotton (1813–1866) was chosen instead.]

6.
CMS C34/39
Henry Venn of C.M.S. to John Venn
25 November 1858
11 Highbury Crescent

My beloved son,

I enclose you 2 or 3 of my sermons when I was a young Preacher. If any of the thoughts can be of use to you or the schemes of others – make any use of them. If I had not thought they might have some day been of use to my sons they w^{d} have been long since burnt.

We go to Stratton Park at 3$^{1/2}$ o'clock to day – & shall certainly return to Highbury on Saturday – but at what hour is uncertain – also whether I can get over to Cheshunt. But let me have a note to Stratton to say how your sermon gets on. I shall aim at paying you a visit on Saturday afternoon – write by tomorrow's post to me at "Sir F. Baring Stratton Park[23] – Micheldever Station".

Ever my beloved son

Your most aff^{t} Father
Henry Venn

7.
CMS C34/43
Henry Venn of C.M.S. to John Venn
22 January 1859
11 Highbury Crescent

My beloved son,

Here I am at 3 o'clock but just starting I hope for St Leonards – Harry went off at 10 – I was obliged to stay & prepare Minutes & write letters &c &c. I must return on Monday by an early train & shall dine on that day at Maley's.[24] I shall be detained till 6 at the office for the African letters.[25]

Do not be discouraged by finding no ideas & texts or subjects. I have formerly labored far more than you under that difficulty. Often have I spend all Saturday in laboring at a dry pump while I saw other pipes like the fountains in Switzerland flowing clearly without "working the handle".

---

[23] [Sir Francis Thornhill Baring (1796–1866), 3rd Baronet, later Lord Northbrook, of Stratton Park – who at one point worked as Chancellor of the Exchequer, 1839–41. Henry Venn knew Baring from Cambridge days, when they both joined a debating society, of which the historian George Grote was also a member. See William Knight, ed. *Memoir of Henry Venn, B.D. Prebendary of St. Paul's, and Honorary Secretary of the Church Missionary Society* (London: Seeley, Jackson, and Halliday, 1882), on pp. 31–32.]

[24] [Not identified.]

[25] [During this period, **Henry Venn of C.M.S.** was actively engaged in writing letters for Africa, for instance to keep the Sierra Leone missionaries up to date about recent developments, such as the arrival of the Bishop. See Knight, *Memoir of Henry Venn*, p. 219.]

But I used to comfort myself in the amount of reading which my barrenness enacted of me - & in the humiliation in w^{h} it kept me which often brought me to my knees & kept me from that degree of self sufficiency & courtesy of popularity by which others were entangled.

I also always found that I could make something of a scripture character when all others failed me – Daniel – Moses – Jeremiah – Barnabas – are all very easy & very interesting.

Be of good cheer: my heart lept for joy when Johnson[26] spoke to me of his own accord of the promise which you gave of being a very effective Preacher. He thought your voice excellent & your thoughts & expressions when use has expanded them likely to make a very interesting & instinctive style of preaching. He is not a man to say one word for flattery's sake.

[…]

Ever most aff^{y} yours
Henry Venn

8.
CMS C40/8
John Venn to Henry Venn of C.M.S.
1859
Church Field, Cheshunt

My very dear Father,

I am sorry I forgot to write to you yesterday till it was too late, still as you did not I believe get home till late you did not I suppose expect me. I think I shall not be able to come this week, I am going to Cambridge tomorrow for the day, as I have a good deal I want to do there such as getting books out of the Library & I suppose too they will have come to some sort of decision about the alteration of the college statutes. All went on very well yesterday; I read prayers in the morning; in the afternoon I went to Gough's Oak & had the service there to a tolerable number about 90, they being almost exclusively the very poorest one has to give them something very plain, I cannot manage a regular extempore sermon even to them that is to say not upon a single text or subject.[27] I have the last two Sundays taken the lines of St. John & St. Peter & can make a sort of running commentary upon them, picking out a number of different scenes in their lives & speaking about them, there is the advantage in this case that when ideas come to a sudden stop one has but to turn to another passage & go on at once, I find I can hold out for about half an hour that way.

Did not a scripture reader apply to you a short time ago from Winkfield? We are in want of one here; several of the parishioners have subscribed together to pay for one for 3 years, he would have to be able to conduct a cottage lecture well as it

[26] [Not identified.]

[27] [As Venn explained in his *Annals* (p. 94), while at Cheshunt, he sometimes switched places with the other curate there, so that he could take the address to the poor at Gough's Oak, of whom he was less afraid.]

would be for that especially for wh. his services would be required. We are going to open two more in different parts of the parish. I am going to write today to Henrietta.[28]

Ever y^{r} most aff^{t} son

J. Venn

9.
CMS C40/10
John Venn to Henry Venn of C.M.S.
1860
Allan Bank, Grasmere

My very dear father,

I believe I did get the note from you that Henrietta spoke of, that is I got one on the Sunday before last, though it did not contain so definite an account of your plans as she seemed to think, in fact it seemed to me that you were still very uncertain. I have been dragging about so much from one place to another that I have only just had time to scribble notes to Henrietta about what I have been doing. I have returned here as you see after a very pleasant week's ramble, the weather having been on the whole of a character not to be grumbled at; yesterday was the [*illegible*], but even then it cleared up in the evening. I said that I had seen Barton,[29] & spent a day or so with him about Wasdale & Scafell; he seemed very well & on the whole cheerful, that is he seemed to take an intense delight in our mountain scramble, & in his botanical researches. He had an aneroid[30] which seemed to give rather more consistent results than ours, would you by the way send ours by Henrietta she cd be safely trusted to bring it I shd think. I met Wigram, B's brother in law,[31] on my way from Keswick on Saturday. Barton showed me a letter from you by the way which I was rather sorry to see in which you spoke about not leaving London before the 14th, is that the case?

As far as I can make out you think it would be best for me to go to Hereford[32] till something definite & permanent in the way of a curacy offers itself, wh I shd be very ready to do.[33] It seems also that if this is to be in the neighbourhood where you & Henrietta can live it must be somewhere near Bayswater, but I shall hope to hear more when we meet. I think of getting to Edinbro' tonight & remaining there for the Sunday. Love to Henrietta.

Your most aff^{t} son

J. Venn

---

[28] [Venn's sister, Henrietta.]

[29] [Probably Venn's Cambridge friend John Barton (1836–1908), Christ's, 1855; B.A. 1859; founder of the Cambridge University Church Missionary Society; afterwards vicar of Holy Trinity, Cambridge, 1877–93.]

[30] [A specific kind of barometer, invented in 1844.]

[31] [Probably Frederic E. Wigram (1834–1897), Trinity, 1852; B.A. 1857; (assistant) curate of St Paul's, Cambridge, 1858–1863; Honorary Secretary of the C.M.S., 1880–95.]

[32] [At his uncle and aunt, John Venn of Hereford and Emelia Venn.]

[33] [Following his time at Cheshunt, 1858–9, Venn returned briefly to Cambridge to be ordained pries before taking a temporary curacy for the winter with William Tilson Marsh at St. Leonard's, Hastings.]

10.
CMS C40/9
John Venn to Henry Venn of C.M.S.
14 June 1860
Allan Bank, Grasmere

My very dear Father,

I suppose you have received a letter from Harry[34] before now telling you of his sudden flight from this place & its cause; Tuesday & Wednesday were certainly very bad, & when there seemed every chance of the same incessant drip today he fairly took fight & has fled to Hereford. I think I shall stay here on Sunday with Ann[35] & then if it does not rain strike off into the Scafell region. I have undertaken, or rather promised to do so, the two services at Langdale on Sunday next, the Vicar[36] has died lately & no successor has been appointed as yet, so that Mr. Jefferies[37] the incumbent here, who is in some way answerable for it, was very glad of my assistance. There have been several fine hours this afternoon & I have been out a walk, but the mountains seem soaked to their very centres. For the present my letters can be sent here & if I go off I shall come back again for them.

Have you heard anything from Holbeton? The letter that came from there for me was from the Teddington Strachan[38] inviting me to his missionary meeting.

Love to Henrietta.

Your most aff^t^ son
J. Venn

Since writing the above I have received [Harry's?] note; it seems very much to confirm my impression about Holbeton, tho' it makes him still more undecided. As it now is, his wish for a month or two of trial entirely does away with its main advantage, i.e. that of offering something <u>settled</u>. I can see nothing else to be done therefore than to take any curacy that may offer itself & to give up the hope of finding a place where you and H. could live.[39] I sh^d^ not so mind mind [*sic*] this as Harry will be free from Cambridge in 6 months. In case of there being any delay I cd fill up the time at Hereford, but I suppose I shall hear from you soon.

J.V.

---

[34] [The Venn family's sobriquet for Venn's brother **Henry Venn of Walmer**.]

[35] [Ann Sykes (1841-?), second wife of Venn's uncle Joseph Sykes. See Venn's *Annals*, p. 98.]

[36] [Probably Stephen Birkett (?–1860).]

[37] [Rev. Edward Jefferies (1840–1893).]

[38] [James Morgan Strachan (c. 1789–1874), of Teddington Grove, Middlesex, who had been affiliated with the C.M.S. since 1814. He was appointed an Honorary Governor for life in 1823, in recognition of his services to the C.M.S.]

[39] [It had been **Henry Venn of C.M.S.**'s strong desire that his son should find a conveniently located curacy that would enable him to share a family home with his father and sister.]

11.
CMS C34/73
Henry Venn of C.M.S. to John Venn
16 June 1860
11 Highbury Crescent

Dearest John,

My interview with Holbeton was a little disappointing. It appears that the stipend for the curate is raised by subscriptions & 2 or 3 years ago he engaged one whom the subscribers did not like & withheld their contributions. He promised therefore that in any other case they should judge for themselves before he engaged any one. This would as you say keep us unsettled & uncertain till too far into winter. I have therefore given up the Surbiton plan.

This morning I went to Sir J.S. Clark[40] to ask him whether he could specify any limits of favorable situations. He said decidedly that he thought Bayswater, the higher parts, better than any situation within 10 miles of London certainly better than Surbiton, Sydenham, Croydon or such places: that for the country he could name no place but Reigate.

Upon fully talking the matter over with Henrietta we have determined to look out at once for a furnished house or apartment for the winter in the higher parts of Bayswater which w[d] be within a walk of all our relatives in those parts to send our furniture to the Pantechnikon[41] – & to sell or let or lease this house.

And now with respect to your own plans. When I thought of giving up St John's[42] 15 years ago the main objections was to my mind that I should have no curacy to offer you when God gave me the privilege of a son in the ministry – the only pang I have since felt in being cut off from parochial work has been the same. The fondest hope of my heart for the future year has been that I might be associated with you in your ministerial charge. The little specimen I had at St Leonard's served to increase this desire to an intensity which would make a disappointment a most severe trial: so that I shall take no step towards a settlement any where till you take a curacy. Released as I now am from Surbiton I shall enquire of all my friends, right & left, to hear of one.

In the mean time I feel very ominous that you should spend 2 months or so at Hereford. I am sure that uncle John's instructions in delivery would be most useful to you. These are just what you want – Strachan[43] said your sermons were delightful

---

[40] [Not identified.]

[41] [The British company which delivered and collected furniture which its customers wished to store, using a pantechnicon van, a furniture removal van drawn by horses.]

[42] [In 1834, Venn'had accepted the living of St John's, Holloway, in the gift of Daniel Wilson. When he resigned St John's in 1846 to devote himself fully to the work of the C.M.S., he 'felt a pang at the thought […] that I might be depriving myself of the future opportunity of initiating my sons into the Ministry, and I could sometimes now [1861] wish I had a church for this purpose'. Henry Venn quoted in Knight, *Memoir of Henry Venn*, on p. 102.]

[43] [Probably J.M. Strachan. See note **38**.]

all they wanted was to be preached instead of being read. I neglected this art till it as too late - & my influence has been consequently sadly hindered till it pleased God to open to me employment of another kind. But the art of delivery is far more thought of now than in my day: & I am persuaded that it will every day become more & more important. God has given you a pleasing voice all you want is management & the "actio, actio &c".

Most pleasant as it w^d^ be to have you at Braemar I w^d^ give up that joy if it interfered with Hereford.

Harry has indeed left you in the lurch: & I hardly know what to advise you to do, whether to go at once to Hereford & then join us 6 weeks hence in Scotland or to go now a tour in Scotland & break off from us to go to Hereford. There will probably be circumstances to guide you beyond those w^h^ are present to my mind.

[…]

Ever your most aff^te^ Father

Henry Venn

Poor Barton[44] (of Christ's) is at Rydal. He has lost his infant as well as his wife. He will be ordained by the Archbp in September – & go to India shortly afterwards. I have had some most touching letters from him.

12.
CMS C34/78
Henry Venn of C.M.S. to John Venn
2 August 1860
Braemar Aberdeen

My very dear son,

Many happy returns of the day on which I hope this will reach you: & may each of them, as hitherto, be commemorated with increasing gratitude to God & parental joy. The last year has given me a large portion of your society which has been an unspeakable satisfaction [to] me. Though it has been a pause in your ministry, yet it has I trust been a health-gathering time & that now we may look forward to a settled sphere of ministerial labour after a few months of "practice" at Hereford,[45] & of observation of a well worked Parish through the blessed wisdom of female ministrations as well as the Pastor's.

My constant prayer for you in that the Lord may furnish you with every needful gift & then place you in the right post of labour, where you may have the highest of all joys that bringing sinners to a vital union with Christ. Much as I long to be near you, this desire is, I thank God, as nothing compared with the desire of your future "crown of rejoicing" & I cannot bear the thought of interfering with that object for

---

[44] [John Barton. See note **29**.]

[45] [Having no regular curacy following his departure from Hastings in March 1859, Venn spent the summer of 1860 at St Peter's, Hereford, with his uncle and aunt. Since they were initially absent from the parish, it was Venn's responsibility to take services.]

the sake of my own comfort or profit. But I have a good hope that our most gracious master may unite both.

[…]

Ever your most aff^{te} Father

H. Venn

13.
CMS C34/80
Henry Venn of C.M.S. to John Venn
11 September 1860
C.M.H.[46]

My dearest John,

I went over by an early train to Mortlake – & have just returned.[47]

The Church is not formidable. It is much less than Cheshunt a modern brick square ugly building with a flat roof & Manley[48] says it is a very easy church to fill – I sh^{d} have judged its capacity at 800.

Every thing which I could see & hear of the Parish is agreeable – Cheshunt upon a smaller scale. The richer people of a higher class – & none of the very poor – no tramps lodging houses &c – more villagers.

The healthiness most satisfactory – a gradual slope of gravel from Richmond Park to the river – walks in all directions.

Mr Manley is a very young looking man – very pleasing – & I hear the highest character of preaching. His views very decided on the right side.

Altogether the curacy fulfills every condition I have ever ventured to desire for you.

As to a residence for me, there are not often houses in the parish vacant – but Putney on one side & a part of Richmond on the other are each within a mile & a half – & far more pleasant walks than from Surbiton to Norbiton.

Mr Manley is now desirous of a personal interview with you – & if you could go to him on Saturday the 22^{d} & spend the Sunday there, I doubt not that you would equally desire & clench the connexion.

But I must not my beloved son interfere farther in the matter. It is too solemn for any consideration but those of a direct kind between the great Head of the Church & yourselves – I told Manley therefore that I regarded him as unfledged in any way till he & you had talked the matter over – & now I shall only wait & pray.

Ever most tenderly yours

H. Venn

---

[46] [Church Missionary House, Salisbury Square, London.]

[47] [After he turned down an opening at Holbeton, Venn decided that he would just take any curacy that might offer itself and to give up the hope of finding a place where he could live with his father and sister. This was more than pragmatism in the face of a lack of options, as Venn also longed for a fresh start away from the expectations of others, especially his father.]

[48] [John Thomas Manley (1830–1901), Trinity, 1848; B.A., 1852; curate of Clapham, Surrey, 1852–55; perpetual curate of Mortlake, Surrey, 1855–64; vicar of Tonbridge, 1864–1890. See pp. 106–107 of the *Annals* for Venn's opinions on Manley.]

14.
CMS C34/83
Henry Venn of C.M.S. to John Venn
2 August 1861
Braemar

My very dear son,

I cannot be sure that this will reach you on your birthday. But it will assure you that I have not passed over that day without notice. The leisure & retirement which I here enjoy enable me to give to you & dear Harry a larger portion of my thought than when I am with you – & to feel more abundantly thankful to my heavenly Father for the blessing of such sons. Such feelings are not unmingled with many regrets that the pressure of my employments when at home has so much interfered with any advantage which you might otherwise have derived from our mutual intercourse: not unmingled also with self reproach that I have not made more effort to secure time & opportunity for your benefit. These feelings quicken my prayers for you & lead me to form many resolutions for the future if we are permitted to spend the future together. You will easily conceive also how constantly my mind turns to dear Harry & his future course – I shall be anxious to hear from you about him. I trust you speak freely together & that you pray much with him as well as for him. Many many prayers have I made with my younger brother, your dear uncle, from the time he went to Haileybury[49]: & at many critical times since: & God has often blessed such spiritual intercourse to both of us. There was a time when uncle John had scruples about remaining in our Church. We had much discussion on the point: but I am confident that far more was gained by our united prayers, & by our joint reference of the matter to our adorable Lord & Saviour.

[…]

With tenderest love to Harry

Ever your most aff^te^ Father
Henry Venn

15.
GCA C12/1
James Fitzjames Stephen to John Venn
19 January 1862
77 Gloucester Place W.

My dear John,

I read your paper[50] through with the greatest interest & admiration. It is long since I have read any thing so clear terse & vigorously thought out. I hope you will publish it & think I can undertake to get it into either Fraser or Macmillan Magazine

---

[49] [The Haileybury Imperial Service College, where John Venn of Hereford had studied, 1818–20, before going up to Cambridge in 1823, where he graduated B.A. (12th Wrangler) in 1827.]

[50] [Anon. [John Venn], "Science of History," *Fraser's Magazine* (May 1862): pp. 651–660.]

as you please. If you do I would give you a few observations on style which I have learnt the force of by long & laborious experience, but as to substance I have little or nothing to add.

If you care to see what my thoughts on the whole matter are you will find them in the following places – two papers on the Study of History in the Cornhill Magazine for June & July last – a review of Austin on Jurisprudence in the Oct number of the Edinburgh (see the end of the article) & a review of Buckle's first volume published in the Edinburgh (I think in April) 1858.[51] In the review of Austin & Maine you will see that I come very near to the point which you handle so well, from another side of the question.[52] The difficulty to which you refer has often crossed my mind but I never could get hold of it fully enough to work it out as you have.

I think however that you do scanty justice to positivism & sociology[53], for often the prediction is not really the object of their system – though I admit they say it is. What they ought to claim, & would claim if they thoroughly knew their own case is that they apprehend more clearly than others the possibility of devising a <u>classification</u> of human actions which will enable them to make contingent & hypothetical predictions. 'If you do so & so then so & so' – (what you excellently call Jonah prophecies).[54] I have said all I have to say on that point in the Cornhill Magazine for July last. It appears & always has appeared to me that if they would lay aside their strange jargon & their arrogance sociologists & positivists would find themselves as orthodox as the rest of the world.

I hope you will some times manage to let us see you, it will always give me great pleasure.

Ever yr affectionate cousin

J.F. Stephen

---

[51] [Anon. [Fitzjames Stephen], "The Study of History," *Cornhill Magazine* III (June 1861): pp. 666–680; Anon. [Fitzjames Stephen], "The Study of History," *Cornhill Magazine* IV (July 1861): pp. 25–41; Fitzjames Stephen, "English Jurisprudence," *Edinburgh Review* 114 (1861), pp. 56–86; Fitzjames Stephen, "Buckle's *History*," *Edinburgh Review* 107 (1858): pp. 465–512. **Fitzjames Stephen**'s articles in the *Cornhill Magazine* were praised by **John Stuart Mill** in the fifth edition of his *System of Logic* of 1862.]

[52] [Both Venn and **Fitzjames Stephen** argued that a science of history, or any social science, is irrelevant to morality since the predictability of human actions does not mean that these actions are not free. Both contrasted social and physical science to make their point, albeit in a different way. Venn argued that it is always possible for humans to change their course of action in light of the prediction made about it. Fitzjames Stephen argued that social laws do not show that humans are not free but only that they tend to exercise their freedom in a regular manner.]

[53] [For Fitzjames Stephen, and others with him, positivists were essentially those who, like Comte in France and Mill and Buckle in England, believed that morals could be divorced from religion. See Colaiaco, *James Fitzjames Stephen*, chapter 7.]

[54] [Jonah or Nineveh predictions prophecies: examples of cases where the prediction of something evil, like a tyranny, will cause that evil to be avoided.]

III. Moral Scientist: 1862–1899

16.
CMS C34/86
Henry Venn of C.M.S. to John Venn
16 October 1862
East Sheen

My dearest John,

I cannot say how much I miss you. The thought of being in the midst of your X[h] labours,[55] & of seeing you at dinner, was a source of pure happiness, of which I was not insensible at the time, but of which I feel now the value by the void it has left. My thoughts are continually turning to your present occupations, & I try to realize them. Let us keep up frequent communications. This will be my best, though a poor substitute, for our daily meetings during 20 happy months. [....]

I hope you will 'lay yourself out' for the freshmen – a work of kindness in the first 10 days is worth 100 during the rest of the term. You will have to take them to be matriculated[56] & that is full introduction enough. Your public lectures will wholly depend upon your private influence – amongst the many regrets with which a retrospect of every portion of my past life fills me, few are more acute than my neglects of the many precious Octobers among the freshmen of Cambridge – I thought I w[d] wait till their characters were ascertained &c &c &c many other idle thoughts beguiled me.[57] But whenever I did have a few & had serious conversation with them – preparing by secret prayers for a blessing upon the intercourse – my gracious God gave me abundant encouragement – infinitely beyond my expectations.

Xavier[58] will be published on Tuesday next. I will send you a copy this week – collect any of the reviews w[h] notice it.

Ever your most aff[h] Father, H Venn

[55] [When Venn had become curate of Mortlake in late 1860, his father moved the family home from Highbury to East Sheen, giving his son a home while retaining sufficient proximity to the C.M.S. headquarters. Venn left Mortlake in July 1862.]

[56] [Anyone studying at a Cambridge college was required to go through a matriculation, or formal registration process, so as to be entered on the University list. This often took place at the start of their residence, but some students did not, or delayed for months or years.]

[57] [**Henry Venn of C.M.S.**, who had graduated B.A. (19th Wrangler) in 1818, returned to Cambridge in 1824, having worked a few years as curate of St Dunstan-in-the-West, to take up the posts of Dean, College Lecturer and Junior Proctor, 1825–26.]

[58] [Among **Henry Venn of C.M.S.**'s occasional publications was the 1862 *Missionary Life and Labours of Francis Xavier*. A biography of a Catholic saint written by a Protestant, the controversial book analyses Xavier's mission, method and achievements from a non-Catholic perspective. The book met with a mixed but generally negative reception, as neither the Catholics nor the Protestants really grasped its ambition.]

17.
CMS C34/87
Henry Venn of C.M.S. to John Venn
22 October 1862
East Sheen S.W.

My dearest Johnny,

On Sunday you were very much in my thoughts. I had to preach a C.M. sermon at Clapham, at St Paul's Ch., which stand in the site of the old Clapham Church in which my grandfather was curate, of which my Father was Rector & used sometimes to officiate.[59] Beneath the pulpit lie the remains of both. It happened to be the first time I had stood in that pulpit. In the congregation I could discover no one who remembered me except the Clerk who had played with me a boy – he being the son of a former Clerk.[60] Thus all my recollections were thrown into the past. The position I occupy as the 3d & which you occupy as the 4th generation of those who have endeavoured to witness to the Truths of the blessed Gospel which bring salvation & peace to the soul of fallen man.[61] Many an earnest prayer did I put up as I knealt [sic] at the Communion Table giving the prayer, that God might make your witnesses bright & strong – that the God of Abraham & Isaac & Jacob might be able the God of Joseph & replenish him with the grace which shall enable him to turn all other acquisitions to his glory & after all to esteem the reproach of Christ greater riches than all the treasures of Egypt.

[…]

Mind you look after Julius Elliott.[62] There has been such a blessed friendship between the Fathers that I shd earnestly desire to hand it down – & you & Julius are about the relative standing of H.V.E[63] & myself. Pray call also on John Thornton just entered at Trinity – he was a high boy at Rugby & is the eldest son of the eldest son

[59] [Henry Venn of Huddersfield (1725–1797) and John Venn of Clapham (1759–1813). The former, elected curate of Clapham in 1754, was one of the founders of the Clapham Sect; the latter was rector of Clapham from 1792 until his death in 1813.]

[60] [Not identified.]

[61] [The Venn family's Evangelicalism had begun with Henry Venn of Huddersfield's conversion around 1785. A central figure in the English Evangelical revival movement of the late eighteenth century, he was the author of *The Complete Duty of Man* (1763) – a popular handbook of Evangelicalism. He was presented as an Evangelical 'saint' in a biography begun by John Venn of Clapham and finished by Henry Venn of C.M.S.: *Life and Selection from the Letters of the Late Rev. Henry Venn* (1834).]

[62] [Julius Marshall Elliott (1841–1869), Trinity 1860; B.A., 1864; curate of Pluckley, Kent, 1865–66; P.C. of St Mary's Brighton, 1866–96; made the second ascent of the Matterhorn, 24–25 July 1868; killed 25 July 1869, by a fall, while ascending the Schreckhorn.]

[63] [Henry Venn Elliott (1792–1865), Trinity 1809; B.A. (14th Wrangler) 1814; Fellow, 1816; Minister of St Mary's, Brighton, 1827–7; a strict Evangelical whose public objection against the proposal for opening the Crystal Palace on Sundays caused much controversy.]

of John Thornton of Clapham.[64] He will know me well. He was an early pet of Margaret Thornton,[65] & no common child from principle. I so long to have an interview with you that I meditate running down for a night or two, as soon as I am free from other engagements. But I have sermons on the 3 next Sunday, for the Soc.[66] away from this place.

I have seen E.B. Elliott[67] & he has sent me a copy of his most severe strictures on Alford's commentary.[68] He says he has refuted the Com^s on the Revelation, & reduced the Com^s on S^t. Mark to very small dimensions.[69] He shewed me a very meek letter from Alford saying strong blows from a strong man were very serious injuries. But E.B.E. did not relent, one said – he needed them.

Ever your most affe^te Father

H. Venn

18.
CMS C34/88
Henry Venn of C.M.S. to John Venn
5 November 1862
East Sheen S.W.

My dearest Son,

I send you the only two letters about Xavier which I have yet received. Nine tenth of the acknowledgments are wisely sent to me before the book has been opened. If you see any notices in print let me know.

---

64 [John Thornton (1843–1918) entered Trinity in March 1862; graduated B.A. in 1866; afterwards curate at various country parishes. He was the grandson of John Thornton (1720–1790), whose eldest son was another John Thornton.]

65 [Probably Caroline Margaret Thornton (née Christie) (?–1881), who was married to Harry Thornton (1812–1885).]

66 [Church Missionary Society.]

67 [Edward Bishop Elliott (1793–1875).]

68 [In 1862, E.B. Elliott, a prominent pre-millenarian, published *Apocalypsis Alfordiana; or, Five Letters to the Very Rev. H. Alford*, which was a refutation of Henry Alford's (1810–1871) commentary on the Apocalypse in his edition of the *New Testament in Greek*, 8. vols. (1841–61).]

69 [Reference is probably to then standard commentaries on Revelation and the Gospel according to St. Mark. The main purpose of E.B. Elliott's eschatological oeuvre was to defend the supernatural inspiration of scripture against rationalist attacks from within the Protestant camp. As part of his historicist approach to Revelation, he also proposed a time-table of the end times of his own, defending the view of a pre-millennial advent of Christ.]

Can you give me any information about Leslie.[70] His mother fears that he is one with Maurice.[71] What is said about Colenso[72] & Maurice? I understand that a strong effort is being made by the Bp of London to induce Maurice to remain in the Church.[73] Another party is trying to get a relaxation of the terms of subscription to the Articles in our Church.[74] I fear that we are on the eve of great convulsions for these questions will touch the foundation of a national church.

The Trustees for St Pauls met on Monday but we could not decide upon a man: much is in favour of Wigram but we all want a man who c[d] take a more forward & firm position in the university.[75]

How are you getting on with lectures – what are your subjects?

---

[70] [Venn's cousin **Leslie Stephen** (1832–1904). The fear was that Leslie Stephen would, would act upon his religious doubts. He initially continued to perform his clerical duties, taking refuge in a Broad Church view of the 39 Articles, but at some point in late 1862 he ceased to take part in chapel services. This obliged him to resign his position as Tutor and Fellow at Trinity Hall. He left Cambridge in 1864 and eventually resigned orders in 1875.]

[71] [**Leslie Stephen** was the son of Sir James Stephen and Jane Catherine (née Venn) Stephen, **Henry Venn of C.M.S.**'s sister. In 1862, **F.D. Maurice** (1805–1872) was working from London, where he had grown in disfavour with all major religious Church parties, many believing that his Christian Socialism implied acceptance of various atheistic and immoral doctrines. In 1860, in spite of controversy, he had been appointed as incumbent of St. Peter's, Vere Street, a position he would hold until 1869.]

[72] [John William Colenso (1814–1883), mathematician, biblical scholar and first Bishop of Natal who had come under the influence of **F.D. Maurice**. His high-minded views about the treatment of African natives, for instance allowing for polygamy among converts, had created a controversy in 1853; several years later, in 1861–62, he again raised an outcry with his commentary on the Epistle to the Romans, dismissing, for instance, the idea that Holy Communion was necessary to salvation.]

[73] [In 1862, **F.D. Maurice** wrote to Archibald Campbell Tait (1811–1882), Bishop of London between 1856 and 1868, to announce his intention to resign his post at St. Peter's, Vere Street. His motive was to be able to distance himself from the views of, among others, Rowland Williams and Henry Bristow Wilson – two of the authors of *Essays and Reviews*. But since his own controversial views were often identified with theirs, the step was expected to be interpreted to mean exactly the opposite – namely, sympathy with the accused. Hence, the Bishop begged Maurice to reconsider his decision, which he did. See the Maurice–Campbell Tait correspondence in Randall Thomas Davidson and William Benham, *Life of Archibald Campbell Tait. Archbishop of Canterbury* (London: Macmillan and Co., 1891), chapter 18.]

[74] [The debate about the authors of *Essays and Reviews* was focused on their alleged honesty or dishonesty. Both orthodox churchmen and the positivists challenged them to be honest in acknowledging the incompatibility between their position and the formularies of the Church and, by extension, the honest retention of their clerical status. Many liberal Broad Churchmen argued that that honest participation in the Church and ministry did not depend on precise and uniform acceptance of its doctrines, but on one's relationship to the worship and doctrine of Church as a whole. **Fitzjames Stephen** defended Rowland Williams by arguing that he was exercising the liberty sanctioned by the 39 Articles: what he had said was perhaps not correct but it was not forbidden.]

[75] [**Henry Venn of C.M.S.** was one of the trustees of St. Paul's, Cambridge. Wigram had been made assistant curate of St. Paul's Cambridge in 1858. The decision was probably whether Wigram should be made vicar.]

You are to have the terminal C.M. meeting on Saturday. I hope you will be there & say a kind word to Col. Martin.[76] He is a man whom I came to honor greatly. He was for several years a most active Missionary in North India, taking a salary to make himself "subservient" but giving us back five times his salary each year.

Your most aff[te] Father

H. Venn

19.
CMS C40/11
John Venn to Henry Venn of C.M.S.
7 November 1862
Caius College

My very dear Father,

Many thanks for your letter, – I return with this the two letters you sent. You never doubted, I should think, that as one of them says, the Guardian & such other papers could make it a sort of party cry or fight.[77]

I have not heard very much about Maurice here, though I have heard something. He was up here last week staying with Kingsley,[78] & two or three of our fellows met him there. It seems that he had fully resolved on giving up his benefice, & had in fact gone so far as to dispose of his houses but he was persuaded to change his plans. His grounds for resigning seem to me strange & almost fanciful. It was not that he felt his position a false one, so they say, but he is going to oppose Colenso, & wished to obviate the taunt "Oh yet you oppose it professionally, because you are paid to". I thought it strange at first, but it seems from what I have heard that such were his motives.

Colenso's book[79] I have not read or seen, but I should not think it would do so much mischief; even those who are inclined that may speak disparagingly of it; they seem to be all little arithmetical difficulties upon which he has fastened.

I am going off next Sunday to take duty for a late fellow of Trinity whose wife has died, – Ellis[80] by name; he has a small living near Baldock; I wish to have

---

[76] [Colonel W.J. Martin. No further information available.]

[77] [Letters which **Henry Venn of C.M.S.** received upon the publication of his *Life of Xavier*. The worry was apparently that his Protestant take on a Catholic saint would be interpreted as a party attack.]

[78] [Charles Kingsley (1819–1875), Trinity, 1838; B.A. (Classical Tripos, 1st Class), 1842; chaplain to the Queen, 1859–75; Regius Professor of Modern History at Cambridge, 1860–69. Kingsley was one of the electors in favour of **F.D. Maurice**'s election as Knightbridge Professor of Moral Philosophy in 1866.]

[79] [Probably Colenso's 1861 *Commentary on the Epistle to the Romans*.]

[80] [Arthur Ayres Ellis (1830–1887), Trinity 1848; Classical Tripos, 1st Class, 1853; B.A. 1852; Moral Sciences Tripos, 1st Class, 1853; Junior Dean, 1858–60; Divinity Lecturer at Christ's, 1853–57; afterwards vicar of Stotfold, 1860–87; Greek testament critic.]

something to do on a Sunday, but that is somewhat further off than I should generally feel inclined to go.

After that I am, I think, engaged every Sunday till the Xmas vacations; there is a Rev. Preston[81] who has the living of Reach & Swaffham near Burwell who had been disappointed in a curate, so I engaged to assist him on the Sundays whilst I was up, that is for the present; when the weather is bad I think I shall go from Saturday till Monday. It seems strange to think that I have got half way & more through the term; I suppose I shall have done work by about the 10th of December.

About Leslie I know nothing[82]; I never see him & we have scarcely any friends in common. I don't know why he gives up the Tutorship; there the tutors lecture, & very possibly he was tired of lecturing. I remember at Llanberis he said he never hoped to give another mathematical lecture. Col. Martin will dine in our Hall tomorrow, so I shall see him there as well as at the meeting. Yr very affte Son,

J. Venn

20.
CMS C40/12
John Venn to Henry Venn of C.M.S.
18 November 1862
Caius College

My very dear Father,

[…] I still go on with my work, & like it very much; the number keep about the same as they are mostly regular, but I cannot get any of them to come at any other time but the regular one in the morning; somehow as yet they seem to keep out of it, but one will be able to see in time.

There is a meeting here tonight for the Xn Vernacular Soc[83]; I think I shall go if I have time; Titcomb[84] is going to speak at it. Carus[85] too is up & about; stopping at

---

[81] [Thomas Preston, vicar of Swaffham Prior and Reach. No further information available.]

[82] [Venn may have overstated in order to reassure his father, for upon Venn's return to Cambridge in 1862 both he and Leslie Stephen were involved in the organization of the Moral Sciences Tripos and had several shared acquaintances. At the same time, Venn would later emphasize that he and his cousin did not fit well together as young men. See letter **122**.]

[83] [The Christian Vernacular Education Society for India, established in 1858. Its aim was to establish training institutions and supply school books.]

[84] [Jonathan Holt Titcomb (1819–1887), Peterhouse, 1837; B.A. (Junior Optime) 1845; popular preacher who held several curacies and was active in numerous organizations; first Bishop of Rangoon, 1878–1882; one time Honorary Secretary of the Bombay Branch of the Christian Vernacular Education Society for India.]

[85] [William Carus (1802–1891), Trinity 1821; B.A. (28th Wrangler and 9th Classic) 1827; Dean, 1832–50; P.C. and Lecturer of Holy Trinity, Cambridge; 1832–37; Lecturer, 1832, and P.C. of Great St. Mary's, Cambridge, 1844–51; Canon of Winchester, 1851–85; Select Preacher at Cambridge, 1854, 1859 and 1866; vicar of Christ Church, Winchester, 1860–70; an eminent Evangelical and friend of Charles Simeon.]

present with our master,[86] he is going to give a lecture about Simeon. I suppose you have not settled about St. Paul's[87] yet; it lies out of the way & so I do not hear much about it. Give my love to Henrietta.

Ever y[r] most aff[te] son

J. Venn

21.
CMS C40/13
John Venn to Henry Venn of C.M.S.
30 November 1862
Caius College

My very dear Father,

I do not think that there has been anything particular going on since I last wrote. I keep a look out at the reviews &c but I have not seen any further notice of Xavier; I will tell you whenever I do. I have been helping Clayton a little in his parish,[88] I mean in visiting; his curate has gone & he has not yet got another. I did not know before what a large extent of out-of-the may streets & small house there were in the parish. Rather curiously one of the first persons I went to see was an old woman who had been most of her life in Islington, with some people who were great admirers of Wilson[89] & his predecessors, Blunt I think their name was.

[…] The term is beginning to show symptoms of coming to an end; the lectures close on Saturday week. Kingsley, the History Professor has nearly come to an end; I have attended all his & liked them, as lectures exceedingly; he manages to fill the largest lecture room, but that may be partly owing to the subject – the history of the United States.[90] I am glad Harry has come back & enjoyed himself so much. Give my love to Henrietta.

Ever y[r] most aff[te] son,

J. Venn

---

[86] [Edwin Guest (1802–1880), Caius, 1819; B.A. (11th Wrangler) 1824; Fellow, 1824–52; Master, 1852–80; Vice-Chancellor, 1854–55; antiquarian, and an intimate friend of Goethe.]

[87] [Probably St. Paul's Church, Clapham. Perhaps it was considered an option as part of Venn's search for more regular clerical duty.]

[88] [**Charles Clayton**, who was then vicar of Holy Trinity, Cambridge.]

[89] [**Daniel Wilson**.]

[90] [Kingsley's annual lectures on the United States in the 1860s made him aware of how little and could be known about the subject. He advocated the creation of an American Lectureship at Cambridge in 1866, a proposal refused by the university. See G. Martin, "The Cambridge Lectureship of 1866: A False Start in American Studies," *Journal of American Studies*, 7:1 (1973): 17–29.]

22.
CMS C40/15
John Venn to Henry Venn of C.M.S.
1 February 1863

My very dear Father,

[...] I have been preaching twice today for Wigram at St. Pauls'[91]; I got a note from him yesterday saying he was not well, & should be glad of help, which of course I was very ready to give as I had nowhere else to go; the body of the church seems small but it runs up an enormous way, as it seems into the tower; but I found it very easy. I have not found any regular duty yet, & it is not a very easy thing to do, unless I were to take it for the vacation as well, which I should not wish to do at present. I found that Mr. Cook at Peterborough[92] was supplied for most of the Sundays of this term & after that he has a curate coming, so I do not think it very likely that I shall be going there again.

I have been thinking of a plan for the summer, which has not been quite matured, but which may come to something. If you go to Switzerland I should certainly go too, but I should very much like to see something of the north of Italy afterward. Do you know anything about the getting of those foreign chaplainries, for instance such a one as that at Florence? Ponton,[93] a friend of mine, had it last autumn; I believe Mr. M. Thomas[94] is the secretary unless he has left it already for his bishopric. If I could get such a place I think I should like it very much for a few months in the autumn. I shall have no occasion to be at Cambridge next October term, if I could find anything to do elsewhere. I lectured last term, but I am inclined to think that to do so every term in such subjects scarcely answers; I doubt if they come so well as when it is made intermittent. I shall be glad to know what you think of it.

Yr very afft. son

J. Venn

23.
CMS C40/16
John Venn to Henry Venn of C.M.S.
11 February 1863
Caius College

My very dear Father,

[...] I have undertaken to help here as secretary to the Jews' Society,[95] not as principal but as a sort of deputy to Jameson who living at Coton finds it

---

[91] [Frederic E. Wigram (1834–1897), who was then curate of St. Paul's Church, Cambridge.]

[92] [Not identified.]

[93] [Alexander Campbell Ponton (1835–c. 1867), Caius 1853; B.A. 1858; curate of Bloomsbury, London, 1864–6.]

[94] [Not identified.]

[95] [The London Society for Promoting Christianity Amongst the Jews. Founded in 1809 by leading Evangelicals, including William Wilberforce and Charles Simeon, it was the first missionary organization devoted exclusively to converting the Jews.]

inconvenient to keep it.[96] In performance of my office I went to the meeting at the Town Hall, sitting on the platform, but as secretary, not saying anything; Emilius Bayley & a Mr. Wright were the speakers,[97] but I thought it very poor. They talked fluently enough, but it was random stuff, & the stories far fitter for a village children's meeting.

I have also taken the secretaryship of the C. & C. Ch. & S..,[98] but that is little more than nominal, – receiving & sending in the money received from one or two charity sermons. There are not many who will take these things, so they get very much lumped into the same quarter. Perowne has a large share for one.

I have been preaching or reading most Sundays; last Sunday I was at Perowne's at the Round Church[99]; Steel,[100] who is acting as his curate, being elsewhere. The Sunday before, as I think I said, I gave two sermons at St. Paul's for Wigram who was not well.[101]

My work is getting on very favorably; the men seem still interested, & a couple more have joined this term. The term is so short that I almost begin to look forward now to returning. The College examinations begin in a month. [...]

Y^r^ very aff^t^ son

J. Venn

24.
CMS C40/17
John Venn to Henry Venn of C.M.S.
24 April 1863
Caius College

My very dear Father,

I shall certainly hope to come up to London for the C.M.S. meeting; it is I believe Tuesday week. I have duty on the Sunday so I am afraid I can't, as Henrietta

---

96 [Francis James Jameson (1828–1869), Caius 1846; B.A. (6th Wrangler) 1850; Fellow of Caius, 1852–55; Fellow of St Catherine's, 1855–1862; vicar of Coton, Cambridge, 1862–1869. Jameson had been one of Venn's mathematical coaches.]

97 [John Robert Laurie Emilius Bayley (1823–1917), Trinity 1846; transferred to the list of 'Ten-year' men, 1847; B.D. 1862; afterwards vicar of St. John's, Paddington, 1867–88; Select Preacher at Cambridge, 1881. Probably B.W. Wright, mentioned once in W.T. Gidney, *The History of the London Society for Promoting Christianity Amongst the Jews. From 1809 to 1908* (London, 1908), on p. 212.]

98 [The Colonial and Continental Church Society, begun in 1823; the name was changed to Colonial and Continental Church Society in 1861. It is known today as 'Intercon'.]

99 [Thomas Thomason Perowne (1824–1913), Corpus Christi 1843; B.A. (31st Wrangler) 1847; Select Preacher, 1861; Hulsean Lecturer, 1880; afterwards Archdeacon of Norwich, 1879–1910. Perowne was curate of Holy Sepulchre (generally known as the Round Church) between 1862 and 1867.]

100 [Anthony W.W. Steel (c. 1837–1885), Caius 1855; B.A. (2nd Wrangler) 1859; Fellow, 1859–85; Tutor, 1880–85; curate-in-charge, 1867–73, and vicar, 1873–78, of St. Sepulchre, Cambridge.]

101 [See letter **22**.]

proposed, spend it at Sheen. Lamb[102] is going away for the week, & I have promised to help him with a sermon. Next Sunday also I have some work found for me, at the Round church. J.J. Perowne has succeeded Jameson there,[103] & is I believe very much liked; he gets a very fair number of University men to go to him. […] I had a visit from the secretary of a sort of little concern which is a kind of Missionary library, & has its books at Hutt's shop. They wanted a copy of Xavier,[104] which of course I gave them, having a spare one of my own. I shall come & claim another when I come home. Ferrers[105] has been making arrangements about the lakes; that is to say he has written to engage lodgings for himself, in a place where there w^{d} be room for me to [*sic*]; but I have not yet made up my mind for certain. […]

Y^{r} most aff^{te} son

J. Venn

25.
CMS C40/18
John Venn to Henry Venn of C.M.S.
8 May 1863 [= 1865]
Caius College

My dearest Father,

I send back with this the Clerical Subs. Commis. Papers,[106] which are very interesting; many thanks for them. It certainly seems from them & from what you were saying at Aunt Jane's[107] last week, that the Commission is coming to something definite; I shall be very glad indeed to see the accounts of some further proceedings, as soon as they are taken down & copied out. […]

I preach this evening at the little Abbey church, which is now held by Thorpe.[108] Valpy,[109] the missionary, used to be there. It is a little bit of a church, but I have been

---

102 [John Lamb (1825–1880), Caius 1844; B.A. (11th Wrangler) 1848; Fellow, 1849–77; Dean, 1852–60; Hulsean Lecturer, 1860. Lamb was vicar of St Edward's, Cambridge, 1858–66 and Rector of Blofield, Norfolk, 1876–80.]

103 [Venn seems to mistake John James Stewart Perowne for Thomas Thomason Perowne, who succeeded Francis James Jameson as curate of St. Sepulchre, Cambridge, in 1862.]

104 [**Henry Venn of C.M.S.**'s biography of Xavier, published in 1862.]

105 [**Norman Macleod Ferrers**.]

106 [By 1862, **Henry Venn of C.M.S.** was a recognised leader of the Evangelical party in the Church of England. This position was acknowledged by his being place on two Royal Commissions, that on Clerical Subscription and Ritual Reform. The Clerical Subscription commission was appointed in 1864 to examine the terms of clerical subscription. It resulted in the Clerical Subscription Act of 1865, which provided a relaxed form of subscription, with honest subscription no longer implying precise subscription]

107 [Venn's aunt Jane Catherine Stephen (née Venn), wife of Sir James Stephen.]

108 [Not identified.]

109 [Thomas Valpy French (1825–1891), Christian Missionary in India and Persia who became the first Bishop of Lahore, in 1877, and founded St. John's College, Agra, in 1853.]

there once before & prefer it to many others. The Sunday after next I go to Barton 4 miles on the Wimpole road.

I have not seen or heard of any place for the summer; If one does not turn up very soon now I sh^{d} be almost inclined to advertise. Robertson[110] was talking about some friend of his who was looking out for a temporary substitute, but he only wants it for a short time. Perhaps he told you something about it, when he was at home with Harry.

I shall have some work to do during the summer as I have been appointed, or rather shall be very soon, one of the examiners for the Moral Science Tripos. There is not any very formidable amount of work required for it, but as some of the subjects or authors I am not very familiar with, I sh^{d} have to brush them up a little during the vacation. I shall have to set papers I presume in Logic & Political Economy.

Love to Henrietta & thanks for her note.

Y^{r} aff^{te} son
J. Venn

26.
CMS C34/91
Henry Venn of C.M.S. to John Venn
31 July 1863
Lausanne

My very dear son,

All the exquisite beauty of the scenery & the endearments of [Looey?][111] & Harry cannot check the current of my thoughts which <u>sets</u> towards you & your employments; & often makes me conscious of a void. At such times I solace myself with the pleasant recollections of 20 months we spent together at Mortlake, & of your Sunday afternoon sermons, & I pray God soon to open the way to your having a sole change of souls, to whom you may again witness for Christ & for the power of his grace. The recurrence of another birthday reminds me forcibly of these things: & I shall offer many prayers on that day, not only for your growth in grace, but that you may have much wisdom & boldness in making manifest the truth of the Gospel, so that the salt may never lose its savour.

I never feel more humbled for my own deficiencies in this respect than when I am travelling. The snare of "relaxation" begets a negligence in ordinary intercourse which at home or in the C.M.H.[112] one is not so liable to. Precious opportunities of saying a word for Christ are thus allowed to pass by unimproved. At home one's reputation for religion will often give point & casual observations: but abroad we meet & part to meet no more as strangers. The salt loses its savour. Pray for me as I do for you that we may not be ignorant of this device of Satan.

---

[110] [Not identified.]

[111] [Probably a family sobriquet for Henrietta, Venn's sister.]

[112] [The Church Missionary House, Salisbury Square, London.]

[…] We have been amused by the intelligence sent us today by Long[113] that Clayton is to be married to a Miss Brown of Cheltenham.[114] How will this affect his position in College? Your Junior fellows when pleading for license to marry could hardly have anticipated Clayton's availing himself of it.[115] […]

Ever most aff^y yours

Henry Venn

27.
CMS C40/19
John Venn to Henry Venn of C.M.S.
17 October 1863
Cambridge

My very dear Father,

I should have written before now, but I have been away the last two days – at Kelshall. Birks[116] came over here for the P. Aid,[117] to preach & speak, & wanted some one to take his duty, so as I had nothing else to do I went. It was somewhat too far to go for the day only, so I went on Saturday & have just come back this morning. It is quite a little parish, wholly of poor, about 300 in number I believe. I was somewhat hurried & therefore took old sermons, but I certainly hope from time to time to write new ones, so as not to get out of practice. […] There was service in the morning & afternoon, & in the evening a sort of lecture like the cottage-lectures we used to have at Cheshunt. […] I shall be glad to get more of that sort of work about here in the neighbourhood, as I have no regular work of my own on Sunday.

I began my work here in College on Thursday last; as I expected very few came, in fact the subjects never will become interesting. Those who could really understand it are already engrossed in Classics or Mathematics, so for the most part the poor remainder fall to my share. I had 5 on Thursday, 2 of whom I think will do tolerably; they both intend to go in for the Civil Service Examination for India next year, & therefore have a purpose in reading. When I came here, of course, I knew scarcely any one, & it takes some time to get acquainted with them, & I don't think I have the knack of doing it well however I shall see in the course of another term or so. The best man for next January is a very nice fellow, Crosthwaite[118] by name, a

---

[113] [Not identified.]

[114] [**Charles Clayton**, who was then Tutor at Caius and vicar of Holy Trinity, Cambridge.]

[115] [Following the abolition of celibacy ruled in the new University statutes of 1860, Caius was the first College to sweep away all bars to matrimony on every member of its foundation. The first to take advantage of the new statutes was Steel (see letter **23**). The Master, Edwin Guest, had married in 1859 when the statutes were being drawn up. See Christopher Brooke, *A History of Gonville and Caius College* (Woodbridge: The Boydell Press, 1985), on pp. 224–226.]

[116] [**Thomas Rawson Birks** (1810–1883). In 1863, Birks was curate of Watton, Hertfordshire, and Rector of Kelshall.]

[117] [Prisoners' Aid.]

[118] [William Hey Crosthwaite (1840–1866), Caius 1859; B.A. (7th Wrangler) 1863; Fellow, 1863–66.]

cousin of the Jarrett's of N. Cave[119]; I should think he is pretty safe for a fellowship. There is a son also of some Mr. Cunningham of Harrow who seemed to know more about me than I did about him; he is a scholar in the 2d year & will also do well I think.[120] I have not been able to see J. Elliott or Thornton yet, [121] but shall go today if possible.

[…]

Yr most affte son, J. Venn

28.
CMS C34/95
Henry Venn of C.M.S. to John Venn of Hereford
10 February 1864
East Sheen S.W.

My dearest John,

You will have seen my name in the new Commission to consider the subject of clerical Subscription in the papers of this morning. […] It is curious that 30 years ago I joined with Scott, King, Dikes &c in a movement to get rid of "hearty assent & consent &c".[122] […]

I am most anxious to have your views of the subject. Draw me up a paper on the subject – & tell me of any persons whom I could apply to, who have thought upon the subject & are capable of throwing light upon it. […] Though I do not expect much good from the Commission I wish to be full prepared in a question touching the Truth of the Gospel continuing with us.[123] […]

Dearest Johnny is I am happy to say dissatisfied with Cambridge. Men will not attend his lectures in any number. This is universally the case in the 'moral' lectures. He longs for parochial work again. Thanks be to God. I shall at all events try to get him a temporary engagement for the long vacation. If we can find a country parish in any county surrounding Hereford I should rejoice beyond measure. […]

Ever most affy yours

Henry Venn

---

[119] [Not identified.]

[120] [William Cunningham (1849–1919), Caius 1869, migrated to Trinity 1872; Moral Sciences Tripos, 1st Class, 1872; B.A. 1873; Fellow of Trinity, 1891; Hon. Fellow of Caius, 1895; curate, 1879–83, and vicar, 1887–1908, of Great St. Mary's, Cambridge; University Lecturer in History, 1883–91; Professor of Economic Science at King's College, London, 1891–97; Archdeacon of Ely, 1907; Lowell Lecturer, 1914.]

[121] [Julius Elliott and John Thornton, see letter **17**.]

[122] [In the early 1830s, **Henry Venn of C.M.S.** was working in Drypool, Hull. The 'movement' referred to has not been identified. The individuals mentioned were members of the local clergy. See Knight, *Memoir of Henry Venn*, on p. 64.]

[123] [**Henry Venn of C.M.S.** saw in subscription 'to all and everything' a tool of a strong national Church. This view was not popular among the other commissioners. See Knight, *Memoir of Henry Venn*, Appendix ('December 11th, 1867).]

29.
CMS C34/96
Henry Venn of C.M.S. to John Venn
10 February 1864
East Sheen S.W.

My dearest son,

Many many thanks for your very affectionate wishes on this my birthday – on which I enter my 69th year. As long as our gracious God gives me strength to work in his cause, & power of mind to lay hold on the hope set before me, continuance in this life is a blessing: otherwise it is rather a dark dispensation. But I cannot sufficiently thank God that He has preserved me to be of some use to my most beloved children, & to see them all that I could wish: – & to fill up a gap in the management of a great society for the advancement of the Kingdom. Yet all these blessings would be causes of despair if there were not a fountain opened, in which my infinite shortcomings & impurities may be obliterated.

I cannot say with what joy I welcome your proposal that you should take a parish for the long vacation. We are constantly receiving applications from incumbents for such help from the C.M.H. & I will have my eye upon them. I could have no greater enjoyment in this life than to listen to your sermons & if I could spend a few weeks with you in such a situation it will be a far brighter prospect to me than Switzerland or Rome. If I can hear of such a place in the midland or western counties it might easily be combined with a visit to Hereford & Welsh mountains.

Upon the general question of your leaving Cambridge I agree in your remarks but we can talk it over at Easter. […]

You will have seen in today's papers the Commission on clerical subscription, & the respectable company in which they have put my name.[124] I shall very much like to have your opinion on what should be done – send me a few notes however brief.

I saw Prof Owen[125] a short time since who told me that he had been to the S. of France to see a cave lately discovered in which there are human fossil remains under circumstances far more important for the solution of the question of the antiquity of the race, than all formal discoveries. The bones were those of males & females, children & infants – mingled with the reindeer & the several extinct species of a very remote date. His inference was that the remains were far beyond the creation of Adam: but he had not yet worked the matter out.

---

[124] [For the list of names, including the Archbishop of York and the Regius Professors of Divinity of Oxford and Cambridge, see *Accounts and Papers of the House of Commons. Volume 44* (1864), on pp. 65–66.]

[125] [**Richard Owen** (1804–1892). Today known as the 1864 de Lastic Collection, this faunal collection was from a French site called Bruniquel, a Late Magdalenian cave in southwestern France explored first by Vicomte de Lastic Saint-Jal and later by Owen, who purchased the remains and Paleolithic artifacts for the British Museum.]

According to my notions of the scripture history the existence of a race or many races of men before Adam may be admitted during the interval between "the beginning" & the modern era when the earth was "without form & void". […]

Your most aff^{te} Father,

Henry Venn

30.
CMS C34/102
Henry Venn of C.M.S. to John Venn
6 July 1864
East Sheen S.W.

My dearest son,

I must send you a few lines to congratulate you upon being in the midst of your favourite mountains & that the weather is more favourable than when you started. I am sorry that your pupils have fallen through. But it is only another proof of the very slight hold which the moral subjects have on the men at college: & how difficult it will be to make them a reality in the university course.

[…] Your proposal for helping Faithfull[126] for a time seems to me very desirable for his sake & a profitable one for your own to renew old associations & revisit old haunts & such a voluntary assistance would leave at liberty to break it off if any other opening should occur & leave you time for a ramble with Harry in August or September. The idea of our all spending several weeks together in some country parish was a most delightful one to me. But there have been providential hindrances, & we must not murmur.

[…] Ever your most aff^{te} Father,

Henry Venn

31.
CMS C34/103
Henry Venn of C.M.S. to John Venn
30 October 1864
East Sheen S.W.

My very precious son,

[…] I was greatly interested by your account of your pupils & of the prospect of your "morals" becoming a reality. Let me hear how things go on. You will have done a good work if you encourage this branch of university studies. I rejoice also to hear of your taking services in the churches of the Town & of the prospect of a sermon in the chapel each Sunday. I greatly desire to get a day or two for a visit to you this term. The only drawback is that so many hours must be consumed in necessary

[126] [James Grantham Faithfull (1817–1873), vicar of Cheshunt, Hertfordshire, where Venn had worked as a curate in 1858.]

calls. If there were covered way from your rooms to the Library I think I should oftener spend a day with you.

[…] Harry [might be?] continuing his connexion with the C.M.S. for a year or two.[127] But I have not yet talked it over even with H. himself. My heart only years after some such arrangement - & the prospect which lies bedded deeper than any other in my heart, that you & may become shining witnesses of the Truth of the Gospel: taking full advantage of the many facilities which God has given you both, of maintaining such a witness in wisdom & in love. There are a Father's fondest hopes & daily prayers.

Ever your most aff^{te} Father,

Henry Venn

32.
CMS C40/21
John Venn to Henry Venn of C.M.S.
7 March 1865
Caius College

My very dear Father,

I told Henrietta a day or two that I could send you a line soon to say how the meeting went off &c. I think Fenn[128] answered very well. The meeting was not a crowded one, indeed not so full as some that I have seen […]; in the first place it was a wet & windy night, & in the second was a meeting convened that same evening about the Bishop of London's fund, at which Rowsell (the present University preacher) was advertised to speak.[129] However there was nothing to complain of. Fenn succeeded very well. There was no attempt at a formal set speech; he just stood up in a sort of conversational way & talked right on, which is a mode that I think the men generally like. Moreover he confined himself to his own subject, speaking about little else than India […]. There was only one drawback, & that was his manner & appearance when one looked at him. I should have thought he had had enough to prevent him from being nervous but he made the most extraordinary face & twisted about as if he was in agony. […]

I was at the dinner which they have this term at the Eagle on the Monday where some of the principal supports of the Society dine together.[130] Martin & Hall & others were there; Martin asked me to dine with him & Fenn at his house tomorrow.

---

[127] [Venn's brother, **Henry Venn of Walmer** ('Harry'), had begun his work as assistant-secretary at the C.M.S. I 1861. He would stay there until 1867, when he became curate of Mortlake, Surrey.]

[128] [Christopher Cyprian Fenn (1823–1913). He had worked as a C.M.S. Missionary in Ceylon in the 1850s. He subsequently became Secretary of the C.M.S.]

[129] [The Bishop of London's Fund, instituted in 1863 "for providing for the Spiritual wants of the Metropolis". Thomas James Rowsell (1816–1894), St John's, 1834; B.A., 1838; a popular High Church preacher who was active for the Bishop of London's Fund and several religious societies in London.]

[130] [Probably the (Cambridge) Church Missionary Society. The names mentioned have not been identified.]

I do not know that there is any thing particular going on in the University just at present except that matter about the examination for girls which is to be voted about on Thursday next.[131] I think I mentioned it in my last letter to Henrietta. It is an extension of the sort of extra-University examinations that have been had for some years past at different places about England. A large number of ladies & others connected with education wished the papers to be set for girls as well; a discussion in the Arts' school was held last week, & it is now to be voted on soon. I don't know which way it will go. But love to Henrietta. Ever yr most affte son.

J. Venn

33.
CMS C34/104
Henry Venn of C.M.S. to John Venn
11 March 1865
East Sheen S.W.

My dearest son,

[…] I was very much interested in your account of the meeting at Cambridge. We are intending to print D. Fenn's sermon & to make him speak at out Anniversary in Exeter Hall at the morning meeting.

I told Watts the printer to send you a proof of a pamphlet on the West African Colonies, which I wish you to correct & let me have again in a day or two.[132] It is somewhat in your way touching on questions of political economy – yet it is for the most part quotation. I am still anxious about the concluding suggestions whether the wording &c is clear enough. It is put together chiefly for the M.P's. who compose the Select Comme on this subject,[133] & is therefore hardly adapted for general readers. Mark in the margin any "hiatus" which strikes you – or any connecting sentences wh may make it more readable. […]

Did you read Dean Stanley'saddress[134] to a clerical meeting in the last Fraser but one? It seems to me simply to say, amidst a glorious profusion of fine writing, that

---

131 [At Cambridge, girls were admitted to the Local Examinations, established in 1858 and today known as the University of Cambridge Local Examinations Syndicate, in 1865. The goal of the examinations had been to fix a standard of instruction for intermediate and grammar school grades. Emily Davies (1830–1921) played a crucial role in its extension to girls. See Rita McWilliams Tullberg, *Women at Cambridge* (Cambridge: Cambridge University Press, 1975], chapter 2.]

132 [Henry Venn, *West-African Colonies. Notices of the British Colonies on the West Coast of Africa* (London: 1854).]

133 [In 1865, a Select Committee of the British House of Commons was appointed to consider the state of the British establishments on the Western coast of Africa. It marked the beginning of a process in which Britain became increasingly committed to the territories adjacent to the British forts. See Richard Olaniyan, "British Imperial Politics and Diplomacy in West Africa, 1860–1865," *The Historian* 34: 4 (1972): pp. 650–669.]

134 [Arthur Penrhyn Stanley (1815–1881), Dean of Westminster, 1864–1881; a prominent Broad Churchman and the leading liberal theologian of his time. The essay referred is entitled 'Theology of the Nineteenth Century', published in *Fraser's Magazine* 71 (February 1865): pp. 252–268. Dean Stanley here defended the kind of German biblical criticism ('new theology') adopted in *Essays and Reviews* which set aside the historical accuracy of the Bible without denying its value as a literary work.]

there always has been scepticism in the church of Christ, & that of late years it has come to us from Germany in a more subtle & imposing form than ever, & will triumph. I could not discover in the paper any one indication of any advance in knowledge of doctrine or of morals in his so called 'new Theology' – but merely that the historical accuracy of the Bible is set aside without denying its authority as a valuable literary work. The Article seems to me a very dangerous one – & yet capable of a very decisive reply. Could you write a paper upon it for the C.O.?[135] [...]

Ever your most aff^e Father

Henry Venn

34.
CMS C34/109
Henry Venn of C.M.S. to John Venn
15 April 1866
East Sheen

My dearest John,

Busy as I am with the Report[136] it give way at once to the interest in your publication & I send you my congratulations that you have gone so far as to put a part in type.[137] I shall rejoice to see it as soon as you can send me a proof sheet & however pressed I may be I shall read it with such intense interest that it will waken me up rather than interrupt me.

Remember a Bookseller has 2 plans.

1. He pays for paper & printing & at when a clear profit arises divided it half & half. Upon this plan you have no trouble or outlay: especially you are saved the bother of decided [*sic*] how much shall be spent on advertising: which, by the bye, is always a calculation of <u>chances</u> so you may like to keep it in your own hands.
2. The 2^d plan is that you pay for paper & printing within 6 months – decide how much to advertise – wait till the work has been out for 12 or 18 months, then receive an account – so many sold – so much expense of advertising – commission of so much due to Author – probably half the outlay. After 6 or 12 months you have a second <u>settling</u> what may happily turn the balance.

[135] [Venn complied with his father's wish to produce Evangelical polemic. He wrote a response to Stanley in which he insisted on Evangelicalism's centrality to nineteenth-century religion and questioned what good could possible come from the questions and doubts of *Essays and Reviews*. John Venn, "The Theology of the Nineteenth Century," *Christian Observer* (May 1865): pp. 371–387.]

[136] [Probably the 1866 paper in which **Henry Venn of C.M.S.** recorded, for the encouragement of C.M.S. missionaries, the progress which had made since 1851 in the development of the 'native Church', churches suited to local culture. Henry Venn of C.M.S. defended a view of indigenous church mission theory focused on indigenization, with foreign missionaries creating churches which they then hand over to local converts. See Knight, *Memoir of Henry Venn*, on pp. 420–421.]

[137] [Reference is to Venn's first book, *The Logic of Chance* (1866).]

If I understood you right McMillan[138] declined the first plan: but perhaps now that he has seen the work he may be willing to take it. If so, let him have it on the half & half profit. You gain by the interest he takes in pushing the book.

But if he adopts the 2^{d} course I shall be well able & delighted to advance the cost of printing & paper.

I think with Macmillan you sh^{d} have 750 copies – of which the cost will not be formidable.

[…]

Ever you most aff^{te} Father,

Henry Venn

35.
CMS C34/110
Henry Venn of C.M.S. to John Venn
2 May 1866
East Sheen

My dearest John,

The 4 proof sheets arrived by post this morning. I came to Sheen last night after the meeting & had therefore my morning free & have had the great satisfaction & delight of studying the proofs. Had the whole work been in my hands I should have read it through without stopping.

I think it so well done & so important I am perhaps too anxious that it should be worked up to perfection of clearness. Hence I should greatly wish to go over it with you & to suggest points on which I think a slight verbal modification or addition may bring your meaning out more clearly.

A writer can never predict (a priori) what mistakes his readers may make – hence the advantage of another eye or eyes – you may be sure that what strikes one will strike others among your readers.

It is chiefly the 2 or 3 first chapters[139] which require the utmost smoothness & clearness – afterwards the reader, if he reads on, gets into the spirit of the book & does not heed little obscurities, w^{h} in the first chapters he w^{d} resent.

I am indeed astonished at the accuracy with which you have written. I have put a few pencil suggestions to shew you the kind of question which occur to me. I do not know what is the rule of the university press – but our London printers will allow such a work to remain 2 or 3 weeks in type. If so you might bring these first proof on Friday & we might go over them together. I will keep Saturday free. If the matter presses you can deal with my suggestions by pen or India rubber & send them off for 'Press'. […]

Ever believe me

Your most aff^{te} Father
Henry Venn

---

[138] [The London publishing firm which published Venn's book.]

[139] [These are the chapters in Venn's *Logic of Chance* in which he discusses the foundations of probability theory and for which the book is probably most remembered today.]

36.
CMS C34/112
Henry Venn of C.M.S. to John Venn
1 August 1866

My very dear son

I must send a line of most tender & affectionate remembrance on your birthday. I shall think & pray much for you – perhaps the more as we are & have been so long separated at least if not so long so widely separated. I am anxiously hoping to spend a Sunday with you & to worship in your church. I look back upon the Sundays at East Sheen when you were officiating as some of the most bright & interesting Sundays I ever spent & the sad blank at present makes the contrast more striking.[140]

I think of your having St Edwards[141] in connexion with the life of Simeon, whose first effective exercise of his ministry was in that church.[142] His remarkable interview with a woman in a house in the churchyard was a type of his whole future ministry. Yet Simeon was only taking the duty for a few weeks in the vacation. Let us recollect that Simeon's God is our God & made him what he was & can make us "fit for the master's use"[143] – if we offer & present ourselves a living sacrifices.

My thoughts also often recur to the remarkable fact of the prayers & labour of your great Grandfather for Cambridge & its students when he used to go over from Yelling.[144] I felt this when I had the evening lectureship at St Mary's[145] – & still more now that you of the 4th generation are called to exercise your ministry in Cambridge. Oh may all the prayers of that righteous man, for his prayers always comprized his children & his children's children to the 3th & 4th gen^{n} be fully filled in your ministry. […]

Ever most aff^{y} yours

Henry Venn

---

140 [In late 1860, **Henry Venn of C.M.S.** had moved from the family home to East Sheen to be near his son, who was then curate of Mortlake.]

141 [There was an opening at St. Edward's, Cambridge, as John Neale Dalton (1839–1931), Clare 1858, was appointed curate in 1866. He left in 1869 to take up a curacy on the Isle of Wight in 1869.]

142 [Charles Simeon had begun his ministerial life as curate of St Edward King and Martyr, Cambridge, in 1782. He was appointed to Holy Trinity Church in that same year.]

143 [Timothy 2:21.]

144 [Henry Venn of Huddersfield had become Rector of Yelling in 1771. Here, he attracted visitors such as Joseph Jowett and Charles Simeon.]

145 [**Henry Venn of C.M.S.** was Evening Lecturer at St. Mary's, Cambridge, in 1826.]

37.
CMS C34/113
Henry Venn of C.M.S. to John Venn
4 August 1866
East Sheen S.W.

My very dear son

I cannot let this day pass over without assuring you how often you have been in my thoughts & how cordially I have framed wishes & prayers for your best happiness in time & eternity. We have not been much in the habit of observing these days – perhaps it w^d^ have been wiser if we had. Time passes so rapidly, slipping away from us, that we need a few stops, for reflection on the past & for looking into the future.

Which I thank God on your behalf for the useful and honourable position in which his providence has placed you: while I bless Him for all the comfort & enjoyment which I have had in you, while I thank Him for having given you talents for the service of his church, which your residence at Cambridge enables you to exercise & enlarge[146]: – yet I pray in respect of the future that He may soon open a way for the more direct exercise of your ministry. When all your talents & advantages may be brought to that best of all works the winning of souls to Christ & advancing his Kingdom.

I look back to the days when each Sunday I sat under your ministrations, as golden days in my life. I never pas your lodgings without a rush of feeling in the recollection of the many hours we spent together & the prayers in which we united. I cannot but think that we may manage in our future intercourse to have more of that spiritual union which is our privilege. I rejoice in the assistance of Mee[147] & in his willingness to take off my shoulder any amount of my work, chiefly in the hope that I may have more time & leisure for enjoying profitable intercourse with my children. […]

Ever your most aff^tle^ Father

Henry Venn

---

[146] [**Henry Venn of C.M.S.** remained committed to the view of Cambridge as being in the service of the Church and of an academic position as an opportunity to exercise Christian influence.]

[147] [John Mee (1824–1883), Cambridge graduate, first Dean of Grahamstown in South Africa, and Secretary of the C.M.S.]

38.
CMS C34/114
Henry Venn of C.M.S. to John Venn
13 October 1866
East Sheen S.W.

My very dear son

I feel quite satisfied that you have done right in giving in your name as a candidate for the professorship.[148] There has appeared to me a providential leading & guidance which encourages you to go forward. Your study of a branch of the subject without the remotest thought of the Professorship – the publication of the work delayed till this precise moment – the advice of such a friend as Todhunter[149] – the absence of any candidate of commanding claims – these are all providential circumstances. But whether they are indications of the divine call must be decided by the event. They seem to me to justify your offering yourself – & having done that we must keep our minds in suspense. For myself I cannot form a wish either way – you have dedicated yourself to the ministry of Christ

In my early days at Cambridge Simeon was helped & his efficiency promoted to an extent to which none of the present generation can conceive by Professors Farish & Jowett when they were lost, Prof^r Scholefield was raised up & Prof Lee.[150] Scholefield united the office of that of a popular preacher with his university standing, but the other three were mainly useful in the countenance they gave to Evangelical truth, through their academical standing.

I have written by this post to Jeremie & Selwyn.[151] It is right that if we offer ourselves for a post we sh^d use all means within our power to place ourselves fairly before the Patrons – do not hesitate therefore to use whatever private influence you can command & let us hear from day to day if any events occur. I presume the foundation must have been materially altered since by the old Calendars a B.D. degree was required & 40 years of age as a qualification.[152]

---

148 [Venn stood as a candidate for the Knightbridge Professorship of Moral Philosophy, which had fallen vacant upon the death of **John Grote**.]

149 [Venn's former mathematics coach **Isaac Todhunter**.]

150 [William Farish (1759–1837), Joseph Jowett (1751–1813), James Scholefield (1789–1853) and Samuel Lee (1783–1852). Farish, Jowett, Lee and Scholefield were Evangelical supporters of Simeon. In retrospect, their appointments to professorships at Cambridge have been deemed noteworthy in a twofold sense; first, as proving the existence of Evangelical intellectualism and, second, as evidence of Evangelicalism's increasing impact on universities in the early nineteenth century. See, for instance, Kenneth Hylson-Smith, *Evangelicals in the Church of England, 1734–1984* (Edinburgh: T. & T. Clark, 1988), chapter 4.]

151 [Two of the electors known to **Henry Venn of C.M.S.**: James Amiraux Jérémie (1802–1872), Regius Professor of Divinity at Cambridge from 1850, and William Selwyn (1806–1875), Lady Margaret Professor between 1855 and 1875.]

152 [The requirement for the Knightbridge Professor regarding their age had been repealed in 1861; the requirement that the holder of the post should be a Doctor or Bachelor of Divinity was only repealed in 1882.]

The Lord direct & guide all to his glory.

Ever your most aff^te^ Father
H.V.

If I can be of any use by visiting Cambridge before the election I shall rejoice to ran down. But I am not aware of any influence I have except with those to whom I have written.

39.
CMS C34/115
Henry Venn of C.M.S. to John Venn
27 October 1866
Brighton

My very dear son

The result[153] of the election has not altered my feeling as far as you are concerned – I feel confident that the Lord has better things in store for you - & I wait till he opens out a path of more public usefulness or ministerial appointment.

I long to know more particulars of the election – probably you will wait till you come to Sheen before you divulge all the gossip connected with events.

I cannot but deplore the election of Maurice.[154] I had occasion to read several of his work with great care some year ago, when one of our Missionaries was said to be infected by them - & this left a most painful impression upon my mind as baffling the grasp of Truth by mysticism & refinement of language.

Cambridge is evidently in a transition state, like the Church at large, & what things are tending to who shall tell? Certainly not as I think to Romanism or to Rationalism.[155] These have become elements of much importance: but the ultimate form of our church & religion will be of a far more composite character. Will that be more or less favorable for the growth of true Christianity? Certainly the present day appears to me more favorable than I ever knew it before. The very opposite opinions which exist apparently in equal force, the absence of any recognized standard of biblical exposition, the liberty of private judgments, these & many other

[153] [**F.D. Maurice** was appointed to the chair, then officially called the Knightbridge Professorship of Casuistry, Moral Theology and Moral Philosophy.]

[154] [**F.D. Maurice**'s appointment was seen by many outside Cambridge, such as Henry Venn, as the vindication of a controversialist. By the younger men at Cambridge, like Venn, it was perceived as a step back in professional dedication to the Moral Sciences.]

[155] [In Victorian Britain, religious rationalism was associated those who argued that the larger part of revealed truth was intelligible and demonstrable by reason. One of the allegedly dangerous implications was that religious truths did not need to be accepted on authority alone. For a discussion see Joshua Bennett, "A History of 'Rationalism' in Victorian Britain," *Modern Intellectual History*, 15: 1 (2018): pp. 63–91.]

things leave a young man more free to follow the guidance of the divine council than was formerly the case. […]

Your letter just arrived – I rejoice in your having stood.

Ever your most aff^te^ Father
Henry Venn

40.
TCL MAYOR B16/135
John Venn to J.B. Mayor
4 November 1866
Caius Coll.

My dear Mayor,

Many thanks for your letters, which I would have answered sooner had I not been absent from Cambridge for two or three days. I shall take the first opportunity of going over to Trumpington & bringing back a book with me as a memorial of Professor Grote,[156] & of some of the happiest hours I have spent in Cambridge. Please do remember me to Mrs. Mayor & give her my sincere thanks for her thoughtfulness in sending me that message.

My book[157] is, to me, full of memories of our meetings at Trumpington. A good deal of the substance of it was read there, & our discussions afterward were of the utmost use in suggesting the ideas & working them out. The last time I saw Professor Grote – about a month or so before his death, – was to talk with him about one of the chapters which he had looked over when in type, & the last letter I had from him was on the same subject; full of sympathy & encouragement, as everything that came from him.

I will call at Kensington[158] when I am in London during the vacation, & take my chance of finding you there.

Believe me
Y^rs^ very sincerely
J. Venn

41.
CMS C34/118
Henry Venn of C.M.S. to John Venn
6 April 1867
East Sheen

My very dear son

I have rejoiced in all the recital Henrietta brought me of your enjoyment of the visit to Cambridge: & now I am anticipating the great pleasure of seeing you & Susie[159] here for a few days at least. May they be days of profit as well as of

[156] [**John Grote** had died on 21 August 1866.]

[157] [Venn's *Logic of Chance*.]

[158] [In 1866, **J.B. Mayor** was working as Headmaster of Kensington Proprietary School, 1864–68.]

[159] [Venn's wife **Susanna Carnegie Venn** (née Edmonstone), whom he had then just married.]

pleasure. Soon your visit to my roof as to home will cease. You will have a home of your own. You will be the head of a family: – an institution appointed of God for the maintenance of his worship & service in any Christian land. Hence arise responsibilities, opportunities, & talents for which we must here after give a solemn account. The question will be put to each of us "what have your friends – your servants, your children – seen in your house?

[…] Ever your most aff^te^ Father

H. Venn

42.
GCA C52/1
John Stuart Mill to John Venn
4 February 1868
Avignon

Dear sir,

I am late in acknowledging the gift of your very valuable work 'The Logic of Chance'.[160] As I perceived at once, as well as heard from others who read it, that it well deserves consecutive study, for which pending occupations left me at the moment no leisure, I laid it by till the time should come when I must necessarily take up the subject in revising my System of Logic for another edition.[161] This time has only recently arrived, and I have availed myself of your criticisms to make several alterations in the corresponding chapters of the book, which alterations however are rather in the mode of expression than in the substance of the opinion expressed.[162]

I think your book very important both as a contribution to the theory of the subject, and as a corrective to prevalent errors. Your general mode of viewing this class of questions is far the best and most philosophical I have met with; and while there is evidence of a great agreement between us in our mode of regarding the great problems of inductive philosophy, you have, in this particular subject, thrown light upon many more points than space and time have allowed me to enter into. Your book is one of the highest compliments which could have been paid to mine; for I have scarcely met with any thinker who seems to have so completely assimilated the best thoughts and principles of my book thereby affording strong indication that your own thoughts had flowed in much the same direction independently of any external suggestion.

There are, however, some points on which we differ, and on which your book has not convinced me. For one thing, you seem to go farther in rejecting the doctrines of mathematicians on the subject than even I do. If I understand you rightly, you attach

---

160 [Venn's *Logic of Chance*, published in 1866.]

161 [Mill's *System of Logic*, originally published in two volumes in 1843; 2nd–8th ed., introducing many changes, 1851–72.]

162 [**John Stuart Mill** mentioned Venn's work at two points in the 7th edition of *System of Logic* of 1868. See John Stuart Mill, *System of Logic*. 7th ed. (London: Longmans, Green, Reader, and Dyer, 1868), p. 77, p. 170.]

little value to the rule for determining the probability by which of several causes a known event has been produced which rule seems to me to rest on solid grounds, and to be quite reconcilable with the principle that all evaluations of probabilities must depend on appropriate statistics. But our chief difference (not unconnected with this last) consists in you not recognizing what I regard as one of the fundamental distinctions in philosophy, that between causal and empirical laws. The root of this difference seems to be the opinion you express that the entire sum of the circumstances which make up the unconditionally invariable antecedents of any effect, is hardly ever repeated.[163] This seems to me, if I may be allowed to say so, a less well-considered opinion than most of yours. If the exact combination of antecedents that is connected with the effect, did not reappear neither would the effect itself ever reappear; except so far as it might be an effect which had several independent modes of production.

If I had time to enter fully into the subject I should perhaps trouble you with a detailed examination of the arguments in your chapter on Causation.[164] But, from the general evidence of your book, I shall be surprised if you do not end by agreeing with me on this point, by the progress of your own thoughts. On my side, I promise myself that when I have more time to devote to the subject, I shall have still more to learn from your book than I have yet learnt from it, and that at some future time it may help me to expand and improve materially those parts of my Logic.

Again thanking you for the pleasure and instruction you have given me, I am, Dear Sir,

Yours very truly
J.S. Mill

43.
GCA 52/2
John Stuart Mill to John Venn
18 March 1868
Blackheath Park, Kent

Dear sir,

I thank you very sincerely for your further remarks in illustration of the matters treated in your book. I shall keep them by me for a further and more deliberate consideration that it is in my power to give them during the session of Parliament.[165] I will mainly, at present, say a few words on a single point; the flaw which seems to you to exist in the theory of the predictability of human history, from the influence

[163] [Venn, *Logic of Chance*, on pp. 334–335.]

[164] [Venn, *Logic of Chance*, chapter 14 ('Causation').]

[165] [**Mill** was a Member of Parliament for City and Westminster, 1865–68, as a member of the Liberal Party. See Bruce L. Kinzer, Ann P. Robson and John M. Robson, *A Moralist In and Out of Parliament. John Stuart Mill at Westminster, 1865–1868* (Toronto: University of Toronto Press, 1992).]

which the foresight itself may have in modifying the parts foreseen.[166] This influence is real; but does not seem to me to affect the theory; for the self-consciousness of mankind, and their foresight of their own future is itself a foreseeable and calculable element. It might have been predicted by a superior being that at such and such an epoch in the development of the human mind, mankind would begin to produce the law of their own future and that their perception would modify the future in such and such a way.

There is an inconsistency though only a verbal one between the two passages which you cite from my Political Economy.[167]

Your examination papers in Political Economy are very thorough and searching.[168]

You are most welcome to use any part of my letter as a testimonial, provided it is so used as to show that it was not written with a view to the present vacancy in the Examinership in Logic[169]; as with regard to that vacancy any influence it is in my power to exercise is pre-engaged in favour of another highly qualified candidate. I am, Dear Sir

Very truly yours
J.S. Mill

44.
CMS C34/135
Henry Venn of C.M.S. to John Venn
29 April 1868
East Sheen, Surrey S.W.

Dearest John,

I must not let the pressure of business deprive me of the happiness of frequent communications with you. Indeed I sh[d] be a loser by that bargain. For I always gain by interchange of thoughts with you: & a subject is now much on my mind which I especially need your help – namely the establishment or disestablishment of a church.[170] I send some detached thoughts & I shall be anxious to receive in return your remarks upon my thoughts & your own thoughts in addition to mine.

---

166 [This had been the main point of Venn's 1862 'Science of History'. The same point was made in *Logic of Chance*, chapter 15 ('On Statistics as Applied to Human Actions').]

167 [Venn's letter to **Mill** has not been located.]

168 [Venn apparently sent **Mill** the examination papers he made for the Moral Sciences Tripos.]

169 [In 1868, there was a vacancy for the office of Examiner in Logic and Moral Philosophy at the University of London. George Croom Robertson (1842–1892) also stood for the position. See *Testimonials in Favour of G. Croom Robertson, M.A., Candidate for the Examinership in Logic and Moral Philosophy in the University of London*. The candidate chosen was Mark Pattison (1813–1884), educated at Oriel College, Oxford, where he graduated B.A. in 1836. Cf. letter **54**.]

170 [Disestablishment became the subject of much public debate in the years 1868–69. **Henry Venn of C.M.S.** believed in the establishment and wanted to see it maintained. The Church of Ireland was formally separated from the Church of England in 1869 through the Irish Church Act 1869. See Shenk, *Henry Venn*, pp. 99–100.]

[…] There is another important subject I have upon my mind to w[h] I alluded in my last – the improvement of the course of instruction in our Islington College.[171] On one point I sh[d] like you to send me a few lines before the end of next week, viz.: How far logic sh[d] be taught – & if so by what text book. Whately?[172] – or any other with a view to equip Missionaries to contend with acute Brahmins or Mohametans: or at least to make them cautious of entering into controversy. I am afraid you will think that I have too many irons in the fire. But that has been my lot since the age of 18 when I first engaged in the edition of my Father's sermons[173] – though Z. Macaulay[174] took 19 out of 20 parts of the trouble – [*illegible*] as soon as I left College & had the curacy of S[t] Dunstan's.[175] I began my attendance at the Ch-Miss-Committee & from that day to this have always had great questions to handle – yet thank God they have always been practical not theoretical subjects & when we shall see things as they are, we shall appreciate the honor & blessing of being in any sense "an instrument for the Lord's use".

With tender love to Susie[176].

Ever your most aff[te] Father
Henry Venn

45.
CMS C34/141
Henry Venn of C.M.S. to John Venn
18 February 1869
East Sheen S.W.

My very dear son,

The intelligence you sent me[177] burst upon my soul like a [*illegible*] from heaven – saying I have heard their prayers.

---

[171] [The Church Missionary Society College in Islington, north London was founded in 1820. Its mission was to prepare C.M.S. missionaries for work overseas. It was closed in 1915.]

[172] [Richard Whately's *Elements of Logic*, first published in 1826. A popular textbook responsible for the renaissance of the study of logic in Britain in the nineteenth century. See James van Evra, "Richard Whately and the Rise of Modern Logic," *History and Philosophy of Logic* 5:1 (1984): pp. 1–18.]

[173] [John Venn of Clapham had died in 1813, when his son, **Henry Venn of C.M.S.**, was seventeen years old. Despite his youth, he was made one of his father's executors. He also edited his father's sermons, published between 1814 and 1818 in three volumes as *Sermons of the Rev. John Venn, M.A., Rector of Clapham.* See John Venn, *Annals of a Clerical Family*, p. 152.]

[174] [Zachary Macaulay (1768–1838). After his sister had married Thomas Babington, a prominent member of the Clapham Sect, Macaulay moved to Clapham where he came under the influence of Babington, William Wilberforce and Henry Thornton and underwent a conversion experience. In 1792, he became governor of Freetown, Sierra Leone, the British colony for freed slaves.]

[175] [**Henry Venn of C.M.S.** was ordained deacon in 1819, and priest in 1820, and soon afterwards accepted the curacy of St. Dunstan-in-the-West, where he stayed for four years.]

[176] [Venn's wife **Susanne Carnegie Venn** (née Edmonstone).]

[177] [Venn informed his father that he had been elected to give the Hulsean Lecture before the University – a mark of distinction, as Venn was preceded in his role by J.J. Stewart Perowne,

For daily have I put up my petition that you may have opportunity & grace to witness for the Truth of the Gospel – yet my thoughts had never anticipated this particular & very special opportunity – a parochial pulpit, or a Select Preacher were only in my thoughts.

Dean Jeremie[178] was at the Ritual Comm^{n}[179] yesterday & told me the particulars of your election: & expressed in the warmest terms his satisfaction, as well as the unanimous opinion of the Electors that you had the first claim: though they had not previously appointed anyone who had not approved himself as a Select Preacher.[180]

But I do not forget the caution – Let not him that putteth on his armour[181] &c.

I now therefore shall daily pray, (with thanksgiving) that you may have the wisdom "profitable to direct"[182] in the choice of subjects: & the [sanction?] that is from above "to teach you all things"[183] necessary to the discharge of your high office. I pray in the confidence of one of old. If the Lord had not designed a favour towards us, "he would not as at this time have told us such things as these".[184] How greatly shall I rejoice to converse with you on the choice of subjects & the way of treating them. How deeply shall I sympathize in every step of your preparation – How often do I thank God that He has given you in dearest Susie one who can share in all these feelings […]. I can hardly wait till Easter to converse with you upon a subject which for the time eclipses the Ritual Commission in C.M.S. A good day talk with you w^{d} set my mind more free for other things. Let me know the day when your lectures end.

[…] Ever my beloved children. Your most aff^{te} Father

Henry Venn

46.
CMS C34/143
Henry Venn of C.M.S. to John Venn
29 March 1869
East Sheen S.W.

My dearest John,

I must send you a line before you leave Cambridge to thank you most cordially for your promise to send some papers for the C.O. I think a short article on the

Bishop of Worchester, and succeeded by F.W. Farrar, Dean of Canterbury. The Hulsean Lectures were established from an endowment made by John Hulse to the University of Cambridge in 1790.]

178 [James Amiraux Jérémie, Regius Professor of Divinity at Cambridge and Dean of Lincoln.]

179 [One of two Royal Commissions for which **Henry Venn of C.M.S.** was selected; the Royal Commission into Ritualism, created in 1867 to inquire into the differences in ceremonial practice in the Church of England.]

180 [The sermons before the University at Great St. Mary's, Cambridge, were preached by Select Preachers, appointed yearly by the Select Preachers' Syndicate.]

181 ["Let not him that putteth on his armour boast himself like him that putteth it off": Kings 20:11.]

182 [Ecclesiastes 10:10.]

183 ["But the Advocate, the Holy Spirit, whom the Father will send in My name, will teach you all things and will remind you of everything I have told you": John 14:26.]

184 [John 16:4.]

Cambridge local & female Exam^ns[185] will be interesting if sent by "an Examiner" – giving your hopes of its development – & mentioning the School in the West Indies.[186] But more especially I shall be interested in J.J.S. Perowne's Hulsean lectures on Immortality.[187] I will get the book & we can talk it over when we meet which will be I hope on Saturday. […]

The C.O.[188] takes up much time. But the importance of the work grows upon me. It takes the lead of all strictly religious serials, in the authority with w^h it speaks. I have a good hope that Elliott & Knox will make good Editors.[189] But I must stick to it for a few months till they are welded together as co-workers. If it puts £100 a year into C. Elliott's pockets it will be a help to him. […]

Give me as much time as you can for I have very much to talk over with you.

[…] Best love to Susie.

Ever your most aff^te Father
Henry Venn

47.
CMS C34/152
Henry Venn of C.M.S. to John Venn
7 September 1869
7 Waterloo Crescent, Dover

My dearest son,

Your precious MS arrived safely by the day mail – & I have read it through with great interest once – but I shall wait & read it several times before I offer any remarks. I think your subject[190] a very appropriate one, & that it may be made very useful. I shall not cease to pray as I have done daily since I heard of your appointment to the Hulsean that in this light you may see light.

[…] Your most aff^te Father

Henry Venn

---

[185] [See note **131**.]

[186] [The Cambridge Local Examinations for girls were imported into India in the late 1860s. See Tim Allender, *Learning Femininity in Colonial India, 1820–1932* (Manchester: Manchester University Press, 2016), chapters 2–3.]

[187] [J.J. Stewart Perowne, *Immortality. Four Sermons. Preached before the University of Cambridge, being the Hulsean Lectures for 1868* (Cambridge: Deighton, Bell, and Co., 1869).]

[188] [*Christian Observer*, a London Evangelical periodical which appeared from 1802 to 1874. Henry Venn of C.M.S. was editor of the *Christian Observer* between 1868 and 1872.]

[189] [Charles Elliott (?–?) and George Knox (1845–1922), afterwards Secretary of the C.M.S. and editor of the Church Missionary Intelligencer.]

[190] [As his subject, Venn chose 'Some of the Characteristics of Belief Scientific and Religious'. For a brief discussion of this choice see Michelle Clewlow, "Intersecting Sets: John Venn, Church and University, 1834–1923," PhD-dissertation, University of Cambridge, 2007, pp. 135–136.]

48.
CMS C34/153
Henry Venn of C.M.S. to John Venn
9 September 1869
7 Waterloo Crescent, Dover

My very dear son,

I must not keep your sermon any longer. I have read it over 4 times, once by myself & then again with Isabel, Harry[191]: & it has been with increased interest & satisfaction.

I am persuaded that it is the most appropriate subject you could have chosen – & I shall long to see the 2d part as well as to read this in its finished state. I have thought over all the suggestions I can send you – not with the idea that they can in themselves be of any use to you: but they may be hints to call forth matter of your own to guard yourself from being misunderstood.

[…] Were I addressing an ordinary congregation I might warn you against exercising yourselves in matters too high for you & refer you at once to the written word of God for the authoritative resolution of your doubts.

But my present position among you, my knowledge of the contents of thought – my sympathy for enquiring minds whose way to holy scripture needs to be cleared from preliminary obstacles, will induce me on this occasion to meet you on your own grounds & to help you to solve your difficulties by methods of strict reasoning & scientific research &c &c.

You cannot possibly use one syllable of what I have written. But you will see that I would not have it said that you ignore a direct appeal to Scripture.

If I have rightly grasped your scheme you propose to hew the unreasonableness of unbelief – upon the very principles of the Rationalists.[192] How far wd this form a running title of the sermons? It has been anticipated by some well known authors whose name I cannot remember.[193]

There is only sentence wh I have marked in pencil, "The knowledge… weapon".[194] It is not necessary to your argument, & is liable to be misinterpreted.

191 [Venn's brother **Henry Venn of Walmer** and his wife Isabel Louisa Venn (née De Butts) (1846–1908).]

192 [On rationalism in the Victorian era see note **155**.]

193 [A central goal of the tradition of Christian apologetics, a branch of Christian theology, is to show the unreasonableness of unbelief. Throughout the centuries, it has taken many forms. Major apologetics of the period were William Paley and Joseph Butler.]

194 ['Everybody is familiar with the fact (explain it how we will) that the approach of death often causes a great change in our religious views. The knowledge of this may sometimes be rhetorically abused by preachers, who yield to the temptation of wielding with little trouble to themselves a terribly potent weapon, but the fact itself is notorious'. John Venn, *On Some of the Characteristics of Belief: Scientific and Religious. Being the Hulsean Lectures for 1869* (London & Cambridge: Macmillan and Co., 1870), pp. 7–8.]

In the quotation from Dean Tucker I should suggest one word of caution e.g. "upon the mere principle of the Light of nature". For the edifice built upon Scripture is built upon a rock.[195]

In the quotations you cite I am not sufficiently acquainted with the characters of the authors – but I need not remind you that a quotation from the University pulpit gives a certain stamp to a name – of which it sh[d] be fully worthy.

In one case I sh[d] suggest the omission viz: in the reference to Coleridge.[196] The point[197] in which you cite him is as you say most obvious & trite – then where is the need of any quotation? It looks as if the name were brought in as an adornment of the prayer, whereas your prayer need no glitter of quotations.

Again I must say how intensely I desire to read your 2[d] part. You have indicated that faith which does not produce action is a non entity? But we must allow for a very feeble faith, dormant under certain circumstances, & yet capable of producing fruit?

[…] Ever your most aff[te] Father

H. Venn

49.
CMS C34/156
Henry Venn of C.M.S. to John Venn
15 September 1869
7 Waterloo Crescent, Dover

My very dear son,

We have read over, separately & together, your second sermon[198]. It is not we conceive so fully worked out as your first sermon, but if we rightly supply, by imagination, the additions you will make we think it will be very interesting & useful.

It strikes us that you have had the chief point of the Sermon so fully in your own mind, that you have failed to fix it prominently enough in the minds of your hearers.

The passage at p. 14 (to be inserted at p. 4) is I conceive the Malakoff of your fortress[199] – against this the attack of opponents will be directed: under shelter of

[195] [Following Abraham Tucker's *Light of Nature* of 1768, Venn argued that the grounds of religious belief were like 'those Dutch buildings which rest, not on well-laid courses of brick or stone, but on an assemblage of many piles driven deep into a soft and yielding bank of mud and sand'. Venn, *Hulsean Lectures*, on p. 13.]

[196] [Samuel Taylor Coleridge (1772–1834), famous poet, critic and philosopher; a notorious controversialist, though, or perhaps because, his philosophy and theology was 'understood by very few'. James H. Rigg, *Modern Anglican Theology* (London, 1859), on p. iii.]

[197] [Venn referred to Coleridge twice, once to his *Friends* (probably *The Friend*) and once to his *Lay Sermons*. See Venn, *Hulsean Lectures*, p. 15 and p. 19.]

[198] ['The Criterion of Truth'. In his first lecture, Venn explored the logic of differences in religious belief; in the second lecture, he asked what was the criterion of truth, that is, on what basis is it possible to decide which opinions were right and wrong.]

[199] [The Battle of Malakoff was a French attack against Russian forces during the Crimean War which resulted in the fall of Sevastopol on 9 September 1855. Venn's 'Malakoff' in his second lecture was his proposed criterion of truth of religious opinions, namely that of 'living by the doctrine in question'. Venn, *Hulsean Lectures*, on p. 35.]

this your adherents must keep their ground. If so it needs to be amplified & to be strengthened fourfold: at present it is hardly discernible from your out-works to common hearers. I will just give you rough hints of the directions in which I think it needs amplification.

1. It needs its connexion with your first sermon to be more clearly pointed out. Is it not indicated in that sermon? Have you not in fact shewn in that sermon that the test you propose is that which men of science rest upon in their department & men of business in theirs &c &c i.e. admitting statements are true – as truths – as convictions on the authority of a sufficient number of competent judges who have investigated the points at issue: & therefore they are bound in consistency to admit religious truth upon the same second-hand testimony. If not, a preconceived prejudice blinds them. They love darkness rather than light. If I have rightly conceived your meaning this will I think [be?] the most telling part of your argument.
2. I do not think this main part of your argument will be understood generally without <u>examples</u> of Truths which have received the stamp of the crucial test you describe. These examples will be I conceive for the most part the great truths of natural religion, but above all the conviction that the Bible is inspired. Without these examples your test will be as vague as the "quod semper, quod ubique, quod ab omnibus"[200] which vanishes into nothing as soon as you ask what dogmas have ever received their sanction. [...]
3. It seems to me that you require in this sermon a re-statement of the distinction between the sharp outlines of scientific Truth & the necessary & indistinct outlines of social, political & religious Truth. Your first sermon touched upon this: but it needs to be re-introduced here. It appears to me just the point you are capable of making clear & are specially called upon to rivet in the minds of young men.

Each of these hints (2 & 3) are attended to in your sermon[201] in Caius Chapel & therefore I conceive you intend to bring them into this course. But them seem to me specially needed in this sermon: which will be in fact the <u>first</u> of the sermons to many young men who will not have come up for the Oct. 3 & which as I have intimated contains the <u>great principle</u> of your lectures.

4. I should earnestly suggest also the caution I expressed in my last letter that you not ignoring the Bible as the Source of Truth; that you should at once distinctly & boldly state that your argument is ad hominem. That God has provided in his inspired word a highway to Truth so clear & lifted up – its own direction – that even the wayfaring man though a fool shall not err therein & that you are only shewing the Positivists & Secularists that upon their own principles if <u>impartially</u> followed out, in a truth loving spirit, – the very principles on which Rationalism

---

[200] ["What has been taught always, everywhere and by all": the famous maxim from the *Commonitory* by Vincent of Lerins.]

[201] [No record of this sermon has been found.]

rests – would lead them to embrace that ready, safe & save road: that all you have said of living by a principle holds in an infinitely higher degree of living by faith that it is richer – fuller of daily & hourly experimental proofs of the reality of our convictions.

If you thus amplify the main position of the sermon a good deal of what you have written might be put into notes.

[…] Ever your most affte Father

Henry Venn

50.
CMS C34/163
Henry Venn of C.M.S. to John Venn
17 January 1870
East Sheen S.W.

My very dear son,

I am very thankful that you have let me see your Preface, because I think there are some important points in which I can save you from being misunderstood, & from an outcry being raised, which might hinder the usefulness of your volume & check its circulation.

We two live in totally different atmospheres. I am surrounded by men who all speak one language & fire up at the shadow of any compromise of the Truth. You live surrounded by men speaking different languages & leaving the Truth to take care of itself.

So that in the good providence of God we may mutually assist each other during the brief space that I may be continued below) by guarding the one against narrowness, the other against a breadth beyond the limits of Truth.

I have also some reputation for putting things clearly – & this I shall try my utmost to do with your Preface, because I am persuaded much depends upon the Preface being as clear as daylight.

[…] I am anxious […] to direct attention to a distinction which I draw between religious belief and Christian Faith. It is not in accordance with scripture to represent faith as merely an intellectual state, as it must be if it were formed solely upon evidence.

[…] This whole passage to the end of the para. on the next page appears to me very liable to be misunderstood – as if you denied the sin of unbelief.[202] The sin you will say lies in the moral element of faith – men hate the light – because the deeds are evil – they are ungrateful & disobedient to their great Benefactor & their Lord & Master. But it appears to me that the great sin of unbelief is that men will not calmly investigate the evidences notwithstanding the probability that God has spoken, & the certain checks of their own consciences. […]

---

202 [Venn, indeed, did not insist on the sin of unbelief; instead, his point was to show that religious faith was reasonable. He did so by showing that there was no categorical difference between scientific and religious belief.]

51.
CMS C34/164
Henry Venn of C.M.S. to John Venn
17 January 1870
East Sheen S.W.

My very dear son,

Besides the Hulsean Preface – my heart is full of anxiety on account of dearest Susie's temporary separation from you,[203] & her present & past anxieties. It appears to me almost unnatural that you too sh^{d} have had such anxieties & that I should not have sympathized with you in them. But perhaps it was wisely ordered in Providence that you sh^{d} be thus called early in your marriage life to the experience that bitter ingredients are infused into the choicest cup of happiness. […]

I have said that I regret that through my ignorance I have not sympathized as I sh^{d} have done. But perhaps I was serving you more effectively by my daily prayer for your growth in grace & that the fullness of the blessings of the Gospel of Christ might be imparted to you. Had I known of your anxieties I might have prayed more earnestly for their removal; I know that they might produce in you the peaceable fruit of righteousness.

[…] The Lord bless & keep you in all your ways -& your precious wife.

Ever your most aff^{te} Father,

Henry Venn

52.
CMS C34/166
Henry Venn of C.M.S. to John Venn
12 August 1870
East Sheen S.W.

My very dear son

[…] I am anxious that you should consider one or two things in your Review of Newman[204] – I sh^{d} very much like to put it into the next month's C.O. But for the C.O. readers it appears to me necessary to make two additions – first to answer the objection of the Stephens[205] that it ignores Truth & justified every man in believing what he likes – I put the matter roughly. Do you think you could handle this matter?

---

[203] [Throughout her life, **Susanna Venn** struggled with poor health. Aelfrida Tillyard later reported that one of the causes of Susanna Venn's suffering had been recurrent problems during pregnancies: 'She had borne him as many children as Queen Anne, and of these, only one had survived'. Aelfrida Tillyard quoted in Clewlow, "Intersecting Sets," on p. 187 (n. 16).]

[204] [John Venn, "Dr. Newman's Grammar of Assent," *Christian Observer* (October 1870): pp. 727–739. On the reception of Newman's *Grammar of Assent* (1870) see, for instance, Geoffrey Rowell, "Anglican Theological Reception," in Frederick D. Aquino and Benjamin J. King, eds. *The Oxford Handbook of John Henry Newman* (Oxford: Oxford University Press, 2017), pp. 517–537.]

[205] [Both **Leslie Stephen** and **Fitzjames Stephen** attacked Newman's *Grammar of Assent* in print. Fitzjames Stephen would publish a negative review of the book in *Fraser's Magazine* in 1872. The objection to which Henry Venn of C.M.S. refers was probably communicated privately.]

The second point is the divine influence promised to the prayerful study of the holy scriptures. Does Newman give any place to this topic?

If you think these points might be introduced into your review consistently with your own plan I will send it to you – or if you prefer it I will wait till we can meet & talk over it. […]

Tenderest love to Susie.

Ever your most aff^{te} Father,
Henry Venn

53.
CMS C34/167
Henry Venn of C.M.S. to John Venn
17 August 1870
East Sheen S.W.

My dearest John,

I very much wish to have your Review of Newman in the next C.O. Elliott also specially pleads for it as able to give the magazine a character which it wants. I send you a part of the Review[206] in the Pall Mall[207] – the rest I unfortunately have not kept but you will see the estimate they form of the book.

I send you under the same cover my article on the Baptismal service[208], which please read over carefully & say whether any link is wanting in the argument – or how it may be made more strong – any suggestions of yours I shall love & value, let me have the proof back by Friday morning.

I have Newman's book[209] if you have left your's abroad. […]

Tenderest love to Susie.

Ever your most aff^{te} Father
Henry Venn

54.
GCA C15/1
Alexander Bain to John Venn
18 March 1871

Dear sir,

I received your letter of 2^{d} March, and some days later, your sets of Examination Papers. I shall be quite glad, if you apply for the vacant Examinership in London

[206] [This review has not been identified.]

[207] [The *Pall Mall Gazette*, an evening newspaper of Liberal stamp which appeared from 1865 until 1921.]

[208] [An opinion piece on 'the proper interpretation of the baptismal service of the Church of England', published as Anon. [Henry Venn of C.M.S.], "The Proper Interpretation of the Baptismal Service of the Church of England," *Christian Observer* (September 1870): pp. 643–653.]

[209] [John Henry Newman's *Grammar of Assent* (1870).]

University,[210] although I feel myself precluded, by their applications, in which I am bound to take an interest, from giving you a recommendation for the office. I studied your work in the Logic of Chance with great interest [...] and made considerable use of it in my Logic Manual.[211] Your Examination Papers are very admirable, in enforcing out of a very cramped and narrow circle of books, a host of pregnant topics.

The candidates that I take an interest in (two in number) are both attached to my special modes of viewing the mind; and, it may be the wish of the Senate, that such views should not be represented in the present appointment. I am desirous that they should have the option of taking such an able logician and Examiner, as you are. [...]

At last election (where Pattison[212] was appointed) your pretentions were warmly supported by at least one influential member of the Senate; and, on that occasion, I should have been well pleased if you had been successful. [...]

Yours truthfully, A. Bain

55.
CMS C34/177
Henry Venn of C.M.S. to John Venn
29 April 1871
East Sheen S.W.

My very dear John,

I congratulate you on your new appointment at the London University,[213] & especially rejoice that it will bring you up occasionally to Town. £80 for 3 years is not in itself despicable – that I trust it may lead to other things after the 3 years expire. [...]

Many thanks for offering me a review[214] on a book on religious belief which I have not seen, but I shall be very thankful for it. I go on from month to month, determining to give up the C.O. but each month something occurs to satisfy me that it has an important & useful influence. [...]

Tenderest love to Susie.

Ever your most aff^{te} Father,
Henry Venn

---

[210] [Cf. letter **43**, when Venn also stood as a candidate. **Alexander Bain** (1811–1877) held this post in 1857–62 and 1864–69. Venn would be External Examiner at London University in the years 1871–76.]

[211] [Probably **Alexander Bain**'s *Logic*, published in two parts in 1870.]

[212] [Mark Pattison (1813–1884). See letter **43**.]

[213] [Cf. letter **54**. Venn would hold the post in the years 1871–76.]

[214] [Not identified. Contributions to the *Christian Observer* were anonymous. The only review from 1871 that can be attributed to Venn on the basis of his correspondence with his father is Venn's 'Bishop Butler reviewed in the light of modern thought', published in September 1871.]

56.
CMS C34/178
Henry Venn of C.M.S. to John Venn
[not dated, but probably August 1871]

My very dear son,

This will reach you on your birthday – & may many days of honourable usefulness & of domestic happiness be given to you. […]

I was rejoiced & most thankful for your offer of two more articles for the C.O. The Article on Butler[215] will be very valuable. You must think of some Title which will be "catchy". What say you to – "Bishp. Butler adapted, tested or reconciled with Modern Thought". I will reckon upon your sending me this in time for the next number. I shall be equally glad of an Art. on Darwin in an expansion of the remarks upon the Abp. of York.[216]

Our circulation has been sadly damaged by a new publication edited by Bickersteth, 'Evening Hours'[217] with his name & the names of writers paraded in advertisements – Bp. Ripon[218], Ryle[219] &c &c. & the subjects similar to the lighter Articles of the C.O. & the price sixpence! The first number came out in April & our circulation dropped 100 that month. I shall however go on to the end of the year, & then either let it drop or turn it into a 2/6 quarterly. There is no quarterly which occupies its ground & many of our Articles are too heavy for a monthly, of which the general type & fashion is lighter reading. […]

Your most aff^te^ Father,

H. Venn

---

[215] [John Venn, "Bishop Butler reviewed in the light of modern thought," *Christian Observer* (September 1871): pp. 681–692.]

[216] [Neither the 'remarks upon the Abp. of York' nor an 'Art. on Darwin' have been identified. There did appear an article on Darwin in the *Christian Observer* in August 1873, entitled 'On the theological import of the Darwinian doctrine of development of species', but since all articles appeared anonymous, this cannot be definitely attributed to Venn.]

[217] [*Evening Hours*, initially edited by Edward Henry Bickersteth (1825–1906, the evangelical Bishop of Exeter. Founded in April 1871, *Evening Hours* was advertised as a 'Church of England Family Magazine'. Under Lady Barker's (1831–1911) editorship from 1874, the emphasis shifted away to become 'A Family Magazine'.]

[218] [Probably Robert Bickersteth (1816–1884), who was Bishop of Ripon, 1857–1884.]

[219] [Probably John Charles Ryle (1816–1900), the Evangelical bishop of Liverpool.]

57.
CMS C34/189
Henry Venn of C.M.S. to John Venn
[7?] January 1872
East Sheen S.W.

My very dear son,

If you are not overpressed you can help me in two C.O. matters. I wish to put in a notice of Arch.[n] Pratt,[220] particularly in respect of his high scientific attainments & his unshaken faith in God's word: the public testimony is found in his work Religion & Science which has just reached a 6th edition.[221]

I hear, but this is what I want you to establish for me, that a work of his on the figure of the earth[222] is of the highest merits – a Mr. Stokes[223] of the Royal Society says that [he] was regarded there as among the very first of scientific men – especially since he vindicated his priority of discovery over Airy[224] in some points in the figure of the Earth. Now I want you to furnish me with what I may safely say on this question of his position as a man of science.

Should you have formed any opinion of his "Science & Religion" I shall be most thankful for any hints.

[…] May every blessing rest upon you.

Your most aff[te] Father,
Henry Venn

220 [John Henry Pratt (1809–1871), student of **William Hopkins** at Caius; B.A. (3rd Wrangler) in 1833; tutor of Harvey Goodwin; author of various works in mathematical physics; appointed Archdeacon of Calcutta in 1850.]

221 [Probably John Henry Pratt, *Scripture and Science not at Variance; or, The Historical Character, Plenary Inspiration, and Surpassing Importance of the Earlier Chapters of Genesis*, first published in 1836; sixth edition 1871.]

222 [The study of the shape of the earth, the so-called 'figure of the earth', was a major mathematical preoccupation in Britain for much of the nineteenth century. In the 1830s and 1840s, earlier studies of the figure of the earth, by Laplace, among others, were extended by Pratt and several others. Pratt's major work, published in 1836, was *The Mathematical Principles of Mechanical Philosophy*; it was revised in 1842, then expanded and republished under the title *On Attractions, Laplace's Functions and the Figure of the Earth* in 1860. Pratt assumed that the Earth behaved as a fluid and showed that the resulting shape would be an oblate spheroid.]

223 [George Gabriel Stokes (1819–1903), Pembroke 1837; B.A. (Senior Wrangler and First Smith's Prize) 1841; Fellow, 1841–57 and 1869–1902; Master, 1902–03; Lucian Professor of Mathematics, 1849–1903; Fellow, 1852, and President, 1885–90, of the Royal Society.]

224 [George Biddell Airy (1801–1892), Trinity 1819; B.A. (Senior Wrangler) 1823; Fellow, 1824; Tutor, 1824–26; Lucasian Professor of Mathematics, 1826–28; Plumian professor of Astronomy and Experimental Philosophy, 1828–36; Astronomer Royal, 1835–81. Reference to Pratt's vindication of 'his priority of discovery over Airy in some points in the figure of the Earth' probably refers to their dispute about their respective hypotheses concerning the thickness of the Earth's crust.]

58.
CMS C34/191
Henry Venn of C.M.S. to John Venn
1 April 1872
East Sheen S.W.

My very dear son,

I have seen the acc^{t} of the death of Prof. Maurice[225] – & also a list of the Electors.[226] You will of course be a candidate – but I see not any way in which I can exert any influence.[227] Therefore I can only wait & pray that the Lord, whose providential leadings it is our glory to follow, many make his will clear, & turn the event to your best interest. I see, most vividly, how much the event may affect your future course of life. I dare not form a wish for you: For He who sees the end from the beginning alone sees what will be for the best. But I shall make it a matter of much prayer, till the matter is decided. […]

Ever your most aff^{te} Father,

Henry Venn

59.
GCA C52/4
John Stuart Mill to John Venn
14 April 1872
Avignon

Dear Sir

I willingly enclose an expression of opinion on your qualifications, such as my knowledge of your writings enables me to give. I am much aided in doing so by the pamphlet[228] which I have had the pleasure of receiving from you, and which shews that you have applied the same clear and vigorous intellect to the subject of ethics which was conspicuous in your logical speculations. Whether your opinions and my own on that subject would altogether agree is in such cases a secondary consideration. I should not fear to defend Kant's maxim[229] against your criticism. He could not mean, nor could Paley[230] mean, that we should so act that the whole human race could with general benefit do exactly what we are doing; they meant that our conduct ought to be capable of being brought under a rule to which it would be for the

---

225 [**F.D. Maurice** had died on that same day, 1 April 1872.]

226 [Upon **F.D. Maurice**'s death, the Knightbridge Professorship of Moral Philosophy fell vacant.]

227 [Cf. letter **38**.]

228 [Venn wrote a pamphlet, entitled 'Practic' – now lost – which contained reflections on ethics, particularly utilitarianism, a version of which Venn also seems to have ascribed to Kant.]

229 [Kant's maxim, or categorical imperative, found in his *Critique of Practical Reason* (1788), which says that 'Act only on that maxim through which you can at the same time will that it should become a universal law'. Venn interpreted it in a utilitarian way.]

230 [William Paley (1734–1805), the Christian apologist who also worked on ethics. His *Principles of Moral and Political Philosophy* of 1785 contained a defence of 'theological utilitarianism', which, like non-theological varieties of utilitarianism, said that rightness depends entirely on the value of the consequences of an action.]

general benefit that all should conform. This rule, in your example of taking orders, would not be that all mankind might with public advantage take orders, but that the choice of a profession should depend (under limitations which could be stated) on the aptitudes and convenience of the individual.

One more remark. I agree with you that the right way of testing actions by their consequences, is to test them by the natural consequences of the particular action, and not by those which would follow if every one did the same. But, for the most part, the consideration of what would happen if every one did the same, is the only means we have of discovering the tendency of the act in the particular case. In your example from Austria, it is only by considering what would happen if everybody evaded his share of taxation, that we perceive the mischievous tendency of anybody's doing so. And that this mischievous tendency overbalances (unless in very extreme cases) the private good obtained by the breach of a moral rule, is obvious if we take into consideration the importance, to the general good, of the feeling of security, or certainty; which is impaired, not only by every known actual violation of good rules, but by the belief that such violations ever occur.

I am,

Dear Sir
Yours very truly
J. Stuart Mill

From my knowledge of Mr Venn's writings, I have been impressed, in an unusual degree, by the clearness, vigour, and precision of his intellect, as well as by his power of putting his clear ideas clearly and forcibly into words. These qualities are apparent as well in what he has written on ethical as on logical subjects; and on both he has shewn decided power of original thought. I am not sufficiently acquainted with his positive opinions in moral philosophy, to know how far I agree or differ with them; but in the case of a public teacher, his actual doctrines are of less importance than his influence in stimulating the exercise of thought in his pupils, and at the same time guarding them by an exact Method, from the evils of vagueness and looseness in thinking: and in both these respects I think it likely, judging from his writings, that Mr Venn would be a highly successful teacher.

J.S. Mill

60.
CMS C34/192
Henry Venn of C.M.S. to John Venn
16 April 1872
East Sheen S.W.

My very dear son,

I have read with deep interest your fragmentary documents on "Practic'".[231] During the first half I almost doubted whether you would improve your chance of

[231] [Cf. letter **59**; a pamphlet that Venn circulated to improve his chances of obtaining the Knightbridge Professorship of Moral Philosophy, which had fallen vacant upon **F.D. Maurice**'s death.]

the Professorship. But after I had read the whole I fully concurred in its being printed & circulated among the electors. I have seen so much of the abominable clanship of the great colleges that I shall not anticipate the result. But I am thoroughly satisfied with a better result than success, namely that you deserve the Office & are prepared to do full justice to the opportunities of usefulness which it will afford. "The lot is cast into the lap, the whole disposing thereof is of the Lord".[232] This is my comfort & my joy. The Professorship or something better I feel assured that your God & the God of your Father will bestow.

[…] God ever bless & keep you both,

Your most affte Father, H. Venn

61.
CMS C34/193
Henry Venn of C.M.S. to John Venn
22 April 1872
East Sheen S.W.

My very dear son,

[…] I think you act very wisely in keeping your own mind in an even balance about the professorship - & I trust dearest Susie will do the same. I heard from the relations of the wife of the eldest of the Perowne's[233] – at Lampeter – that a prof.rship at Cambridge had been offered to him & he was considering of it. I heard from another quarter that the Master of Peterhouse[234] had said he would have a high churchman. Putting these two things together I suspect that the Master has got two of Electors to stand with him. If so, as he has a second vote, the election would be in his hands. Nevertheless all may be over ruled – & will be to your best interests.

Ever your most affte Father

Henry Venn

62.
GCA C45/1
William Stanley Jevons to John Venn
9 October 1874
Parsonagehead, Wittington, Manchester

Dear Mr Venn

I have been reading your review[235] of the "Principles of Science" in the Academy with the greatest interest & gratification, & have to thank you warmly for the careful & impartial way in which you have treated my poor volume.

---

[232] [Proverbs 16:33.]

[233] [J.J. Stewart Perowne (1823–1904), Corpus Christi 1841; B.A. 1845; Fellow, 1849–64; Hulsean Lecturer, 1868; Praelector in Theology at Trinity, 1872–75; Vice-Principal of St. David's College, Lampeter, 1862–72.]

[234] [Henry Wilkinson Cookson (1810–1876), Peterhouse 1828; B.A. (7th Wrangler) 1832; Master of Peterhouse, 1847–1876.]

[235] [John Venn, "Review of *The Principles of Science: A Treatise on Logic and Scientific Method.* By W. Stanley Jevons," *The Academy* (October 1874): pp. 381–384.]

Passing over the points where you indicate more or less agreement I should like to notice briefly the objections which you raise, not with the object of taking up your time in controversy but simply to explain the difficulties under which I lay.

Mill's so called inductive methods are certainly given in my Elementary Lessons[236] but that work was only intended as a small introductory text book in which it was impossible to discuss the exact value & nature of doctrines commonly accepted. What Herschel[237] & Mill[238] treated as the special methods of induction are by me treated under the head of experiment. It seems to me, rightly or wrongly, that they are rather rules for observing or experimenting so as to gain facts from which hypothetical reasoning may afterwards extract laws & principles. The process of induction proper is, on this view, what I have treated in chapters XXII to XXVI.[239]

I quite agree with you however that much vagueness attaches to the name induction; and I think it very likely that I have not used the word always in exactly the same sense but my excuse can be that even Mill seems to me to have used the word loosely, indeed he wavers very much & calls geometry for instance sometimes deductive & sometimes an inductive science.

To the want of a psychological analysis of the basis of reasoning I plead guilty. On the point the Spectator[240] raised almost exactly the same objection as yourself at the same time. No doubt to a considerable extent I have avoided the true difficulties of the subject but this does not preclude me from attempting to remedy the defect at some future time, if I live long enough, and can feel that I see my way to a more settled state of opinion. My efforts have been directed painfully to analysing in order the more formal and mechanical parts of logical method, which may be useful in itself, though only a preliminary task to a more profound discussion of the bases of knowledge.

The main point in which I should venture to differ from your criticisms refers to the symbolic method & its usefulness. It is a matter which cannot be adequately discussed in writing, but I would remark that the principal question is whether or not the symbolic processes correctly represent the relation of classes of things & the course of our thought about them. If not the symbols must be given up, or modified until they do. In short I venture to look upon them as an essential mode of expressing a true system of logical forms, not meaning of course merely the general letters A.B.C., &c, but any corresponding use of words or signs for expressing the like general relations of terms.

---

[236] [William Stanley Jevons, *Elementary Lessons in Logic: Deductive and Inductive* (London: Macmillan, 1870).]

[237] [John F.W. Herschel (1792–1871) in chapters 6 and 7 of his *A Preliminary Discourse on Natural Philosophy* (1830).]

[238] [**John Stuart Mill** in Book III ('On Induction') of *System of Logic*.]

[239] [William Stanley Jevons, *The Principles of Science: A Treatise on Logic and Scientific Method* (London: Macmillan, 1874), pp. 105–241.]

[240] [The *Spectator*, 19 September 1874, p. 1178; a second portion of the review appeared there on 26 September, 1874, p. 1204.]

No doubt I have not adequately noticed Mill's objection to 'Quantification',[241] but I felt that to enter into discussion & criticism would add too much to the length of an already heavy & costly book.

With regard to your example of possible confusion at bottom of 3rd column p. 382, it seems quite possible that a student might make the mistake suggested, but it would be by a breach of the rule of substitution,[242] but substituting 'consequence of gravity' for 'consequence of gravitating matter' there being no warrant whatever for this substitution. Thus I do not see that the strict scientific generality of the principle of subn is impeached or indeed intended to be impeached.

In the next column you have given a very nice example of a logical question, simple yet perplexing without some method of symbolic analysis.

I have to thank you for pointing out oversights about Encke's comet[243]; it is obvious now that I have committed a blunder in the Elementary Lesson which no one had before pointed out. The angular velocity of the comet is increasing, so that of course it turns each time a little sooner than it wd without resistance. The resisting medium produces an effect which wd in the absence of solar attraction produce retardation.

This & some of the requisite corrections which you point out shall receive the closest attention if ever the time arrives when a second edition becomes possible.

Believe me

Yours very faithfully

W. Stanley Jevons

63.
JRUL JA6/2/360
John Venn to William Stanley Jevons
11 October 1874
Claremont, Holdenhurst Rd, Bournemouth

My dear Prof. Jevons,

I was very glad to get your letter this morning, & to find that you were not altogether dissatisfied with my review. I had been thinking of writing myself a few days

---

241 [In his *System of Logic*, **Mill** had objected to the introduction of the quantification of the predicate into syllogistic reasoning. This innovation, which led to new propositional forms, had been proposed by Augustus De Morgan and William Hamilton in the 1850s. See L.M. Laita, "Influences on Boole's Logic: The Controversy Between William Hamilton and Augustus De Morgan," *Annals of Science* 36:1 (1979): pp. 45–65.]

242 [The central rule of **William Stanley Jevons**'s logical system, first elaborated in detail in his 1869 with the same title. 'All acts of reasoning seem to me to be different cases of one uniform process, which may perhaps be best described as the *substitution of similar*. This phrase clearly expresses that familiar mode in which we continually argue by analogy *from like to like*, and take one thing as a representative of another'. William Stanley Jevons, *The Substitution of Similars* (London: Macmillan, 1869), on p. v.]

243 [Comet Encke, first recognized as a periodic comet in 1819 when its orbit was computed by Johann Franz Encke.]

ago, but when one has just been throwing stones at a man's windows one is shy of ringing the bell & calling on him. The review was the best I could do, in the allotted space, & was at any rate the result of several weeks almost exclusive study of your book.

There were several points about which I should like a good talk, & much hope to get it when you come to Cambridge this term. I should of course have hoped to have got you to stay at our house then, but, owing to my wife's weak state of health, we have had to break up our home for the winter, & I don't expect to be more than a partial resident at Cambridge for the next 6 months. When I wrote to you in the Spring,[244] I think you undertook to appear once at any rate, & as I shall be more or less there till the beginning of December, we may have an opportunity of meeting.

With regard to the Academy, I felt that your work was too extensive to be properly discussed in a weekly journal, & that the 30 or 40 pages of a Quarterly were wanted to give any adequate discussion of the many points raised. It seems to me too that the Academy plan of signatures has a slight drawback to its advantages; one is (commercially speaking) writing cheques instead of issuing bank notes; & therefore has to keep strictly within the limits of one's own private solvency. Hence since blame is easier to support by reasons than praise, the latter becomes somewhat difficult to any who have not a commanding right to be heard, when merely expressing their judgment. In particular I should have liked to express my high admiration of the concluding part of your 2ᵈ Vol., where you are speaking of the infinitude of science.

I hardly know what has been said by any of your other critics, as I avoided reading any before writing my own; this was written more than two months ago, but press of matter from Associations & Congresses, prevented the Edition from putting it in earlier.

I suppose that on the matter of the [*del*: real] value & place of symbolic reasoning there is a real difference of opinion between us; I have never been able to feel the extreme admiration which some high authorities (like yourself & Todhunter[245]) express for Boole's work,[246] wonderfully profound & ingenious as it is in its own way; – but I admit that I have never thoroughly worked through it. One point occurs to me which seems illustrative of what I think his needless complexity of process;

---

[244] [This letter has not been found.]

[245] [**Isaac Todhunter**, one of Venn's mathematics coaches as an undergraduate.]

[246] [George Boole's work on algebraic logic in his *Mathematical Analysis of Logic* (1847) and *Laws of Thought* (1854). Todhunter had recommended Boole's *Laws of Thought* to Venn in 1858. In his *Annals* Venn admits that it was not until the late 1870s that he was able to make sense of it. See *Annals*, p. 96.]

he actually gives a proof of the 'Law of Contradiction', employing more than one mathematical step in so doing.[247]

By the way, in your next Edition I wish you would touch upon the question whether when we employ symbolical processes in logic it is necessary that every step of our process, or only our final result should be intelligible. Boole w[d] doubtless employ impossible quantities: – I forget if he actually does so.

I had had rather a good illustration in practice of the sort of difficulty which want of familiarity with symbolic methods, though, may cause. The example about the Board & bondholders (originally suggested when reading your 'Dichotomy' plan in the Elementary Lessons) was set at London University & college exercises &c to I suppose between 200 & 250 men, without my getting in a simple case the correct answer, though they tortured it to fragments with syllogisms.

On the question of the Special Methods of Inductive Enquiry, there is little more than a difference of degree in our opinions. I have for some time past expressed the opinion in lecturing, that the value & logical completeness of Mill's 4 Methods[248] was much overrated, & shall find in future much assistance from your statements & examples.

On the question of Least Squares I suppose you have not seen Leslie Ellis' remarks on Herschell's [*sic*] proof &c.[249] I can't detect exactly where the latter goes astray, but it seems to me no more than a mere juggle. There is a paper also by J.W.L. Glaisher[250] which you would like to see on the same subject. It was in the Astronomical Soc. Reports, Vol XXXIX, p. 75; but has been republished separately at Strangeways, Castle St. Leicester Square.

Believe me

Y[rs] very sincerely
J. Venn

---

[247] [Venn's point seems to be that, in making the law of contradiction a consequence of the law of duality ($x^2 = x$), Boole took away the fundamental character of the law. See George Boole, *An Investigation of the Laws of Thought* (London; Macmillan and Co., 1854), chapter 3 ('Proposition 4').]

[248] [**Mill**'s 'four methods of experimental inquiry', from his *System of Logic*: method of agreement, method of difference, method of residues and method of concomitant variations.]

[249] [Robert Leslie Ellis (1817–1859), Trinity 1836; B.A. (Senior Wrangler) 1840; Fellow, 1840–49. Ellis had returned to the topic of least squares in 1850, when he wrote a critical response to John F.W. Herschel's widely-read article on Adolphe Quetelet's work on probability, published anonymously in the *Edinburgh Review*. Ellis's short publication appeared in the *Philosophical Magazine* in November 1850.]

[250] [James Whitbread Lee Glaisher (1848–1928), Trinity 1866; B.A. (2nd Wrangler) 1871; Fellow, 1871–1928; Tutor, 1883–93; Lecturer in Mathematics, 1871–1901; President of the Royal Astronomical Society, 1886–88 and 1901–03. The paper referred to is James W.L. Glaisher, "On the Law of Facility of Errors and Observations, and on the Method of Least Squares," *Memoirs of the Royal Astronomical Society* 39 (1872): pp. 75–124.]

64.
JRUL JA6/2/358
John Venn to William Stanley Jevons
23 March 1876
Petersfield House, Cambridge

My dear Jevons,

I am telling Macmillan to send you a copy of my Logic of Chance.[251] If any remarks occur to you in reading it please let me have the benefit of them. It is to the criticism of competent friends in this way that one finds one must look for hope of improvement; for, from the ordinary literary critic (according to my experience) not a particle of benefit is to be got, – at least in works of a somewhat special & technical kind.

Have you any speedy prospect of a 2[d] ed[n] of your 'Principles of Science'?[252] If so there are some points in it to wh I sh[d] like to call your attention. They are mostly in connection with Boole, whom I have been reading this winter preparatory to lecturing on him; it does not seem to me that you have quite done him justice.[253]

About Disjunction (pp. 84, 85; 130) you speak as if Boole were confined to one way of representing them.[254] It appears to me that your way is defective. There are of course two perfectly distinct questions here (1) What sense does ordinary usage put on the disjunctive "or"? (2) How is logic to express these senses, if more than one? Much of your discussion (pp. 84, 85) is occupied with (1), & it does not seem to me that you have very clearly distinguished it from (2). Coming to (2) it is obvious that there are two senses of "or", as you say. Boole expresses both with perfect precision, thus (a) $x(1-y) + y(1-x)$ = "X or Y but not both" (b) $x+y(1-x)$ = "X or Y or both". [With?] these meanings being perfectly distinct it seems to me that no symbolic language in logic can be admitted wh does not improve upon the confusion of common language by distinguishing sharply between them. (Boole p. 60 also notices the form [*del*: analogous somewhat to yours] for the disjunctive "$x+y$", wh as he says means "X or Y", with the assumption that X & Y are exclusive, for otherwise we sh[d] be counting the common part twice over) a & b are in the two regular forms wh. Boole adjusts, according to the meaning intended to be conveyed, & I can't find any such tacit assumption as you refer to on p. 84, 130 (see e.g. Boole p. 170; numerous other instances might be given). It seems to me, I confess, that

---

[251] [Venn's *Logic of Chance* had originally appeared in 1866. The second edition, then just finished, appeared in 1876, with many additions and changes.]

[252] [The second edition of Jevons's *Principles of Science* would appear in 1877; it was reprinted with corrections in 1879, 1883, 1887 and 1892.]

[253] [Venn introduced Boole's work at Cambridge in his 'Advanced Logic' course. For the discussion between Venn and Jevons on algebraic logic see, for instance, Lukas M. Verburgt, *John Venn: A Life in Logic* (forthcoming at The University of Chicago Press), chapter 9.]

[254] [Where Boole insisted on an exclusive interpretation of disjunction when applied to classes, Jevons interpreted it inclusively, such that $a + a = a$ would count as a legitimate law of logic. Venn initially followed Boole, but changed his mind and adopted inclusive disjunction in the second revision 1894 edition of his *Symbolic Logic*, recognizing the practical advantage of the notation.]

your "X• | •Y"[255] is a sort of concession to the laxity of common language wh. the logician ought to avoid.

Again in explaining your form A=AB (for All A is B) you speak as if Boole's $x = vy$ (v an indefinite class symbol) were different from yours, when Boole's method & form <u>includes</u> yours. Boole always (or rather whenever required) [?] this symbolic v, & gets the $= x(1-y) = 0$. This gives at once $x-xy = 0$ viz <u>x = xy</u> wh. is your form (see Boole p. 104).

Where is your Logical Machine to be got? & does it entirely supersede the Abacus?[256] As I say, I am going to lecture on these symbolic methods next term, & want to find out as much as I can about them meanwhile.

Yrs very truly,

J. Venn

65.
GCA C45/2
William Stanley Jevons to John Venn
25 March 1876
Withington

My dear Venn,

Thanks for telling Macmillan to send me a copy of your new edition which shall have my best attention. You admire Mill so much more than I do, that to a certain extent it makes us look from different points of view, but there is nothing like the free expression of opinion for getting towards the truth.

There is no immediate prospect of a second edition of my Principles being required, but possibly in a year or two I may have an opportunity of correcting errors & omissions.[257]

I am greatly obliged to you for stating your objections to my remarks on Boole. I did not want to take up much space in that book with controversy, more especially as I had said what I had to say in my earlier essay 'The Pure Logic"[258] of which I send you a copy. My logical notation stands or falls with the substantial accuracy of what I have said about exclusiveness of alternatives.[259] The point is certainly a nice

---

[255] [Jevons introduced a new symbol for disjunction, which he used instead of the addition sign, +, that Boole had used and that he himself had also adopted in his *Pure Logic* of 1864.]

[256] [Jevons was one of the first to see that Boole's methods were suitable for mechanization. He devised a number of labour-saving devices of which the logical abacus (1866) and the logical machine (1870) were the most important. He himself used them for classroom instruction purposes at Queen's College and Owen's College, Manchester. See Harro Maas, *William Stanley Jevons and the Making of Modern Economics* (Cambridge: Cambridge University Press, 2005), chapter 6.]

[257] [See note **252**.]

[258] [William Stanley Jevons, *Pure Logic: Or, The Logic of Quality Apart from Quantity: With Remarks on Boole's System and on the Relation of Logic and Mathematics* (London: Edward Stanford, 1864).]

[259] [As Jevons wrote in *Principles*: 'I discuss this subject [i.e. of disjunctive propositions] fully because it is really the point which separates my logical system from that of the late Dr. Boole'.

& debatable one, but 15 years of consideration lead me to maintain what I have said. Doubtless Boole can express whether the alternatives are exclusive or not –

$$x(1-y)+y(1-x)(\alpha)$$

$$\&xy+x(1-y)+y(1-x)(\beta)$$

and so can my notation equally well

Ab. / .aB

AB. / .Ab. / .aB

but the point is that his notation necessitates the expression, so that as he himself told me in a short correspondence,[260] the terms are uninterpretable rules upon this condition. Thus x+y has no consistent meaning in his system except it be explained as (α) or (β). Until you decide which it is, it cannot be used. But one general result of my logical cogitations is that you always use terms subject to subsequent restrictions or explanations.

I cannot see that I am wrong on p. 130. By "a condition concerning the exclusive nature of alternatives" I mean simply that terms whenever joined by + in his system must be held to be exclusive since

$$x+y=(xy+x(1-y))$$
$$+(yx+y(1-x))$$
$$=2xy+x$$

would be actually uninterpretable, in fact a contradiction in terms.

In answer to your remark that my "X• | •Y" is a concession to the laxity of common language, I can only refer to pp. 82–85 in Principles or to the latter part my Pure Logic of which I send you a copy.[261]

Jevons, *Principles of Science*, 83.]

[260] [The correspondence between Boole and Jevons was first published in 1913, but without notes. In the 1970s edition of Jevons's papers and correspondence, the correspondence is omitted. The texts have been transcribed and annotated in Ivor Grattan-Guinness, "The Correspondence Between George Boole and Stanley Jevons, 1863–1864," *History and Philosophy of Logic* 12:1 (1991): pp. 15–35.]

[261] [On pp. 82–85 of the *Principles of Science* **Jevons** wrote, for example, the following; 'In his 'Law of Thought' (p. 32) [Boole] expressly says, 'In strictness, the words "and"," "or," interposed between the terms descriptive or two or more classes of objects, imply that those classes are quite distinct, so that no member of one is found in another.' This I altogether dispute. In the *ordinary use* of these conjunctions we do not necessarily join distinct terms only; and when terms so joined

It is not easy to establish my view but I am convinced of its truth. The best proof perhaps is to be found in the extreme simplicity and generality of my notation compared with the intricate & almost incomprehensible system of Boole. I am quite convinced that Boole's forms 1/1 0/1 1/0 & 0/0 have no real analogy to the similar mathematical expressions. In logic they merely indicate that a term agrees or is contradictory to one or other on both or neither side of an equation. But they have nothing to do with the values 1, 0, ∞ &c.

As regards the form A = AB, I say on p. 49 that in Boole's logic expressions of the kind A = VB were freely used, but that I found indeterminate symbols only introduced complexity.

I do not say that Boole exclusively, but only freely used A = VB that is the form corresponding to x = vy. I am of course aware that he also used x = xy. My point on p. 49 is that the introduction of v or V at all is a mistake and in no way an advantage.[262] I think my remarks are correct though very concise, but you should understand that in view of the extent of the book I felt obliged to restrict myself to a simple exposition of my own notation avoiding all controversy.

I am not sure whether you mean that my acknowledgment of dependence on Boole were not ample enough. I should be sorry if this were so, but you will find the Pure Logic abundant recognition of the fact, and in my "Elementary Lessons" which is the only one of my books widely read the whole subject is put down under the heading of Boole's logic. The fact is that a friend well acquainted with the matter remonstrated with me for representing my notion as being that of Boole.

If I ever come to a second edition of the Principles I shall be happy to amplify a little upon the points of agreement & difference with Boole.

You will gather from the last chapter of the 'Pure Logic' that I hold Boole's system to be absolutely perfect within itself, entirely self consistent but I deny that it is logic pure & simple. The relation to logic is one of so difficult a kind that I cannot say more than I have said in the 'Pure Logic'.

My machine is rather a large & awkward thing, about 3 feet high. There is only one in the world & probably there never will be another. I would offer to send it only it will not bear travelling, & I fear it w[d] be useless to send it you. It is moreover in bad order. I enclosed a photograph & a paper giving full details of it.[263] The Abacus[264] will do all that the machine can & more, as it takes in five terms, but it is rather troublesome to

---

do prove to be logically distinct, it is by virtue of a *tacit premise*, something in the meaning of the names and our knowledge of them, which teaches us they are distinct'. Jevons, *Principles of Science*, on pp. 83–84.]

262 [Boole had regarded *v* as an indeterminate class-term ('all, some, or none').]

263 [William Stanley Jevons, "On the Mechanical Performance of Logical Inference", *Proceedings of the Royal Society of London* 18 (1870): 166–169.]

264 [As **Jevons** wrote in *Pure Logic*: '[W]hile the logical slate or printed abecedarium is convenient for the private study of logical problems, the abacus is peculiarly adapted for the logical classroom. By its use the operations of classification and selection, on which Boole's logic, and in fact any logic must be founded, can be represented, and the clearest possible solution of any question can be shown to a class of students [...]'. Jevons, *Pure Logic*, on pp. 151–152.]

use. If you like to shew it to your students I shall be happy to send it with or without the black board & you could return it to me after we are settled in London. It does the problems in a very short time if you get the knack of moving the combinations.

Believe me,

Yours very faithfully
W.S. Jevons

66.
JRUL JA6/2/359
John Venn to William Stanley Jevons
28 March 1876
Petersfield House, Cambridge

My dear Jevons

Many thanks for the book & paper which you have so kindly sent me. What I wrote referred purely to the 'Principles'; & probably I shd have altered somewhat what I said if I had read the 'Pure Logic'. As explained in your letter the point is more subtle, & I sh^{d} like to think it over a good deal more before writing again. But I still think that what you say in the Principles (p. 84) is misleading, especially the paragraph, "Take, for instance…".[265] It left on me, after several readings, the impression that you regarded Boole's system as <u>unequal</u> to expressing any but mutually exclusive alternatives, not merely in the sense (as explained in your letter) that <u>his</u> alternatives are always mutually exclusive, but in the sense that he cd not express in his fashion any common statement in which the alternatives were not so. That is, your example about the peers seemed to me to imply not merely that Boole has a peculiar way of expressing alternatives, but that he can't even express at all on his system any common statement except of the 'exclusive' kind. I still think, if I may venture to say so, that you would make these pages much clearer if you were to indicate that there are two very distinct questions: – (1) What, as a matter of fact, does the particle "or" express? (purely philosophical) & (2) How ought the logician to express the two distinct meanings that it possesses? (logical)

With regard to your objection in the 'Pure Logic', p. 78,[266] I can't admit the objection that Boole has to appeal to the matter in an illogical manner. It must surely

[265] ['Take, for instance, the proposition 'A peer is either a duke, or a marquis, or an earl, or a viscount, or a baron.' If expressed in Professor Boole's symbols, it would be implied that a peer cannot be at once a duke and marquis, or marquis and earl. Yet many peers do possess two or more titles, and the Prince of Wales is Duke of Cornwall, Earl of Dublin, and Baron Renfrew. If it were enacted by parliament that no peer should have more than one title, this would be the tacit premise which Professor Boole assumes to exist. Nor is the restriction true of more common terms'. Jevons, *Principles of Science*, on p. 84.]

[266] [Venn's reference is probably to the following passage: 'He [Boole] says, 'that before attempting to translate our data into the rigorous language of symbols, it is above all things necessary to ascertain the *intended* import of the words we are using.' This simply amounts to consulting the matter, and Professor Boole's symbols thus constantly imply restrictions not expressed in the forms of language, but existing, if at all, as tacit or understood premises'. Jevons, *Pure Logic*, on p. 78.]

be always postulated that language is unambiguous, & if language in any case unfortunately is ambiguous, we must either decline to treat the example or look to the matter to see what is meant. 'All these troubles are too much for me', 'All these troubles are unexpected'; no one w^{d} blame a logician for saying that since 'all' is sometimes collective & sometimes distributive we must gather from the context or matter wh. is meant in the case in question. So Boole w^{d} say (I suppose) that since 'or' sometimes means one thing & sometimes another we must somehow come to an understanding in such case what it does mean which in reference to meanings so distinct as these is important. But to fix accurately the meanings of your premises surely is not to 'appeal to the matter'.

Don't trouble to reply to this: I will write again when I have thought more on the matter.

Many thanks for your kind offer about the 'machine'. I had no idea it was unique in these days of logical reading & lecturing & examining. I did not indeed suppose that such things were to be had at country stationers, but I thought that there might be a few perhaps somewhere in London.

Yours very truly,

J. Venn

67.
GCA C12/2
Leslie Stephen to John Venn
14 June 1876
8 Southwell Gardens, South Kensington

Dear John,

I have just finished reading your book[267] & the reading has given me very great pleasure. I am really impressed both by the ingenuity & clearness of the thought & by the skill wh which you have compressed your matter into so small a space. At the same time, and rather for the same reason partly, I am very glad that I did not undertake to review it. I would not have done you justice. You get beyond my depth in some very important parts as e.g. in the discussion of "modality"[268] – a diabolical concern of wh I scarcely know the name. And further my brain is at present really in an enfeebled state. I am about always headachy & puzzled – not that I am exactly ill, but what we used to call in the old athletic slang "stale". Various causes wh you can guess at have brought me into this & though I hope that it will disappear when

[267] [The second edition of the *Logic of Chance*, published in 1876.]

[268] [Venn added a chapter on 'Modality' – the logic of necessity and possibility – in which he defended the view that modal propositions can only be satisfactory treated in the logic of probability, instead of in formal logic. For a discussion see Ilkka Niiniluoto, "From Possibility to Probability: British Discussion on Modality in the Nineteenth Century," in Simo Knuuttila, ed. *Modern Modalities. Studies of the History of Modal Theories from medieval Nominalism to Logical Positivism* (Dordrecht: Kluwer Academic Publishers, 1988), pp. 239–274.]

I get my holiday, I can't think about a logical problem without getting confused after a little.

I agree with you in all your first principles & agree as far as I can follow in almost everything – though I would suggest some difficulties. Most of them I suspect are due to my own want of clearness. But I always feel that there is some perplexity wh I have never succeeded in unravelling to my own satisfaction & wh still bother me dimly even in reading your book. If I tried to state it now, I should only get into a mess; & I imagine that it too is due to a want of mathematical capacity or sufficient thought. In an equal way, I may say that there is a point at wh induction & the calculation of probability seem to get muddled together & here I always stick.[269] However I won't try to unravel my own muddles.

The moral objection to gambling is, as it seems to me, that it means a simple love of excitement wh ought to be turned to better account; whereas the advantage of insurance is that it diminishes excitement by enabling one to feel superior to chance.[270] Therefore though you may in one sense compare insurance to betting, it seems to me to be really the antithesis. I did not feel sure whether this was properly brought out in your chapter.

I suspect too, a fallacy in the last chapter,[271] ingenious as it certainly is. I don't quite see why the effect of prophecy should not be so calculable that I should make my prophecy coincide with the event. E.g. if I say to a man that path leads to a precipice I may prophesy that he can't walk over it.

But, as aforesaid, I won't try to argue. I may make a muddle out of a bit of your book when my head is better; but I don't quite feel up to it at present.

I wonder if C.J. Monro is a man in the year below me (I think) to whom I once took a great liking but who somehow disappeared altogether from my circle & whom for some reason, I fancied to be dead.[272] His name, I think, was Cecil. If so, I should like to hear of him again – I remember being much struck by his abilities – long ago –

Yours affect,
L. Stephen

I hope that your wife is pretty well

---

[269] [In the 1876 edition, the relation between induction and probability theory was discussed by Venn in chapter 8 ('Induction'). Venn's view was that probability was founded on induction, rather than vice versa. As a result, he simply rejected the application of probability to inductive inference., as proposed by **Jevons** and others.]

[270] [In chapter 15 ('Insurance and gambling') of the second edition of the *Logic of Chance*, Venn captured the contrast between insurance and gambling in the following terms: insurance is a means of diminishing uncertainties whereas gambling is a means of increasing them.]

[271] [Chapter 18 ('Statistics as Applied to Human Actions').]

[272] [This was, indeed, **Cecil James Monro** (1833–1882). On account of ill-health, he afterwards lived largely abroad, where he devoted himself to letter-writing and criticism of proof-sheets for friends.]

68.
GCA C45/3
William Stanley Jevons to John Venn
5 October 1876
Withington

My dear Venn,

I got your letter in London 2 days ago but I was not able to see your article[273] in Mind until this morning after my temporary return home.

I thank you much for your reference to my logical notation so far indeed as it is mine. Your whole article is full of interest to me and it seems one of the best notices of Boole that I have seen. One of the professors of U.C.L. expressed himself as quite pleased with it.

Actually I do not agree with some of your remarks but now when I am tired with travelling & in the thick of an approaching removal I do not feel able to discuss the matter adequately.

What you say at the bottom of p. 488 does not exactly correspond to my view of the matter but on turning to the page of the Principles quoted I find I cannot complain of your view.[274] Of course you observe that I make no claim to any exclusive use of the form X = XY as at the bottom of p. 50 I distinctly disclaim this. I myself prefer the form for reasons which I will refer more clearly when I have an opportunity.

My successor Adamson[275] has found that Boole's two laws xy = yx & x^2 = x are most distinctly stated by Leibnitz together with some cases of substitution!

[...]

I will send you my London address in a week or two. Yours very faithfully

W.S. Jevons

69.
GCA C32/1
Francis Galton to John Venn
7 July 1880
42 Ruthland Gate, London

My dear Venn,

Your kind words written on May 31 reached me abroad (as no doubt my sister who forwarded letter told you) & though we have been back a fortnight I happen to have been laid up & all my letters are in arrear so pray excuse my long silence.

---

[273] [John Venn, "Boole's Logical System," *Mind* 1:4 (1876): pp. 479–491.]

[274] [Venn, "Boole's Logical System," pp. 488–489, where Venn criticizes Jevons's rejection of Boole's use of $x = vy$ for 'All X is Y' and Jevons's proposal to express is as X = XY.]

[275] [Robert Adamson (1852–1902). In 1876, Adamson had come to England from Germany as **Jevons**'s successor in the Chair of Logic and Philosophy at Owens College, Manchester. He was afterwards Professor of Logic at the University of Glasgow, 1895–1902. After Robert Leslie Ellis noted Leibniz's anticipation of Boole's 'Index Law' (1847) or 'Law of Duality' (1854) in 1854–55, Adamson was given credit by Jevons, in the second edition of his *Principles of Logic*, for pointing to Leibniz's anticipation of the substitution of similars.]

I am very glad you continued to re-make a satisfactory logical machine.[276] I told Jevons of its existence and excited his cupidity & whetted his eagerness. I dare say that you have long since been in correspondence with him. It certainly is a beautiful little thing & I could fancy it a lovely affair with ivory areas. I see we are joint contributors to 'Mind' in this month's number.[277] I just saw the volume for the first time yesterday but will read your article carefully.

I fear that even if you are now in Cambridge all thoughts of a visit must be postponed, thank you much for thinking of us. With our united kind remembrances to yourself & Mrs Venn

Sincerely yours,
Francis Galton

I send by book post an advance copy of my "Visualized Numerals"[278] in case you like to look at the lithographed "Forms".

70.
GCA C45/4
William Stanley Jevons to John Venn
18 August 1880
Hampstead NW

Dear Mr Venn

On my return home from my usual summer tour I am much pleased to find a copy of your paper on the Mechanical Representation of reasoning awaiting me.[279] It must have come just after I left home. I have not yet had time to read the whole of it but I have already got an idea of your diagrammatic machine,[280] of which I had previously heard some account from Mr Galton.[281] It is exceedingly ingenious & seems to represent the relations of four terms very well.

---

[276] [Cf. letter **65**. Venn, who was critical of the practical usefulness of Jevons's mechanical inventions – writing, for instance, that what logical machines did was but a very small part of the entire process of reasoning – did come up with one or two proposals for a logical machine himself. See John Venn, "On the Diagrammatic and Mechanical Representation of Propositions and Reasonings," *The London, Edinburgh, and Dublin Philosophical Magazine* 10:59 (1880): pp. 1–18, on pp. 16–18.]

[277] [John Venn, "On the Forms of Logical Proposition," *Mind* 5:19 (1880): pp. 336–349; Francis Galton, "Statistics of Mental Imagery," *Mind* 5:19 (1880): pp. 301–318.]

[278] [**Francis Galton** (1822–1911) published two articles with the title 'Visualized Numerals' in 1880. Galton put forward the thesis that many people possessed or utilized visual number systems that he called number forms. Number forms were the spatial or diagrammatic systems by which people imagined the organization of numbers. Galton reasoned that, if number forms could be pictorial, then statistical thought itself could be pictorial.]

[279] [Venn, "On the Diagrammatic and Mechanical Representation of Propositions and Reasonings."]

[280] [Cf. note **276**.]

[281] [**Jevons**'s letters from and to **Galton** are not included in the volume with papers and correspondence published at Macmillan in the 1970s.]

In the next week or two I hope to finish off the proofs of my 'Studies in Deductive Logic',[282] the whole of which is now I think in print. It ends with what I propose to call the Logical Index,[283] a complete statement of the 256 cases of four terms.

I see you refer on p. 16 to the use of stamps. About a year ago it occurred to me to have the Logical Alphabet for 3 or 4 terms made into the India rubber stamps now so much used, & I find these answer admirably, so as quite to supersede the Logical Slate I had previously used.[284] I only [*illegible*] our impressions.

Believe me

yours very faithfully
W. Stanley Jevons

71.
GCA C5/7
Alexander Bain[285] to John Venn
16 December 1880

My dear Mr Venn,

Aberdeen[286] is a bad place for rare old works in Logic. Edinburgh is more hopeful. I shall be there in about three weeks, and shall be glad to prosecute an enquiry for any special items that you may desiderate.

[…] Of course you know that Jevons[287] has been hunting out, at least the names of old logical writers, so as to make up an exhaustive bibliography.

If you will give me titles of your desiderata, I will try all the sources that I have access to.

With best wishes,

yours truly
A. Bain

---

282 [William Stanley Jevons, *Studies in Deductive Logic: A Manuel for Students* (London: Macmillan and Co., 1880).]

283 [Jevons, *Studies in Deductive Logic*, chapter 28. 'Four' is a slip; Jevons provided an index of all 256 logical relations between three terms.]

284 [Cf. Venn, "On the Diagrammatic and Mechanical Representation of Logical Propositions," p. 16; Jevons, *Studies in Deductive Logic*, p. 285. As Jevons explained here: 'In my previous logical books [i.e. *Pure Logic* (1864), *Substitution of Similars* (1869), *Elementary Lessons in Logic* (1870) and *Principles of Science* (1874)] I described a Logical Slate with five series of the combinations of the Logical Alphabet engraved upon it. I first made such a slate in May 1863'.]

285 [**Alexander Bain** (1818–1903).]

286 [**Bain** had been appointed Regius Professor of Logic at the newly united University of Aberdeen in 1860. In 1871, he had received the honorary degree of Doctor of Law from the University of Edinburgh.]

287 [Cf. letter **72**.]

72.
GCA C45/5
William Stanley Jevons to John Venn
11 May 1881

Dear Mr. Venn,

Accept my best thanks for your volume on Symbolic Logic,[288] received yesterday. I have been looking forward to the interest & pleasure of reading the book, of the progress of which I have heard from time to time.

I have already looked through the volume with a good deal of care & made myself acquainted with its purpose & contents. I fear must defer a detailed reading until I get again into the logical vein. At present I am much involved in Economics.[289]

It is natural that I should not agree with all your remarks but I cannot omit to thank you for the evident care with which you have treated my opinions especially where you see reason to dispute them.

I am surprised to find how many books you cite which I have never heard of or at least not been able to see. Your list will assist Adamson[290] & me in the task we are engaged upon of forming a bibliography of logic probably for the Lond. Society.[291]

Much as I may differ from you on minor points I rejoice at the publication of your book as I feel sure it will establish & ensure the recognition of the still greater points in which we agree.

Believe me

Yours very faithfully,
W.S. Jevons

73.
GCA C17/2
T.S. Baynes[292] to John Venn
14 June 1881
University of St Andrews

My dear Venn,

I have just returned home after my weeks absence & find your "Symbolic Logic" on my table. Prey accept my best thanks for the volume, which, if you will allow me

---

[288] [John Venn, *Symbolic Logic* (London: Macmillan and Co., 1881); 2nd ed. 1894.]

[289] [**Jevons** had resigned his post as Professor of Political Economy in University College, London, in 1880. Around this time in 1881, Jevons was working on his *The State in Relation to Labour* (1882).]

[290] [**Jevons**'s collaborator Robert Adamson (1852–1902), who had succeeded Jevons in the Chair of Logic and Mental and Moral Philosophy at Owens College, Manchester in 1876.]

[291] [This (possibly unfinished) bibliography has not been found. It was probably compiled for the Statistical Society of London, of which **Jevons** had become Vice-President in 1880.]

[292] [Thomas Spencer Baynes (1823–1887), pupil and advocate of the logical work of Sir William Hamilton; appointed Professor of Logic and English Literature at St Andrews University; superintended the ninth edition of the *Encyclopaedia Brittanica*, begun in 1873.]

to say so, appears to me a most solid & valuable contribution both to the science & history of logic.

I hope it is not too late to thank you for the quotation from Hoffbauer referring to Ploucquet's recognition of the Categorical predicate.[293] It is quite new to me, as I have never seen Hoffbauer's work.[294]

In glancing through your volume I almost sigh for the learned leisure essential to its production and also, to the [*illegible*] at least, to its adequate comprehension & intelligent use. […]

Very truly yours,

Mr. Thos. Baynes

74.
JRUL JA6/2/361
John Venn to William Stanley Jevons
27 July [1881]
Innsbruck

My dear Prof. Jevons

I must just send you a few lines to express my sense of obligation to you for your review in Nature.[295] It was not merely friendly, but (considering that I had had to express my differences from you more than from any other writer) I ought rather to call it generous. However I know that you are one of those excellent combatants who, whilst they will hit their hardest against any authority in defence of what they consider the truth, will also (what does not by any means follow) have no objection to being hit at in return, provided it be fairly done. I can honestly say that I tried to conform to the latter.

Thanks for your reminder about Garden.[296] I have the book, & now you mention it I seem to remember having seen his diagrams: but it must have slipped my memory at the time.

There is one thing I am sorry for, viz. that it should seem to you that I had been unfair or snubbing to Maccoll.[297] I have been greatly bothered by him. I had a very

[293] [The German logicians Johann Christoph Hoffbauer (1766–1827) and Gottfried Ploucquet (1716–1790). In logic, a categorical statement is a proposition that asserts or denies that all or some of the members of the subject term are included in the predicate term. Baynes's reference is probably to Ploucquet's development of a new syllogistic logic where the categorical forms are interpreted as (set-theoretical) identities, or diversities, between the full extension, or a non-empty part of the extension, of the subject and the predicate. Plouquet, thereby, arguably anticipated the quantification of the predicate. See Wolfgang Lenzen, "Ploucquet's "Refutation" of the Traditional Square of Opposition," *Logica Universalis* 2 (2008): pp. 43–58.]

[294] [Hoffbauer's 1792 *Analytik der Urtheile und Schlüsse*.]

[295] [William Stanley Jevons, "Symbolic Logic," *Nature* 24 (1881): pp. 233–234.]

[296] [Francis Garden, *An Outline of Logic, for the Use of Students* (London, 1867). See *Catalogue of a Collection of Books on Logic Presented to the Library by John Venn* (Cambridge, 1889), p. 115.]

[297] [Hugh MacColl (1837–1909), Scottish mathematician and logician, today remembered for his innovative contributions to modal and non-classical logics. See, for example, Shahid Rahman and

long correspondence[298] with him some 2 years ago, which was dragged out by constant complaints that I "must in fairness acknowledge" and such things about his method[299]; – (things which I could not see my way to acknowledge). However, to tell the truth, I thought that my opinion about the value of his general work was higher than yours: e.g. you spoke of his having "thrown the science back into ante-Boolian confusion",[300] – I should not say this, & I thought that my general judgment of him (in a note) was much more favourable than this.[301] The really distinctive characteristic about his papers (though he did not seem to see it), – I mean in contrast with Boole, – was his a+b as against a + *not*-ab notation. But this of course had been adopted & urged by you before Macoll. He has been persistently trying to get me into a correspondence in Nature or elsewhere which I have as far as possible avoided, as I think that rather intricate principles cannot possibly be thus decided. The first two of his letters I took no notice of, but the third[302] long one (in Nature) contained a charge[303] which I was bound to notice. I thought it was one which he had no business to make, as my long correspondence with him must [*page torn*] that I never dreamed of denying his having written his papers before reading Boole. But I declined (as you did in your letter[304] in Nature some months ago) to be dragged into a sort of examination by working out examples to see which could do it quickest & shortest.

I certainly had no intention of undervaluing his work, which indeed I believe I rate decidedly higher than such a first-rate judge as C.J. Monro[305] does.

However an independent person like yourself is in a better position than I am to say what the general effect of my words was; though I feel sure that if that general effect was depreciatory it was unintended.

Yours very truly

J. Venn

Juan Redmond, "Hugh MacColl and the Birth of Logical Pluralism," in Dov. M. Gabbay and John Woods, eds., *Handbook of the History of Logic. Volume 4: British Logic in the Nineteenth Century* (Amsterdam & Boston: North Holland, 2008), pp. 533–604.]

298 [This correspondence has not survived.]

299 [In a series of three papers appearing in 1877–78, MacColl had put forward a propositional logic based on the meanings of sentence connectives, rather than on the algebra of classes. Whereas MacColl claimed that his system differed from that of Boole in several important respects, Venn and Jevons agreed that it was merely a development of one part of Boole's system. For context see Lukas M. Verburgt, "The Venn-MacColl Dispute in *Nature*," *History and Philosophy of Logic* 41:3 (2020): pp. 244–251.]

300 [Jevons, *Studies in Deductive Logic*, on p. xv.]

301 [John Venn, "Symbolical Logic," *Nature* 24 (1881): pp. 140–141.]

302 [Hugh MacColl, "Symbolical Logic," *Nature* 24 (1881): pp. 124–126.]

303 [In his *Symbolic Logic*, Venn suggested that, if he had read Boole before writing his three papers of 1877–78, MacColl would not claim originality. MacColl responded by accusing Venn of not having read his work properly.]

304 [William Stanley Jevons, "Symbolic Logic," *Nature* 24 (1881): pp. 233–234.]

305 [**Cecil James Monro** (1833–1882.]

75.
GCA C54
C.J. Monro to John Venn
22 January 1882

My dear Venn

I have delayed writing a very long time. The excuse is that indisposition my brother spoke of; it is, among others an indisposition to write, at least to take the consequences of writing. But it has been better in this rather cold weather, so I write now before the rumours of change to mild are confirmed.

I am sorry Mrs Venn has to go abroad, but glad you can arrange it in the way you describe; this satisfaction not being quite neutralized either by the consideration that the arrangement seems to be for the good of lecturing so well as of lecturers.

I am sorry again that I shall not meet you: I keep to the west side of France if I go at all. I may do so, but later.

Lastly, on the chapter of sorrow, I was sorry at first to see you were going back to common logic,[306] but in the way you describe your aim there is room for originality as well as thoroughness. Anyhow I am glad you told me, because I am merely in time to ask you what you think the best book on the subject, big or little, in any language, at present. I will send back Tiverton[307] presently.

I never did anything more to z/y = z/w[308]: I meant and mean to recast & publish, but I have been horribly idle and intent has rather stopped the way, – not that I have done much to that either. It is a serious discouragement though that so large a fraction of my public has "forgot it all".

I am glad you in any manner or degree thought well of my "critical notice"[309]: it gave me a great deal of trouble to try & do justice to the subject both in detail & as a whole, and I fear I have only proved that I could not do this.

I have had no communication with MacColl[310] since I forget when, but it was something I mentioned to you; and I am glad of it.

Of the equiangular spiral[311] it only occurs to me to say that the question is not in [form?], geometrical I mean of course not logical. "Have an end to it" is not a geometrical notation I fancy, unless you make it one by disposition. I should say the spiral has an end much as the infinite straight line has: the only difference is that in

---

306 [Around this time, Venn started work on what would become the *Principles of Empirical, or Inductive Logic*, published in 1889, which aimed to give an overview of the whole of logic from a quasi-Millian perspective.]

307 [Not identified.]

308 [Context unknown. Possibly a contribution to the *Educational Times*, to which **Monro** had contributed before – a periodical which often contained mathematical questions and solutions.]

309 [**Monro** reviewed Venn's *Symbolic Logic* for *Mind*. C.J. Monro, "Review of *Symbolic Logic*, by John Venn," *Mind* 6:24 (1881): pp. 574–581.]

310 [Monro's correspondence with MacColl has not survived.]

311 [The equiangular spiral, logarithmic spiral or growth spiral is a self-similar spiral curve, first described by Descartes and extensively investigated by Jacob Bernoulli. The context of Monro's reference to it is unknown]

the case of the spiral you can show a point which unlike the intersection of the straight line with the line infinity is at a finite distance, not indeed along the locus but along the paths; and how does this matter? But I dare say I miss the point.

Yours sincerely

C.J. Monro

76.
UCL 3/3/21/2 F1–F4
John Venn to Francis Galton
1 May 1887
Petersfield House, Cambridge

My dear Mr Galton

I am engaged on a new edition of my Logic of Chance,[312] & naturally want to work through again what you have written upon the subject of 'Law of Error' during the last few years. I have a copy of your (1877) Royal Institution paper[313]: What you could then only test on sweet peas, you have I think tested since by the result of the answers to your prize questions.[314] Can you give me a reference to any other of your papers; as I don't want to miss any of them?

What you say in your Royal Institution [paper?] seems to me very important, & I must take rather more shame than some others for not having seen the necessity for such conditions, because I have rather insisted upon the effect where they are absent. Might we not illustrate what comes in that case (i.e. when 'reversion'[315] &c are absent) by a consideration of the following result. Suppose a large number of persons with equal fortunes to begin with, incessantly & indiscriminately betting for small & equal sums, say shillings. If we were to look at their fortunes at the end of a certain time we should find these to be differentiated according to our familiar curve. If they had had time to gamble 100 times we should find most of the fortunes

---

312 [The third edition of Venn's *Logic of Chance* would appear in 1888. It contained many additions, mostly relating to statistical topics, such as error laws.]

313 [Francis Galton, "Typical Laws of Heredity," *Proceedings of the Royal Institution* 8 (1877): pp. 282–201. This is the paper where Galton for the first time used "normal" to refer to the distribution which was then still variously called the law of error, the mathematical law of deviation or the exponential distribution. He noted that the production of peas of the same class deviate normally on either side of their own mean weight. See Stephen M. Stigler, *Statistics on the Table* (Cambridge, MA: Harvard University Press, 1999), p. 412.]

314 [Venn's reference is to **Galton**'s circulation of a fly-leaf in 1884 offering '£500 in prizes to those British subjects resident in the United Kingdom who shall furnish him before May 15, 1884, with the best Extracts from their own Family Records'. Galton used to answers as part of his promotion of the cause of eugenics: they resulted in the publication of two books, *Record of Family Faculties* (1884) and *Life-History Album* (1884). For context see Theodore M. Porter, *Genetics in the Madhouse: The Unknown History of Human Heredity* (Princeton & Oxford: Princeton University Press, 2019), chapter 10.]

315 [The notion, introduced by **Galton**, of reversion toward the mean. See Stephen M. Stigler, *The History of Statistics: The Measurement of Uncertainty Before 1900* (Cambridge, MA: Belknap Press, 1986), pp. 283–293.]

near to old mean, [*illegible*] down such way till we found one with +£5 & one with – £5 on his old fortune. Examine these again a little later & the 'dispersion' will be greater; & so on until some have been ruined. This represents the pure results of theory, supposing many small & independent influences; & having seen & enforced this, I ought to have asked why it does not take place where natural classes are concerned: – the question you <u>have</u> asked & answered.

But then this seems to raise a further difficulty. In the case of the gamblers we see that the 'curve' being a straight line to start with, it closes up more & more as the modulus alters: i.e. the equal fortunes are more & more differentiated. Now your theory seems to involve perfect stability without any alteration, & then the question arises how the differentiation we actually observe was brought about? I suppose you would, (what one always must allow where questions of fixity or variability of species are concerned) that so far as any possibility of observation goes the species <u>do</u> remain absolutely unchangeable. That is, your remarks in no respect preclude an inappreciably small tendency towards further dispersion. If the present variations have resulted from equality or indistinguishable similarity to start with, then it is probable that in the course of ages our giants will really be bigger & our dwarfs smaller, unless new controlling influences are brought into play.

I was sorry to miss you here in the winter: but, as you know, I was severely ill: I am getting all right now.

Yours very truly

J. Venn

77.
UCL 3/3/21/2 F5–F7
John Venn to Francis Galton
4 May 1887
Petersfield House, Cambridge

My dear Galton

Many thanks for your kindness in sending me all these pamphlets: – I have somewhat of a collection of short papers on such subjects & am going to have them bound up together. They will be given to the University some time: &, as I need not say to you, there is a great convenience to any future reader in having them in one volume.

Thanks also for your kind enquiries about myself. I am getting all right again now, though I still have to take precautions.

As regards what I said about "gambling", I don't think there is the slightest difference of opinion between us. All that I meant was that such an example represents the outcome of the pure chance theory, & illustrates that gradual further dispersion which as you very well show (& were I think the first to show) does <u>not</u> prevail in regard to the vital phenomena, or natural species: Or rather its influence is counteracted by <u>Reversion</u> &c there.

I was interested in the paper about extinctions of family names which I had not before seen.[316] Like yourself I had often thought about the subject (having been dabbling in genealogies during the last 2 or 3 years) & I had poked at it a little in an empirical way, but Watson's paper was quite new to me.

I hope if you come here again to stay at Trinity Lodge you will let me know so that I may have the pleasure of calling on you there.

Believe me

Yours very truly,
J. Venn

78.
UCL 3/3/21/2 F8–F11
John Venn to Francis Galton
23 September 1887
Petersfield House, Cambridge

My dear Galton

Many thanks for yours just received. It will be a very great pleasure to see you here, & go over some of the results of those Cambridge measurements with you.[317] I worked (after my last letter) through the breathing statistics, & find the same marked superiority here again over your S. Kensington results (i.e. an average of 253 against 219).

I think it will be best now to wait until the clerk at the Philosophic Library has indexed the 1440 cards. I think I am right in concluding that you wished him also to undertake such work as making separate lists for each college: I can then send these to the various tutors & get the men sorted into 3 classes as regards "honours". After this is done the real interest of the Cambridge results will commence in the comparison between these different results.

[*Pages missing*]

I am sorry to hear you got a fever in Italy. When I did so, to a slight extent, some 10 years ago, I found English air a complete reformative. And, though subsequent

[316] [Together with Rev. Henry Watson, **Francis Galton** had set out to develop a mathematical theory of the extinction of surnames, which gave rise to the Galton-Watson branching process. Henry William Watson and Francis Galton, "On the Probability of the Extinction of Families," *The Journal of the Anthropological Institute of Great Britain and Ireland* 4 (1875): pp. 138–144.]

[317] [Following up on **Galton**'s successful laboratory at the International Health Exhibition held in London in 1885 and his well-attended Rede Lectures at Cambridge, in 1886–87, the Cambridge Anthropometric Laboratory had been set up in the Library of the Philosophical Society. With a set of instruments presented by Galton himself, measurements were made of keenness of eye-sight, strength of "pull", strength of "squeeze, head size, breathing capacity, height without shoes, and weight. Venn was asked by Galton to undertake an analysis of the data, and to compare them with the data obtained at South Kensington. See John Venn, "Cambridge Anthropometry," *Journal of the Anthropological Institute of Great Britain and Ireland* 18 (1889): pp. 140–154; and Susannah Gibson, *The Spirit of Inquiry: How One Extraordinary Society Shaped Modern Science* (Oxford: Oxford University Press, 2019), chapter 7.]

returns of it were freely prognosticated by my friends, I had but one & that for a few hours.

Believe me

Yours very truly
J. Venn

79.
GCA C62
J.S. Reid[318] to John Venn
3 November 1887

My dear Venn,

I did not know when you left that you intended to make such a decided change as that described in your letter.[319] For the sake of the College, I am heartily sorry. You were in a position to give us very important help. However, I hope we shall see you often at the meetings of the Governing Body.

I sincerely hope the change will secure the principal purpose you have in making it – better health for Mrs. Venn and for your little boy.

The Governing Body will, I am sure, cordially agree with you in wishing that your most valuable work on the history of the members of the College should go forward.[320] Already the records you have got together from a collection such as no College possesses. The information will be valuable for all time, and there can be no hesitation about supplying you with the means for carrying on further investigations, so long as you feel able to direct them. Not only we, but our successors will be your [*illegible*] for all the labour and thought you have expended on the work. I will take care that the matter is put before the Governing Body at the next meeting.

Roberts[321] agrees with me in thinking it very undesirable that you should formally give up your position as Catechist. It would be a great pity to uproot the old office[322] altogether, and I see no reason why you should not continue to hold it. When you cease to hold it, I trust that it will be attached still to some one of the outlying studies of the College which can be [repr…ted?] by one of the Fellows.

The best thing for our men would be to get some one (Johnson[323] perhaps) to see them for an hour a week and advise them generally, giving them sometimes papers. The fee for this is about £3 per term. I have already got Johnson to make this

---

[318] [J.S. Reid, (1846–1926), Christ's, 1865; B.A. (Senior Classic) 1869; Fellow, 1869–72; Fellow of Caius, 1878–1926; Tutor, 1885–99; University Lecturer in Classics, 1883–99; Professor of Ancient History, 1899–1924.]

[319] [Venn had expressed his wish to resign the post of Catechist, which he had held since 1862.]

[320] [John Venn, *Biographical History of Gonville and Caius College, 1349–1897* (3 vols.) (Cambridge: Cambridge University Press, 1897–1901).]

[321] [Ernest Stewart Roberts (1847–1912), Caius 1865; B.A. (6th Classic) 1869; Fellow, 1870–1903; Junior Proctor, 1875–76; Tutor, 1876; Senior Proctor, 1884–85; Senior Tutor, 1885; President, 1894–1903; Master, 1903–1912; Vice-Chancellor, 1906–08.]

[322] [The office of Catechist had formally existed since 1508. See Venn's *Annals*, p. 117.]

[323] [Possibly W.E. Johnson (1858–1931), King's, 1878; B.A. (11th Wrangler) 1882; Moral Sciences Tripos, 1st Class, 1883; Fellow, 1902–31.]

arrangement for Holman,[324] who is prevented by his engagement with Dr Luard[325] from attending lectures regularly. With this kind of arrangement the cost for tuition of each man would amount to about £5 per term. There is I am afraid no prospect that we shall have more than two or three men at a time studying Moral Science.

The tuition of the College is working, on the whole, well just now. Gallop[326] is developing considerable power as a teacher. Such friction as we have just now in College is mostly connected with the Chapel. With all his pleasant qualities, Wallis[327] is often a difficult colleague to work with.

Yours very sincerely

J.S Reid

80.
UCL 3/3/21/2 F12–14
John Venn to Francis Galton
7 November 1887
Petersfield House, Cambridge

My dear Galton

We have got an alphabetical M.S. list of our "measures" here, & a rough list of the men of each college. These last I am getting the tutors to sort out for me into 3 classes: viz. High honours, Low honours and poll or pluck. This last job will, I hope, be got through in a few days now.

I am afraid I should not have time to go through the original cards, with the view of selecting the measurements of the men belonging to each of these classes. But if you would authorize the clerk to do this mechanical part of the work, I will undertake all the consequent calculation & comparison of the separate classes with each other & with the general average.

I worked through the general lot of papers in respect of 4 characteristics, viz.: height; pull; squeeze; breathe.

Are you in hopes of being able to get down here this term?

Yours very truly
J. Venn

---

324 [Henry Holman (1859–1919), Caius 1889; Scholar, 1890; B.A. (Moral Sciences Tripos, 1st Class) 1890; Professor of Theory of Education, Aberystwyth College, 1893.]

325 [Edward Gurner Gallop (1862–1936), Trinity 1880; B.A. (2nd Wrangler) 1883; First Smith's Prizeman, 1885; Fellow, 1885–89; Fellow of Caius, 1889–1936; Mathematical Lecturer, 1886; Tutor, 1903–09.]

326 [Probably Henry Richards Luard (1825–1891), Trinity 1843; B.A. (14th Wrangler) 1847; Fellow, 1849; Junior Bursar, 1853–62; Assistant Tutor, 1855–65; University Registrary, 1862–91; Lady Margaret's Preacher, 1881; President of the Cambridge Antiquarian Society, 1864–65; Vicar of Great St. Mary's, Cambridge, 1860–87.]

327 [Frederic Wallis (1854–1928), Caius 1872; B.A. (Classical Tripos, 1st Class) 1876; Fellow, 1878; Dean, 1878; Senior Proctor, 1892; afterwards Archdeacon of Wilts., 1911–1912 and Canon of Salisbury, 1913; described in the *Alumni Cantabrigienses* in the following terms: 'scholarly, humorous and intensely lovable, he left a name in Cambridge which was never mentioned save with abiding affection. As the guide, counsellor and friend of youth, Wallis was one of the best known and best loved of Cambridge dons'.]

81.
UCL 3/3/21/2 F17–19
John Venn to Francis Galton
16 February 1888
Fern Mount, Suffolk Road, Bournemouth

My dear Galton

I set to work on my arrival here & have worked out the results in a form which you will be able to interpret & make use of. But of course a good deal more will have to be done before a report upon them can be drawn up. It has involved a good deal of work. There are, as you see, 1051 men, each of whom undergoes 11 measurements. As every column was added up twice, for accuracy, it involves some 25,000 elements. I thought it better to calculate the mean in each case by the full arithmetic process, as the numbers in each subdivision were mostly too small for it to be safe to substitute the "median".

Without the help of 'Crelle',[328] which you so kindly sent me, the work would have been vastly more laborious; but I have used it in all the multiplications and divisions.

By the way I am not sure of the nature of the head-measures. I presume that they represent, in a way, the three elements of a solid measure, & that therefore the product of the three is a fair criterion of the size of the head.

I will wait for your comments before drawing up anything in the way of a Report. But some facts strike the attention at once: –

(1) There is practically no physical difference between the reading &non-reading men; – except, –

α. The eye-sight deteriorates slightly with study
β. The head is slightly larger
γ. The "Pull" seems a trifle less strong

(2) In all 3 classes alike the eye-sight seems to deteriorate slowly after 19 (& perhaps long before). But the apparent improvement in each case at 24 is either a chance or a puzzle.

The strength seems in each case to decline after 23 or 24.

Yours very truly
J. Venn

[328] [August Leopold Crelle's *Rechentafeln* (Berlin: G. Reimer, 1875). Among other things, the book gave the products of every two numbers from one to one thousand. The first English-language edition appeared in 1885: *Tables for Facilitating the Operations of Multiplication, Division, and Evolution. Abridged from Dr. A.L. Crelle's* Rechentafeln. By Lieut. J.A. Borris. (Washington: Government Printing Office.]

82.
UCL 3/3/21/2 F21–23
John Venn to Francis Galton
26 February 1888
Fern Mount, Suffolk Road, Bournemouth

My dear Galton

Many thanks for your kind comments: – if any criticisms occur to you in reference to my new edition[329] I shall be most thankful for them.

I have sent you off a sort of 'Report upon those Cambridge measurements'.[330] But you must regard it as offered for your criticism & suggestion. So pray send it back to me if there seem any serious errors or omissions. If any minor points occur to you you might perhaps like to add them in the shape of a comment. Pray communicate it to the Anthropological or any other Journal you think best. If it is printed I should like to have a fair no. of copies as of course I must send one to each of the tutors (25 or so) who helped me with information.

If the attendance keeps up there will probably be some 1500 more measures for discussion in a couple of years. This may serve to clear up some doubtful points.

Yours very truly
J. Venn

P.S By the way I am not sure whether (in comparing the Cambridge and Kensington measurements) I have called attention to the fact that I have reckoned mean squeeze (R. and L.) & that you reckoned extreme squeeze (R. or L.). This makes an important distinction when comparing the absolute strength.

83.
UCL 3/3/21/2 F25
John Venn to Francis Galton
1 March 1888
Fern Mount, Suffolk Rd., Bournemouth

Would it be possible to test the average discrepancy between the 'Pull' and 'Squeeze' results at Kensington & Cambridge? If your former instruments are in existence & physically uninjured they might be sent to Cambridge & tried in 40 or 50 cases upon the same men as those who use our present ones. If it was in London one would like to have your former superintendent[331] to see that his instruments were used according to his method for fair comparison. But perhaps when you are next in Cambridge you would see Mr. White[332] & ascertain exactly what measure of

[329] [The third edition of Venn's *Logic of Chance*, published in 1888.]

[330] [Venn, "Cambridge Anthropometry,"]

[331] [A Sergeant Randall.]

[332] [The Librarian of the Philosophical Society.]

difference can be allowed for difference of usage. (I hope you took no injury from your cold journey to Cambridge).

Y^rs truly J. Venn

84.
GCA 7339/229
John Venn to Joseph Foster[333]
28 March 1888
Petersfield House, Cambridge

Dear Sir,

I have just been looking at the 1st vol. of your Oxford Matriculations,[334] which we ordered for our Library, & it prompts me to ask a question of you. I am engaged, in spare time, with a complete M.S. list of all the persons (towards 8000 in no.) who are known ever to have belonged to the College[335]: adding a few notes about their subsequent career, where this can be ascertained. Is there any present prospect of your List of Admissions to Inns of Court being published?[336] Such a work would clear up a large number of my present blanks. Of course I have consulted your Men at the Bar[337] for the later names, & your list (of A & part of B) in the Misc. Genealogica[338]: but the large help I have found from these makes me very desirous to consult the remaining entries. Have you got them in any form in which they could be consulted? Of course if they are likely to be published within any reasonable time I should hope to see them in print.

Of course if I can ever be of any help in reference to former members of my own College I shall be most glad, & if you have not got a copy of our recently published Admissions to Gon. & Caius Coll.[339] I will send you a copy.

Believe me,

Yours very truly,
J. Venn

333 [Joseph Foster (1844–1905), genealogist whose major works – transcriptions of the admission registers of the Inns of Court (published in 1885 as *Men-at-the-Bar*) and of the matriculation registers of the University of Oxford for the period 1500 to 1886 (published as *Alumni Oxonienses*, 4 vols.)– are still used today.]

334 [Joseph Foster's *Alumni Oxonienses: The Members of the University of Oxford, 1715–1886, etc. Being the Matriculation Register of the University*, 4 vols. (Oxford, 1888).]

335 [Venn had then already published *Admissions to Gonville and Caius College in the University of Cambridge, March 1558–9 to Jan. 1678–9*. Edited by J. Venn and S.C. Venn (London: C.J. Clay and Sons, 1887).]

336 [In 1889, Joseph Foster published *The Register of Admissions to Gray's Inn, 1521–1889*, which was privately printed.]

337 [Joseph Foster, *Men-at-the-Bar: A Biographical Hand-List of the Members of the Various Inns of Court* (printed for the author by Hazell, Watson, and Viney, London, 1885).]

338 [Probably Foster's *Collectanea Genealogica*, which was published by subscription in 1881.]

339 [See note **335**.]

85.
UCL 3/3/21/2 F26–F28
John Venn to Francis Galton
8 May 1888
Petersfield House, Cambridge

My dear Galton

My only complaint of your Notice in Nature[340] is that you should have given our readers the impression that the investigation was mine, instead of my merely working loosely on your suggestions.

As regards a Psychological Laboratory[341] on any small scale here I think I could guarantee its being carefully looked after: – in fact (failing more skilled & experienced supervision) I would do so myself. But whether it would be adequately utilized & appreciated is another matter.

At present I am at work on a somewhat systematic attempt to write a Treatise on Inductive Logic (something between Mill & Jevons in respect of scope & general treatment) which has occupied much of my time for many years past.[342] As soon as this is done, – & I hope it will be published before this time next year, – I should like nothing better than such a complete change as that involved in turning to Psychologic measurement & calculations &c.

As regards W.E. Johnson,[343] I know the man well. In fact he was for a short time a pupil of mine. He was a good mathematician to begin with & has studied Logic & Psychology very carefully since. His notice of my Logic of Chance[344] seemed to me very careful & acute. But as with most of our Moral Science men here, few as they are, there is much difficulty in finding any reasonable, not to say lucrative, employment for him.

I am not able to come to the R.S.[345] tomorrow, having an engagement here; but those crowded meetings are not the best occasion of getting a quiet talk, at least not with a person like yourself who knows everybody there. I may be in London some

---

[340] [Francis Galton, "On Head Growth in Students at the University of Cambridge," *Nature* 38 (1888): pp. 14–15.]

[341] [On the complex history of experimental psychology at Cambridge in the 1880s see Michael M. Sokal, "Psychology at Victorian Cambridge: The Unofficial Laboratory of 1887–1888," *Proceedings of the American Philosophical Society* 116:2 (1972): pp. 145–147.]

[342] [Venn's *Principles of Inductive or Empirical Logic*, which would appear in 1889.]

[343] [W.E. Johnson (1858–1931), King's 1878; B.A. (11th Wrangler) 1882; Moral Sciences Tripos, 1st Class, 1883; Fellow, 1902–31; University Teacher of Theory of Education, 1893–8; University Lecturer in Moral Sciences, 1896–1901; Sidgwick Lecturer, 1902–31; played a leading role in the development and teaching of the moral sciences (especially logic and psychology) at Cambridge; frequent contributor to *Mind*.]

[344] [W.E. Johnson wrote a review of the third edition of Venn's *Logic of Chance* for *Mind*. W.E. Johnson, "Review of *The Logical of Chance*. By John Venn," *Mind* 50 (1888): pp. 268–280.]

[345] [The Royal Society. Venn had been elected a Fellow in 1883.]

day the week after next; if so I will write to know whether, & when, you will be free.

Thanks for the reference to those papers by Dr Meyer of Göttingen.[346]

Yours very truly
J. Venn

86.
GCA C69
Frederick Ryland[347] to John Venn
9 September 1888
St. Hubert, Montserrat Road, Putney S.W.

Dear Mr. Venn,

I have to thank you very much for your kind permission to use the testimonial and your particularly kind note, in which the permission is given.

Believe me, I often look back with a kind of regretful longing to the hours spent in the lecture room at Caius, when Ward[348] and Keynes[349] were among the auditors and actually ventured to discuss points with the lecturer; which the lecturer, who seemed the most terrific don I had seen, was always ready to discuss difficulties with anyone and everyone.

Yours sincerely

F. Ryland

87.
CMS C41/5
Albert Venn Dicey to John Venn
10 January 1889
All Souls College, Oxford

Dear John,

I had meant to have written to you at once in answer to your question, but I had not the statutes by me & cd not then answer them. Thus I forgot each day till too late.

Your letter is at home but I remember the questions.

---

346 [Georg Hermann von Meyer (1815–1892), famous German anatomist. See Karl Pearson, ed. *The Life and Letters of Francis Galton. Volume I: Birth 1822 to Marriage 1853* (Cambridge: Cambridge University Press, 1914), p. 132.]

347 [Frederick Ryland (1854–1902), St John's 1873; Moral Sciences Tripos, 1st Class; B.A. 1877; Assistant Lecturer in Philosophy at University College, London; and a private tutor.]

348 [**James Ward** (1834–1925).]

349 [John Neville Keynes (1852–1949), Pembroke 1872; Moral Sciences Tripos, 1st Class, 1875; B.A. 1876; Sc.D., 1891; Fellow, 1876–82; Hon. Fellow, 1911; University Registrary, 1910–25; University Lecturer in Moral Science, 1883–1910; Secretary of the Local Examinations and Lectures Syndicate, 1892–1910; father of J.M. Keynes.]

1. I consider you admirably suited for the Professorship[350] and both on public & private grounds I heartily wish you might get it;

2. The Statute certainly does not require you shd lecture on Aristotle. The Prof. is required to teach Logic, & its connection with mental philosophy, the laws of evidence, & natural science.

3. The Electors are (i) the Vice-Chancellor (that is Bellamy,[351] the Head of St. John's Coll.), (ii) The Sedleian Professor of Nat. Philosophy, i.e. Bartholomew Price,[352] who is in fact Prof. of Mathematics, (iii) Wallis,[353] the Prof. of Moral Philosophy, (iv) A person nominated by New. Coll. to act as elector on the particular occasion (A. Robertson)[354], (v) A person nominated as permanent elector by the Hebdomadal Council.[355] I don't know who this is, but I will ascertain (D.B. Monro).[356] The Board is not in my judgment a very good one. Few of our Boards are.

4. I don't think any one is standing with strong claims, but I fancy that as regards Logic some persons would feel a preference for an Oxford man.

5. The Professorship, like my own, involves I believe a residence of 6 months. Non-Oxford men have pretty often been appointed to Oxford Professorships.[357] I cannot see any harm whatever that it wd do you to stand.

Y[r] aff[te] cousin A.V.D.

88.
UCL 3/3/21/2 F29–F32
John Venn to Francis Galton
10 January 1889
1 Durrant Villa, Avenue Road, Bournemouth

My dear Galton

I am very unfortunate to have missed you again at Cambridge, as I should have much enjoyed a talk on the subjects you mention, & on others as well. It is kind of you to think of me in respect of your "Natural Inheritance"[358]: I shall be delighted to

---

350 [Venn sought to apply for the Wykeham Professorship in Logic, established at the University of Oxford in 1859. Thomas Fowler (1832–1904) had held the chair between 1873 and 1889.]

351 [James Bellamy (1819–1909), President of St. John's College, Oxford, from 1871 to 1909.]

352 [Bartholomew Price (1818–1898), Sedleian Professor of Natural Philosophy from 1853 to 1898, for a long period took on a large part of the mathematical teaching at Oxford.]

353 [William Wallace (1844–1897), White's Professor of Moral Philosophy at Oxford from 1882 to 1897.]

354 [Not identified.]

355 [The Hebdomadal Council was the chief executive body for the University of Oxford from 1854 to 2000.]

356 [David Binning Monro (1836–1905), elected Provost of Oriel College, Oxford in 1882 and Vice-Chancellor of Oxford from 1901 to 1904.]

357 [In this case, however, Venn lost to an Oxford man: John Cook Wilson (1849–1915), who had graduated from Oxford with a rare double double first, gaining a 1st in Mathematical Moderations (1869), 1st in Classical Moderations (1870), 1st in Mathematical finals (1871) and a 1st in Literae Humaniores (1872). Cook Wilson held the chair from 1889 to 1915.]

358 [Francis Galton, *Natural Inheritance* (London: Macmillan and Co., 1889).]

see it, & it can be safely sent here, as we have taken this house for two months certain from now.

In respect of the "correlations"[359] you speak of, if I were in Cambridge, – or rather, in Granchester, for we have moved out there now, & given up our house on Parker's Piece, – I would look up a little volume which was sent me a few years ago by the author. It is the only discussion on the subject (at least one application of the subject) which I can remember to have seen. The author was a French officer, much interested in the applications of the Theory of Probabilities. The particular problem which he mainly discusses is what I understand to be this 'correlation', in respect to uniforms as required for the vast numbers of men in modern armies. If all men were "similar", as mathematicians say, you would only need, say, 10 sizes, the numbers in each size being determined by the Law of Error. But of course each particular length of body gives rise to the same Law of Error in respect of the width of chest; & therefore the problem before the military clothiers becomes more complicated when they have to decide the allotment of a million uniforms. Is this analogous to your problem? Unfortunately I can't recall the name of the author, & you know how impossible it is to describe a foreign book (especially a little one in pamphlet form) so that a servant can identify it. But I do not suppose that the author could teach you anything on that subject. If I remember right he introduces the conception of a solid for the Law of Error, the usual plane curve not sufficing for the additional element of variability.

If I come to London next month I will try to see you for a talk; I may want to come for a day or two.

We are very fairly flourishing, dislike, rather than dread, of the cold & fog having brought us here.

Yours very truly

J. Venn

P.S. It haunts me that my Frenchman's name is Capt. LeGros.[360]

89.
CMS C41/6
Albert Venn Dicey to John Venn
11 February 1889
The Orchard, Banbury Road, Oxford

Dear John,

I trust you are making every possible effort about the Chair. The chance of success is not I fear great, but it is worth making a great exertion for. Get the very strongest testimonials you can & send in your name as soon as you can. The reason

[359] [**Francis Galton**'s invention of the notion of (statistical) correlation dated from late in the year 1888, It arose when he recognized a thread in different scientific problems he was studying, one of which was anthropometry. See Stephen M. Stigler, "Francis Galton's Account of the Invention of Correlation," *Statistical Science* 4:2 (1989): pp. 73–79.]

[360] [Not found.]

I recommend this is that as you are not much known here it is most desirable that the electors, some of whom I am sure intend to act conscientiously, should have a full opportunity of enquiring into your claims. I earnestly beg you to send copies of the Logic of Chance & the Formal Logic to all the electors but above all to Munro, the Provost of Oriel & to Bartholomew Price, the Sedleian Professor of Natural Philosophy.[361] I mention them because they are mathematicians & also I think less tied to particular candidates than some other electors. Also take care to call attention to a very eulogistic note in, I think, the last edn of Mill's Logic,[362] which I remember your showing me years ago.

I have intimated to these two electors that you are likely to stand. If you find a difficulty in referring to Mill, any of your friends, or I myself could do it, if you think it well which I am not sure it is to have a testimonial from a person so out of the regular logical line.

[…] Above all things lose no time. Of course I assume that people send in testimonials – in fact I have no doubt it is so.

I cannot tell you how strongly I feel about the benefit of your appointment to the place. Please show this to your wife, as I think she is more likely to show energy about the matter than you are. In fact I have been hesitating very much whether I had not better address her instead of you.

I dictate in order to prevent all possibility of misreadings.

Your affec. cousin,
A.V.D.

P.S. Better get a testimonial from Sir J.F. Stephen.[363]

90.
CMS C41/4
John Venn to 'Sir'
[not dated, but early 1889]

Sir,

I beg to offer myself as a candidate for the Wykeham Professorship of Logic. I graduated in Mathematical honours (Sixth Wrangler) in 1857, & became a Fellow of Caius College in the course of the same year.

Having already devoted some attention & study to the Moral Sciences, I was appointed College Lecturer in such subjects in 1862, an office I still hold. With the exception of occasional years of absence I have lectured continuously since that time, at first to Members of the College only, but since about 1868 to students of this University generally. In this capacity, I believe I was the first to introduce the now common practice of Inter-Collegiate lectures.

---

361 [See notes **352** and **356**. 'Formal Logic' refers to Venn's *Symbolic Logic* (1881).]

362 [The eighth edition of **Mill**'s *System of Logic*, published in 1872, reproduced the passage, which first appeared in the seventh edition of 1868, in which Mill praised Venn's *Logic of Chance*. See note **161**.]

363 [Their shared cousin **James Fitzjames Stephen**.]

My principal subject throughout the above period has been Logic: the accessory subjects being Political Economy, Ethics, & Mental Philosophy. I was an Examiner in Logic & Mental & Moral Philosophy in the University of London from 1871 to 1876: & examiner in similar subjects for the Indian Civil Service in 1875. I have also examined 12 or 13 times in the University of Cambridge for the Moral Sciences Tripos, & for the Special Examination in Moral Science for the Ordinary Degree.

I was elected a Fellow of the Royal Society in 1883; & admitted to the degree of Doctor in Science in Cambridge in 1884.

Besides various articles in "Mind", & other philosophical periodicals, I have published as follows, on the subject of Logic: –

The Logic of Chance: Ed. I. 1866, Ed. II. 1876, Ed. III. 1888

Symbolic Logic: 1880

Empirical or Inductive Logic pp. 590 1889 (now completed at the Press).

I send copies of these works, in evidence of the last, the Table of Contents will serve to show the range & scope.

The two pamphlets, which accompany or follow these, will serve as a brief introduction to one application of the Symbolic Logic.

I may add that I have given considerable attention to the subject of Psychophysics,[364] having both knowledge of, & considerable interest in, the requisite mechanical appliances. If elected, I should propose to give some instruction in this department, as it is connected at several points with the principles of Inductive Logic, but is apt to fall through in the curriculum between Philosophy & Physiology.

The works mentioned above will serve to indicate that I have not neglected any branch of Logic, though my own study & tastes have induced me to approach it especially from the scientific & modern side. I have continued to keep up my knowledge of mathematics & physical science as a reader, though not as a lecturer. I gather that this (i.e. the connection of Logic with Natural Science) is one of the qualifications desired in the Professor.

I do not enclose any testimonials, as I understand these are not definitely asked for, & as my works will probably be already known to those who are interested in Logic. But in case the Electors wish for an outside opinion I would refer them to any of the following (to whom however I have not suggested any such application):[365]

---

[364] [The study of quantitative relations between mental and physical events or, that is, the study of perceptual processes by determining the effect of a subject's experience or behavior through varying the properties of a stimulus. The subject was very much *en vogue* in Cambridge in the 1880s, especially due to the activities of Venn, **Ward** and G.F. Stout. See Colin Crampton, "The Cambridge School: The Life, Work and Influence of James Ward, W.H.R. Rivers, C.S. Myers and Sir Frederic Bartlett," unpublished PhD-dissertation, University of Edinburgh, 1978.]

[365] [**Alexander Bain** (1818–1903), then Professor Regius Professor of Logic and Regius Professor of English Literature at the University of Aberdeen; George Croom Robertson (1842–1892), then Professor of Mind and Logic at University College, London; **Henry Sidgwick**; Christoph von Sigwart (1830–1904), Professor of Philosophy at Tübingen; **Francis Galton**; **Leslie Stephen**; **Albert Venn Dicey**; Alfred Marshall (1842–1924), St John's, 1861; B.A. (2nd Wrangler) 1865; Fellow, 1865–77 and 1885–1908; College Lecturer in Moral Science, 1868–77; Professor of

viz. Prof. A. Bain of Aberdeen
Prof. G.C. Robertson, University College, London
Prof. H. Sidgwick, Cambridge
Prof. Sigwart, Tubingen
Mr. F. Galton, F.R.S.
Mr. Leslie Stephen
Prof. A.V. Dicey, Oxford
Prof. Marshall, Cambridge
John Venn

91.
UCL 3/3/21/2 F36–38
John Venn to Francis Galton
25 May 1889
Riversdale, Grantchester, Cambs

My dear Galton,

There are two little points in your "Natural Inheritance" that I want to ask you about.

1. On p. 70, I think the formula $\sqrt{\frac{c^2}{c^2+b^2}}$ must be wrong. For one thing, as it stands, the dimensions are wrong. I suppose it was meant to be printed $\frac{c^2}{\sqrt{c^2+b^2}}$. But in the second place it ought to introduce as a factor the observed lateral distance. If this be x I make the most probable distance from M of the red mark = $\frac{c^2}{c^2+b^2}$◆x. (I had worked it out so myself, & now I see that Mr Dickson seems to have given the same expression on p. 224.)[366]

2. p. 98 "As the two parents contribute equally…Therefore the average regression from the parental to the mid-filial stature must be the one-half of two-thirds…."

Perhaps I do not quite understand this. But it seems to suggest that the average regression from the single parent to the mid-offspring will be less than that from the mid-parent to the same. But the single parents will of course diverge from each other more than the mid-parents do, since the latter are got by averaging the former in pairs?

Yours truly

J. Venn

Political Economy at Cambridge, 1884–1908; President of the Economic Section of the British Association, 1890, and one of the founders of the Royal Economic Society.]

366 [James Douglas Hamilton Dickson (1849–1931), Peterhouse 1870; B.A. (5th Wrangler) 1874; Fellow, 1874–1931; Tutor, 1877–97; assistant to Lord Kelvin and Francis Galton. For Dickson's expression of Galton's formulae see Francis Galton, *Natural Inheritance*, Appendix B. ('Problem by J.D. Hamilton Dickson, Fellow and Tutor of St. Peter's College, Cambridge. Reprinted from Proc. Royal So., No. 242, 1886, p. 43.), pp. 221–224.]

92.
FGP 3/3/21/2 F39–F41
Francis Galton to John Venn
24 May 1889
42 Rutland Gate

My dear Venn

The first error is gross, and basely perhaps, I assign it either to a misprint in part, or in part or wholly to a blind & careless writing out a formula from memory.

My excuse is that when the formula is turned to account in p. 127*, though the confounded $\surd$ reappears there, it has not been used in the calculation and thence the words "of the lateral distance" do not appear at the end of the paragraph in p. 70, their equivalent does appear and is used in p. 127.

(*Here F_n is the equivalent of the a, in p. 127: $c^2 + b^2 = (1.7)^2$ $\quad \frac{c^2}{c^2 + b^2} = \frac{2}{3}$ whence the value of b is 0.98 which is the value given there),

The paragraph in p. 70 would have run

"…lateral distance a of a…. It is $\frac{c^2}{c^2 + b^2} a$."

In the second point I think I am right – what I meant to express was –

$$\frac{2}{3}\left\{\frac{M+F}{2}\right\} = \frac{M}{3} + \frac{F}{3}$$

(the $\frac{M+F}{2}$ being the mid-parental deviation). It would have been better had I put it in that form.

I am so much obliged to you for pointing out these blunders & faults of expression.

There is another one in p. 157, 8 lines from bottom. The figures there given are those in the bottom line of Table 9 B p 207 and which compare the frequency of contrasted to non-contrasted marriages…So far they are carried – but on analysing the non-contrasted there is a tendency of art to marry art though overborne by the tendency of non-art to refrain from marrying non-art.

Here however I spoke with much reservation due to the finallness of the figures, which made me clump together instead of analyzing, as I should have done.

93.
UCL 3/3/21/2 F42–F43
John Venn to Francis Galton
4 August 1889
Riversdale, Grantchester, Near Cambridge

My dear Galton

I was just thinking that I ought to write & say how glad I was that you were satisfied with my attempt to summarize your 'Natural Inheritance', when I saw the comments in Nature of the last week.[367] I don't think there is any occasion for me to add

[367] [John Venn, "Natural Inheritance," *Mind* 14:55 (1889); pp. 414–420; Francis Galton, "Head Measures at Cambridge," *Nature* 40 (1889), p. 643.]

anything to what you have said. Of course, as you say & as I have always felt, it is most important to take a second batch of statistics for comparison with the first. I should hope that, in the course of this year, the accretions will about equal the original numbers. If so I will take them in hand in the winter or spring, & see what they have to say.

Our critic,[368] though he does date from 'Trin. Coll. Camb.' does not seem to have appreciated the security which gathers about a 'mean'.

Yours truly

J. Venn

94.
UCL 3/3/21/2 F45–F48
John Venn to Francis Galton
15 October 1889
Caius College, Cambridge

My dear Galton

I delayed answering your letter as I wanted to see Mr White (the library clerk, & our 'measurer'): also we have been in rather a bewilderment about our house, which has been giving serious indications of unwholesomeness, culminating just now in the illness of one of the servants, which has caused us to leave in hurry somewhat. I don't think we shall go back to Grantchester but our future is very doubtful.

I will gladly undertake the reduction & comparison of the last set of statistics; – do I understand that you propose to pay Mr White the same moderate charge for clerical work as last time? I will readily undertake all the computation. He has already begun the first stage, viz. that of making the separate list of men for each college. The second stage consists in my taking these lists to the tutors, & getting them to table the men as A, B, & C, respectively. When Mr White begins again, & copies out all the details (from the original cards) referring to these 3 classes respectively. Then I can operate upon them in any way desired.

As regards Mr White's qualifications for the actual measuring, H. Darwin[369] would be a much better judge than I. His work in respect of copying, &c, seems to me clear & careful; but as he has not much spare time, he does not work quickly.

I doubt if it would be easy <u>here</u> to get much copying work done quickly. If you want this done it would probably be best to have all the cards sent up to you in London & get it done there. If you decide upon this please ask H. Darwin to see them sent: I shall only be here from time to time during the winter. Of course the

---

368 [**Galton** had responded to criticism voiced in 'H.J.P.', "Anthropometric Measurements at Cambridge," *Nature* 40 (1889): pp. 593–594.]

369 [Charles Darwin's son Horace Darwin (1851–1928), Trinity 1868; B.A. 1874; co-operated with A.G. Dew Smith in the making of scientific instruments (this business became the Cambridge Scientific Instrument Co. Ltd., in 1895); designed a number of anthropometric instruments for the laboratory at Cambridge; Mayor of Cambridge, 1896–97.]

sending the papers away from Cambridge would put a stop to Mr Whites copying them, & prevent me from making as early a start as I otherwise could.

We expect to go to Bournemouth, probably in a week or 10 days.

Believe me
Yours very truly
J. Venn

95.
UCL 3/3/21/2 F49–F51
John Venn to Francis Galton
20 October 1889
Caius Coll.

My dear Galton

I see that 'F.M.T.' has got a backer up in this weeks Nature.[370] I am writing to Glaishier[371] [*sic*] to ask if he can put a name upon the anonymous. I glanced at the list of names in the Calendar, but P's and T's are too common for ready identification. If you make a reply you might perhaps add that I have got a second batch of results in hand, equal in number to the first, so that even if the same individual errors are prevalent here, the result on the averages will be reduced. Also, that I shall take care to examine carefully the alleged inaccuracies of measurements upon the same men, by reference to the original cards.

As regards this new source of error (i.e. that of the imperfect measurer) is it not a case of the superposition of 'independent errors', & to be estimated in the same manner as regards its influence on the accuracy of our results? I mean that if c_1 is the 'prob.er.' of the men's own dimensions, & c_2 the 'prob. error' of Mr White in measuring them, we ought to enlarge our standard of accuracy in the ratio $\sqrt{c_{1^2} + c_{2^2}}$: c_1? Of course any scheme of allowing for errors is a makeshift for removing them, but I only suggest this for your consideration.

I am beginning to collect my materials for the A, B, C, division of my men, from the college tutors; &, as soon as that is done, Mr White will begin to sort out the cards & copy the items as before.

Believe me

Yours very truly
J. Venn

---

370 [Another criticism of the accuracy and inferences drawn from Galton's and Venn's Cambridge measurements appeared in *Nature*: 'F.M.T.', "Head Growth in Students at the University of Cambridge," *Nature* 40 (1889): pp. 317–318. The anonymous F.M.T. may have been Frederick Meadows Turner, a physician trained in London and Cambridge with an interest in statistics who was measured several times in the Philosophical Library.]

371 [James Whitbread Lee Glaisher. See note **250**.]

96.
UCL 3/3/21/2 F56–F59
John Venn to Francis Galton
27 November 1889
Apthorp, Owls Road, Boscombe

My dear Galton

Thanks for your letter: I wish I had been with you in your inspection of the instruments. I will duly investigate, as far as I can, the instrumental error, & will forward the cards to you for examination afterwards. The little I saw of Mr White's work gave me the impression of a careful man.

As regards the "Squeeze" results of the last set. I see, by my tabular results, that very few in proportion of the men reached 112; I doubt if the mean would have been altered by more than a fraction of an ounce by the extension upwards of the graduation. The eye-sight was different: I had already noted that the 'median' rather than the 'mean' was the test there.

There are two other points I wanted to speak to you about. Of course you know Wallace,[372] the naturalist. Have you any objection to asking him if I may call on him? He lives, I find, at Parkstone, a suburb at the opposite end of Bournemouth. I understand that he is in weak health & does not go out much, if at all; but if he would allow one to call I should very greatly value the privilege.

The other point is this. I am anxious to put up F.Y. Edgeworth's[373] name for the Roy. Soc. I have long had a very high opinion of his work; &, since his presidency at the Brit. Assn this year, he must be known to a wider circle. Very likely you have seen some of his many contributions to the theory of statistics, & the Laws of Error, &c; if not, I can send you a list of them. Should you have any scruple in adding your recommendation to this proposal?

We are getting on here very well, as regards health & weather.

Believe me
Yours very truly
J. Venn

97.
UCL 3/3/21/2 F60–F63
John Venn to Francis Galton
15 December 1889
Apthorp, Boscombe Spa, Bournemouth

My dear Galton

I have worked out the new set of results for comparison separately, & combination jointly, with the old ones; & send you the tabulated statistics. I am going to

---

372 [Alfred Russel Wallace (1823–1913), today chiefly remembered for independently conceiving the theory of evolution through natural selection.]

373 [**Francis Ysidro Edgeworth** (1845–1926). In 1889, Edgeworth was Professor of Political Economics at King's College, London.]

Cambridge tomorrow for the inside of the week, so if you like to keep the papers during that time, & return them to me with any suggestions, I can work at them again on my return. When I go there I am intending to collect the cards of all the men who were measured five times or upwards (I told Mr White to select them for me, & have them ready on my visit there). This will enable us to form some idea of the amount of "error" to be assigned to this cause of inaccuracy.

The papers I send will explain themselves. No. (I) is our first set of statistics, of 2 years ago. No. (II) is the new set. No. (III) is the combined result of the two, (with of course due 'weighting' for the relative numbers of each). No. (V) is the diagram of the "Head-products" of the A, B, & C, men separately, & of all three together.

The general conclusions seem in entire accordance with our former results. That there is head-growth after 19 & 20 is surely indisputable, – for look at the last diagram, – but I doubt the evidence of a different degree or rate of growth as between A, B, or C. But perhaps your more experienced eye may detect such things better than mine.

When I return I will draw up a sort of 'Report' on the results. Would it not be well to send it to Nature (if they will take it), as the former results were criticized there?[374]

Yours very truly

J. Venn

No. (IV) of the enclosed papers is to show the effects of age merely, irrespective of mental distinctions. A, B, & C, are here all thrown into one group.

98.
GCA C25/2
Albert Venn Dicey to John Venn
30 December 1889
The Orchard, Banbury Road, Oxford

Dear John,

I much wish for his sake no less than yours, that would call upon Arthur Cohen,[375] who is staying, I don't know for how many days, at Kildare House, Bournemouth. You will find him most interesting to talk to on the connection between logic & law especially in regard to mathematical questions connected with the law of wagers than any man I know.

I shall probably send you a paper[376] I have written upon the law of wagers, though it deals but very cursorily and from your point of view most superficially with the problem about wagers & insurance.

The conclusion I have arrived at is on the whole this.

---

[374] [See letters **82**, **85** and **94**.]

[375] [Arthur Cohen (1830–1914), English barrister and Liberal Party politician; Magdalene 1849; President of the Union Society, 1853; 5th Wrangler but as a Jew could not take his B.A. degree until after the passing of the Cambridge Reform Act of 1856.]

[376] [Not found.]

I believe both on your authority, which is really sufficient for me, & for the reasons you give as far as I follow them, that logically a wager includes both a bet in the ordinary sense of the term & a contract of insurance, and that they can hardly be distinguished on logical grounds. So far Anson is I presume technically right.[377] On the other hand I think that for legal purposes it is possible to distinguish between two classes of wagers by looking at their purpose: 'bets' tend on the whole to increase, insurance tends to diminish sudden and what one may popularly call chance transferences of property, or in other words fluctuations from wealth to poverty & vice versa. The one are discouraged as being injurious to general happiness or prosperity, the other are encouraged as promoting it.

It is very noticeable from this point of view that English law tends as far as possible to invalidate insurances which are not contracts of indemnity. The fundamental principle of all but life insurance being that they must be contracts of indemnity & nothing more. Hedging bet no doubt is in its character (as regards object) a contract of insurance, but the law does not aim at cutting matters fine, it treats it from its connection with real betting, as a bet, or in legal phraseology, a 'wagering contract'.

People in general much overrate the subtlety of law. It decides matters in a rough & ready manner, as I have said to a pupil, the seals of justice are used for weighing not diamonds, but butcher's meat.

A happy new year both to you and Susie,

Your affectionate cousin,
A.V. Dicey

99.
UCL 3/3/21/2 F64–F66
John Venn to Francis Galton
5 January 1890
Apthorp, Boscombe Spa, Bournemouth

My dear Galton

I have finished my paper on the Cam. measurements, & am going to try to get the editor of Nature to insert them. If he won't have them I will send them to you for the Anthrop. Journal. I rather prefer the former, as being the more popular, since

---

[377] [Sir William Anson (1843–1914), British jurist and Liberal Unionist turned Conservative; educated at Balliol College, Oxford, 1862–66; appointed Vinerian Reader in in English Law at Oxford; Vice-Chancellor of Oxford, 1898–99. The reference is probably to Anson's *Principles of the English Law of Contract* of 1879, where Anson discussed so-called wagering contracts, where wagers were understood as 'a promise to pay money, or transfer property upon the determination or ascertainment of an uncertain event'. His goal was to show which wagering contrasts were and which were not permitted by law – apart from this, there was no logically strict distinction between, say, a bet and a contract of insurance. See Sir William R. Anson, *Principles of the English Law of Contract* (Oxford: The Clarendon Press), pp. 165–172.]

anything we can do to stir up a little more general interest in these schemes seems desirable.[378]

As regards the inferences to be drawn from the fresh batch, it seems to me that every previous conclusion gets more or less further justification: there is no opposite indication.[379]

When I wrote to you, some weeks ago, about Edgeworth,[380] you said you would kindly look over the form of Application, & make suggestions. I enclose it in the form in which it first suggests itself to me.

Believe me
Yours very truly

J. Venn

P.S. As the 'Recommendation' may need alteration, I send only a rough copy: – will you send & make suggestions?

The introduction of the fresh "prob. error", (i.e. that springing from the imperfection of the head-measuring instrument) has, I find, very little effect. The "prob. error" of the <u>observed</u> Head-volumes is about 17 cubic inches; that arising from the <u>instrument</u> is about 5 inches. But $\sqrt{17^2 + 5^2}$ only equals 17.7.

100.
UCL 3/3/21/2 F68–F71
John Venn to Francis Galton
23 February 1890
Apthorp, Boscombe Space, Bournemouth

My dear Galton

I have been thinking about what you were saying on the subject of some scientific-popular Introduction to the theory of statistics. I think, with you, that there is a good opening for such a treatise. In some respects I should much like to attempt something of the kind, as I want to prepare (for this or next year) some lectures in that way: – the subject "Theory of Statistics" is now formally introduced into the schedule of the Moral Sciences Tripos, in the department of Scientific Logic.[381] But it would require a good deal of search in the volumes of the Statistical Society,

---

[378] [The paper would be published as John Venn and Francis Galton, "Cambridge Anthropometry," *Nature* 41 (1890): pp. 450–454.]

[379] [The main conclusion remained the same: there was no correlation between mental ability the shape or size of the head of Cambridge students. The data were later re-examined by Karl Pearson whose more fine-grained results showed the same. See Karl Pearson, "On the Correlation of Intellectual Ability with the Size and Shape of the Head," *Biometrika* 5:1–2 (1906): pp. 105–146.]

[380] [Venn had proposed to put **Francis Y. Edgeworth** forward for election as a Fellow of the Royal Society. See letter **96**.]

[381] [Venn would teach the first lecture course on statistics in Britain. The course outline of 'Theory of Statistics' (1891) – which, as in the case of his three logical books, might have functioned as the basis for the book referred to – can be found in the Venn Papers at the Gonville and Caius College archives. 'Notebook: "Theory of Statistics – J. Venn. Caius Coll. Cambridge"; 1891; 48 bound pp; 20 loose pp.', GCA D4.]

Philosophical Magazine, &c, &c, to say nothing of special treatises on Probability. If it were desirable to go into the mathematical foundation of the various rules no one would do it better than Edgeworth.

I am sorry to hear that you have been troubled with Bronchial weakness. We have been by no means free from the Influenza down here: it did not come in a sudden gust, as in London & many other places, but has been dropping in upon people from time to time during the last 6 weeks. We have just sent home one of our servants, whose mater lives close by, & who seems to have got it, though not badly; so we must wait for some days yet to know whether we have got through this last sphere of infectious influence.

When I next have a few hours in London I will try & see you.

Believe me
Yours very truly
J. Venn

101.
GCA C53
William Minto[382] to John Venn
10 May 1890
8 Westfield Terrace, Aberdeen

Dear Sir

I have to thank you for a copy of the catalogue of your most interesting collection.[383] It makes the logician's mouth water. I presume that if one were to make the pilgrimage to Cambridge, one would be allowed access to the collection, otherwise the sending of so rich a catalogue would indeed be a bitter mockery. But after all it would [*illegible*] be instructive as showing one what books there are. I can see already from a hurried glance through the titles that it will be of great service to me, and I thank you most sincerely for sending it.

I have been re-reading with great interest lately your article in the first volume of Mind.[384] I know nothing [*illegible*] keeps us [*illegible*] in clearing up the prevailing confusion of aims in Logic. I have been trying for some time at disentangling them

382 [William Minto (1845–1893), Scottish academic, critic, editor and journalist; he became assistant professor under **Alexander Bain** when he held the Regius Chair of Logic and the Regius Chair of English Literature at the University of Aberdeen; Minto succeeded Bain to the chair in 1880, holding the post until his death in 1893.]

383 [Venn had donated his collection of 1100 books on logic to the Cambridge University Library in 1889. A catalogue was printed in 1889 under the title *Catalogue of a Collection of Books on Logic Presented to the Library by John Venn, Sc.D.* (Cambridge, 1889).]

384 [John Venn, "Consistency and Real Inference," *Mind* 1:1 (1876): pp. 43–52. Venn opened his first contribution to *Mind* with the following statement: 'It would not be going too far to say that the principal difficulty in the way of a student in Logic at the present day (at any rate in England) consists not so much in the fact that the chief writers upon the subject contradict one another upon many points [...] as in the fact that over a large region they really hardly get fairly within reach of one another at all'. Venn, "Consistency and Real Inference," on p. 43.]

by tracing their historical evolution: hence you will understand what a welcome find your catalogue is to me.[385]

Very truly yours
W. Minto

102.
GCA C76/2
John Veitch[386] to John Venn
10 May 1890
The Loaning, Peebles, New College, Glasgow

Dear Sir,

I beg cordially to thank you for the presentation of "Catalogue of a Collection of Books on Logic",[387] just received. I am very pleased to find that any one in these days has cared to make such an excellent collection of logical treatises as the catalogue shows. I have not all of them myself, but I have some which are not in the catalogue.

I have asked 3 publishers the Mess. Blackwood to forward to you a copy of my "Institutes of Logic"[388]– published so far back as 1885 – of this you do not appear to possess a copy. You will take the trouble to had it, I think you will find references to certain of the books included in your Catalogue – and you will also find [*illegible*] of the doctrines of some of those books.

Yrs truly

J. Veitch

103.
UCL 3/3/21/2 F72–F74
John Venn to Francis Galton
13 June 1891
Summerleigh, Chine Crescent R^{d}, Bournemouth

My dear Galton

I think I could, without difficulty, compress into a 15-minute paper some remarks about our Cam. Anthropological results which would be of interest. The only doubt

[385] [In 1893, Minto would publish *Logic, Inductive and Deductive* (London: John Murray, 1893) which opened with a chapter on 'The Origin and Scope of Logic'. Minto approached the question 'Where should we begin in Logic?' in terms of another question: 'Where did the founder of Logic begin? Where did Aristotle begin?'. Minto, *Logic*, on p. 1.]

[386] [John Veitch (1829–1894), Scottish philosopher and historian; began his career as assistant lecturer to Sir William Hamilton and Alexander Campbell Fraser; in 1860, he was appointed to the Chair of Logic and Rhetoric at the University of St. Andrews and in 1864 to the corresponding Chair at the University of Glasgow.]

[387] [See note **383**.]

[388] [John Veitch, *Institutes of Logic* (Edinburgh & London: William Blackwood and Sons., 1885). The various parts of this book were interspersed with 'Historical Notices', for instance on 'Logic Since Aristotle' and 'Quantified Predicate – Historical Notices'.]

is about my own presence to read them, as it is possible that I might find it very inconvenient to be in London on Aug. 14 – if it happens that I were absent I suppose the paper could be read by some one else. I would send you a sketch of the paper some time before: the tables of figures could, I presume, be appended for subsequent publication, & taken for read at the time.

A'propos of your recent paper[389] at the R.S., I was reminded of some experiences of my brother last year. His vicarage at Sittingbourne was cleverly 'burgled' one night. On looking at the windows next morning they saw the finger marks of the professional performers on the glass outside, as they had tried to press up the windows. It was too late when I heard this, or I should have been inclined to say at once, Have the panes cut out & sent up to Scotland yard for the chance of any suspected man being subsequently identified. The police were certain, from the signs of operation, that the men had belonged to some high branch of their profession; – but it seems that even burglars do not always keep abreast of current science. If they have read your paper they will wear gloves next time.

What do you think of my seal?[390]

Yours truly
J. Venn

104.
UCL 3/3/21/2 F75–F77
John Venn to Francis Galton
21 August 1891
Hope Bank, Colwyn Bay, N. Wales

My dear Galton

I duly got your letter: it was very good of you to take the trouble to write, amidst all the tearing hurry of your closing meeting.

My only feeling is that I got a good deal more than my deserts in such a full summary in the Times.[391] I had all the week an uneasy sensation such as the shirk feels, – or at least should feel, – at not being in London. But it would have

---

[389] [Francis Galton, "Method of Indexing Finger-Marks," *Proceedings of the Royal Society* 54 (1891): pp. 303–311. The paper, received 30 April 1891, built upon a memoir, sent to the Royal Society in November 1890, entitled 'The Patterns in Thumb and Finger Marks: On their Arrangement into naturally distinct Classes, the Permanence of the Papillary Ridges that make them, and the Resemblance of their Classes to ordinary Genera'. Galton was not the first to propose the use of fingerprints for identification, but he was the first to place their study on a scientific basis, thereby laying the groundwork for their use in criminal cases. His studies were based on a large sample of fingerprints obtained through his anthropometric laboratories. See Gertrud Hauser, "Galton and the Study of Fingerprints," in Milo Keynes, ed. *Sir Francis Galton, FRS* (London: Palgrave Macmillan, 1993), pp. 144–157.]

[390] [The seal has not been preserved. It probably carried the Venn family's coat of arms. See Venn, *Annals of a Clerical Family*, pp. 6–7.]

[391] ['The Congress of Hygiene', *The Times* (15 August 1891.]

absorbed three days from my stay in Wales, which was a large expenditure for a short paper.[392]

I hope if you pick up any experiences or hints, about intercomparison[393] of results & processes such as ours, you will let me have them. Within the next 12 months the Cambridge statistics ought to bear cropping again, & one would like to have some references for comparison if possible.

I am gradually putting together my ideas for my lectures on "Theory of Statistics"[394] next term. But there are difficulties. Our Cambridge scheme, admirable as it [is] for those who study along the main routes, does certainly not give much encouragement to those who want to turn down what is considered a bye-lane, however important the places to which it leads. So the existence of an audience or class becomes somewhat doubtfull [*sic*]. But the benefit to the writer of the lectures is mainly in the fact of his having to put his thoughts in order, whatever he do with them afterwards.

Hoping you are having a pleasant holiday.

Believe me
Yours very truly
J. Venn

105.
UCL 3/3/21/2 F78
John Venn to Francis Galton
31 December [1891]
Hastings

My dear Galton

I shall be delighted to sign for Edgeworth,[395] who, I have long felt, ought certainly to be F.R.S.

Reciprocation of all good wishes,

Yours very truly
J. Venn

---

[392] [John Venn, "Results of Anthropometry at Cambridge," *Transactions of the Seventh International Congress of Hygiene and Demography. London, August 10th–17th, 1891. Volume X. Division II. Demography* (London, 1892): pp. 308–314.]

[393] [**Galton** had first introduced the notion of 'statistics by intercomparison', the idea that statistical comparisons may be made strictly in terms of the interior variation in the data, in an 1875 paper, entitled 'Statistics by Intercomparison'.]

[394] [See note **381**.]

[395] [**Francis Ysidro Edgeworth** (1845–1926). Cf. letters **96** and **99**.]

106.
GCA C85/1
Karl Pearson[396] to John Venn
28 February 1893
University College, London, Gower Street, W.C.

Dear Sir,

You will perhaps be surprised at hearing from a perfect stranger with a request of the kind that I am putting before you. I am unable owing to the need for entire rest to give my lecture course[397] of four lectures at Gresham College in April, and have to seek for some deputy to do the work. The two courses this session have been on the laws of chance & their applications were to be dealt with in subsequent courses. It has struck me that if I could persuade four specialists each to take a lecture and illustrate in some manner the branches of chance in which they are interested the course would be a very valuable and interesting one to the Gresham audience. That audience is an appreciative, if popular, one and numbers about 300 persons.

I am writing to ask you, whether you could give them a lecture on 18 April, on some subject, such as probability curves, means, double lumped curves, etc. in which you are specially interested. I should leave, of course, the subject to your choice, merely remarking that the audience does not know much about the geometrical side of probability at present, but if you should prefer the philosophical side, your position thereon is already understood by a good many of the audience.

I fear the honorarium (£7.7.) can hardly be an inducement, but it might be a help should there be need to illustrate the lecture in any manner.

I trust you may be able to help me in this matter. I need hardly say that I should consider it a great favour if you can do so. I am, dear sir,

Yours faithfully

Karl Pearson

107.
GCA C58/2
Karl Pearson to John Venn
2 March 1893
7 Well Road, Hampstead N.W.

Dear Sir,

I am very grateful indeed for your ready compliance with my request for a lecture on April 18. I know the paper[398] in the Statistical Society's journal and feel quite sure that you would find material for an interesting popular lecture therein. Of

---

396 [**Karl Pearson** (1857–1936). In 1891, Pearson had been appointed Professor of Geometry at Gresham College.]

397 [A series of free public lectures. Gresham College, founded in 1597, still hosts over 140 free public lectures every year.]

398 [John Venn, "On the Nature and Uses of Averages," *Journal of the Royal Statistical Society* 54:3 (1891): pp. 429–456.]

course the Gresham audience is a less educated one than that at Adelphi Terrace,[399] but they readily catch at ideas, especially if they are graphically illustrated in any way. My last lecture was to have been devoted to the probability curve, but I got no further than showing the mechanical construction of one approach to it by a machine for tossing a practically infinite of times 20 coins, i.e. a machine for finding $(1+1)^{20}$ or the coefficients rather.

Will you let me have in about a fortnight's time, a few lines, say 20 to 30 if put in type for the purpose of the syllabus? I enclose a copy of my last syllabus, which will give you an idea of the sort of sketch the committee like of each lecture. Perhaps you will let me have the title of your lecture at the same time.

Again thanking you for your kindness in giving me this aid, I am

Yours very truly
Karl Pearson

108.
GCA C58/3
Karl Pearson to John Venn
24 March 1893
7 Well Road, Hampstead N.W.

Dear Dr Venn,

Your syllabus seems delightful and promises a most interesting lecture.

Sir Robert Ball is going to take the application of chance to astronomical measurements, Prof. Weldon to biological measurements and Mr Whitworth some kindred topic of which he has not yet sent me the particulars.[400] Would you object to my entitling your lecture "On Frequency Curves; their Nature, Variety, Use etc."? I have used the word 'frequency curve' for any curve in which the frequency of any divergence from the mean is plotted up to that divergence. I had found the current term 'curve of error' rather confusing with a popular audience, who did not first reach the subject from the standpoint of errors of observation. Please do not hesitate to say if you prefer to retain the title 'Curves of Error'.

If it would be of any service to you for illustration, I have an unused diagram of the probability curve with the points obtained from the binomial $(1+1)^{20}$ marked upon it.

The lectures begin at 6. P.M. and last an hour. Many of the audience come on their way home from the City & I have found from experience that 7. P.M. is a practical limit to their powers of staying – i.e. of postponing their dinners or suppers. I will have a cup of tea ready for you at 5:30 on April 18, and will send you a syllabus or two, when they are in type.

---

399 [Adelphi Terrace (originally known as Royal Terrace), comprised a block of 11 houses, four storeys in height, over double basements. It was demolished in 1936.]

400 [Robert Stawell Ball (1840–1913) who in 1892 had been appointed Lowndean Professor of Astronomy and Geometry at Cambridge, and director of the Cambridge Observatory. Frank Raphael Weldon (1860–1906), joint founding editor, with Francis Galton and Karl Pearson, of *Biometrika*; he had been appointed Professor of Zoology at University College, London in 1889. William Allen Whitworth (1840–1905), author of the popular *Choice and Chance* (1901).]

Yours very truly,

Karl Pearson

P.S. In the majority of cases the lectures are not read, but there is no rule on the point.

109.
GCA C26
C.L. Dodgson (Lewis Carroll)[401] to John Venn
11 August 1894
7 Lushington Road, Eastbourne

Dear Sir,

You are quite welcome to make any use you like of the problem I sent you, & (of course) to refer to the article in "Mind".[402]

Your letter[403] has, I see, crossed one from me, in which I sent you "Nemo's" algebraical illustration. I hope you may be able to find room for it in your book.[404] Perhaps you could add it, as a note, at the end of the book, & give, at p. 442, a reference thereto?

I shall be grateful if you will not mention to any one my real name, in connection with my pseudonym.

I look forward with pleasure to studying the new edition of your book.

Very truly yours,
C.L. Dodgson

---

[401] [Charles Lutwidge Dodgson (1832–1898), better known as Lewis Carroll; writer of children's fiction and an eminent logician. His work on algebraic logic – notably his two-volume *Symbolic Logic* (1896) – attracted renewed interest in the late twentieth century. See Robin Wilson and Amirouche Moktefi, eds., *The Mathematical World of Charles L. Dodgson (Lewis Carroll)* (Oxford: Oxford University Press, 2019).]

[402] [In 1894, Venn was preparing the second edition of his *Symbolic Logic*. The article referred to is the note published as Lewis Carroll, "A Logical Paradox," *Mind* 3:11 (1894): pp. 436–448. The paradox – dealing with the ambiguity of the phrase "if…then" – became known as the 'Barbershop Problem'. Venn was the first to discuss the problem in print, in the 1894 edition of *Symbolic Logic*, where he referred to it as the 'Alice Problem'. (See John Venn, *Symbolic Logic*, 2nd ed. (London: Macmillan, 1894), p. 442.) Venn here confirmed that the problem had previously been circulated and discussed among logicians of the time. For context see Amirouche Moktefi, "Lewis Carroll and the British Nineteenth-Century Logicians on the Barber Shop Problem," *Proceedings of the Canadian Society for the History and Philosophy of Mathematics* 20 (2007): pp. 189–199.]

[403] [There must have been other letters between Venn and Carroll over several years, but these are now missing. The only other traces of their correspondence available are twofold: a form, entitled 'Logical Nomenclature. Desiderata (1895)', that Carroll sent to Venn to fill in, and a letter with a diagrammatic solution to a logical problem. This letter has not survived but Carroll reproduced part of it in his *Symbolic Logic*. See Lewis Carroll [Dodgson], *Symbolic Logic. Volume I* (London: Macmillan and Co., 1869), p. 179. I am grateful to Amirouche Moktefi for providing this information.]

[404] ['Nemo' was the pseudonym that Carroll adopted for Cook Wilson (see note **357**) in their dispute on the 'Barbershop Problem'. For details see Amirouche Moktefi and Francine F. Abeles, "The Making of 'What the Tortoise Said to Achilles': Lewis Carroll's Investigations Toward a Workable Theory of Hypotheticals," *The Carrollian: The Lewis Carroll Journal* 28 (2016): pp. 14–47. Venn did not include it in the 1894 edition of his book.]

110.
UCL 3/3/21/2 F79–F81
John Venn to Francis Galton
25 April 1896
Trinity Lodge, Dean Park, Bournemouth

My dear Galton,

Your letter has followed me down here, where I have had to come for a week or 10 days, after a turn of influenza. It was not much, but, as usual, it left something else which the doctors will not give the same name to, but call a consequence of it: – in this case a certain amount of bronchitis. However, I am mending fast, & expect to return to work at Cambridge in a week.

I am sorry to hear what you say about Edgeworth,[405] but I think your remarks exactly his the point. I have a very high opinion of the aggregate of his contributions to the theory of statistics, & have felt the help of his work in a number of directions; but it is difficult to point to any single essay of his which, like K. Pearson's,[406] is decisive at once in the way of power & originality. I feel strongly that Edgeworth ought to be on the R.S., & I can't but think that if I had a role it would go for him, but the more I think about it the more difficult I feel it to say anything quite definite which ought to decide others in his favour.

I am replying at once to your letter, as it has been delayed in coming here; but I will think over the matter carefully. If I can recall anything, of that thoroughly definite character which you rightly ask for, I will write again.

With kind regards to Mrs. Galton.[407]

Believe me,
Yours very sincerely
J. Venn

111.
GCA C82
Andrew F. West[408] to John Venn
30 April 1896
Princeton, New Jersey

My dear Sir,

Your letter of April 19th came whence yesterday. I write without delay to inform you of our sincere regret that the chance we shall see you in Princeton this October

---

[405] [Cf. letters **96, 99** and **105**. **Edgeworth** received many academic honours (e.g. twice President of Section F of the British Association (1889 and 1922) and President of the Royal Statistical Society) but he was never elected a Fellow of the Royal Society.]

[406] [**Karl Pearson** was elected a Fellow of the Royal Society in 1896.]

[407] [Louisa Jane Butler (1822–1897), daughter of George Butler (1774–1853), Headmaster of Harrow School for 1805 to 1829.]

[408] [Andrew Fleming West (1853–1943), Giger Professor of Latin at Princeton University from 1883 to 1928 and the first dean of the Graduate School at Princeton University. He received the honorary degree Doctor of Letters (D.Litt.) from the University of Oxford in 1902, in connection with the tercentenary of the Bodleian Library.]

is so small.[409] We are also sorry to hear you have suffered under the assaults of our common foe, – the influenza, and that this and other considerations affecting your own health and the health of your family, together with university engagements, seem decisive in the matter.

But we cannot refrain from pressing our invitation, begging to assure you that present decision is not necessary, so far as we are concerned. Our invitation is open and will remain open up to the very day of our sesquicentennial Festival,[410] – so that, if the present [*illegible*] circumstances change, nothing will please us more than to hear you can come, even if we learn this at the last moment.

It will be such a pleasure to us to be permitted to entertain you with such degree of quiet or festivity as may best agree with your health and comfort. Then the sea voyage itself will surely be beneficial.

We earnestly hope the outlook may change and that you may be with us in Princeton in those lovely autumn days – the finest of our whole year – the time of our Festival.

I am with great respect

Andrew F. West

112.
GCA C80/1
James Ward to John Venn
4 January 1897
Selwyn Gardens, Cambridge

My dear Venn,

Since our walk the other day several things have happened that force me to reconsider what I then said, which was that if you were a candidate for the new professorship,[411] I should not be.

I said this assuming that if you thought fit to enter for the chair your election was practically certain, and I thought it likely from remarks of yours that I should still have a chance in a few years time. Personally as I have explained to you, I should much prefer this. To be elected now means a very serious pecuniary sacrifice to me: not to be elected at all a serious disappointment.[412] So on the whole I

---

409 [Venn had been offered an Honorary Degree at Princeton University.]

410 [The celebration of the one hundred and fiftieth anniversary of the founding of Princeton University, held October 20–22, where Woodrow Wilson, then Professor of History, Politics and Jurisprudence there, was the keynote speaker.]

411 [In 1897, the University of Cambridge finally created a second professorial chair in philosophy, to be named Professor of Mental Philosophy and Logic. It was lobbied for, and financially backed, by **Henry Sidgwick**, then Knightbridge Professor of Moral Philosophy. From the Venn–**Ward** correspondence it becomes clear that the original plan was to have two new chairs, one in Mental Philosophy and one in Logic. It was probable due to financial constraints that this plan did not materialize. See Byron E. Wall, "John Venn, James Ward, and the Chair of Mental Philosophy and Logic at the University of Cambridge," *Journal of the History of Ideas* 68:1 (2007): pp. 131–155.]

412 [The problem was that the new Professorship carried a much smaller stipend than what **James Ward** – who had to support three generations of his family – was then earning as a Tutor at Trinity College, Cambridge. He had been appointed to this post in 1881, thanks to **Henry Sidgwick**'s support.]

was content to take my chance of succeeding you when you resigned. I asked you accordingly to let me know if you were not going in, feeling that in that case unless I stood now some younger man would be elected and I should be excluded altogether.

But it has now become clear to me that these two sets of considerations cannot be kept distinct. I have heard nothing definite, but I "gather" a good deal from the talk of people in general and from letters that I have received. The line taken up is this and there is nothing personal in it at all: – The chair is one of *Mental Philosophy and Logic*. No man will take both and "mental philosophy" is wanted most. We have a line of Lecturers in Logic already,[413] but we have nobody to take charge in Psychology and Metaphysics. If the chair were exclusively a chair of Logic the course would be a simple one for the electors: as it is, Logic is put in the second place. In brief, before personal claims are discussed at all, yours will be unfavourably handicapped: as a result it seems clear that there is a risk of your not being elected simply on the ground that a representation of psychology and metaphysics is the first consideration. And I hear that very strong names in these departments are among the candidates. I am quite aware, and everyone allows it, that my claims in my department are not comparable with yours in logic. But if the contest is likely to be primarily among the psychology field, it really looks as if I ought to go in as I am strongly urged to do by many who have spoken and written to me about it. Thus what would have been for me the ideal solution – your election at once and five or six more years of my tutorship for me seems as little likely to be the fact as ideals usually are.

As I had heard nothing from you I concluded that you were going in and so had thought very little about the matter till to-day, when the case was put to me and I have tried to state it here. I only learnt yesterday that applications have to be sent in by Saturday next. I start on Wednesday morning for Aberdeen and have therefore not much time left in which to make up my mind. But I feel that the first thing to do is to let you know the changed light in which the *pros* and *cons* now appear. If you *are* meaning to send in your name, I can sincerely say that I hope devoutly you will be elected. But it would be some disappointment to me to see our young friend Stout successful.

Forgive this troublesome letter and pardon me if I have failed to paint the situation in befitting words.

Ever yours sincerely

James Ward

---

[413] [E.g. Thomas Woodhouse Levin (1834–1904), St Catherine's 1857; B.A. (1861); worked for many years as (Intercollegiate) Lecturer on Logic and Philosophy for the Moral Sciences Tripos.]

113.
GCA C80/2
James Ward to John Venn
9 January 1897
Aberdeen

My dear Venn,

I have got your letter of the 6th and sincerely thank you for it. I have sent in my name to the V.C.[414] at the last minute. Stout,[415] I find had sent in his, but I know of no others; though I gather there are some Oxford men.

I have done my best to put my view of the situation before the three or four electors, whom I have had an opportunity of seeing or of writing to, and who had invited me to say my say.

I fully realize the propriety [of] the practical justice, of your being made the first occupant of a chair in Cambridge to which the subject of Logic is attached and I still hope the electors will take this view. And as I said before, I am satisfied that unless they proceed to weigh the subjects before they weigh the men, the presence of Stout and myself will make no difference to you.

I saw Bain[416] a few days ago. He has aged seriously since I saw him a year ago and is quite unequal to a journey to Cambridge just now. He gave me his views with his usual candour: if he were there I do not think he would vote for either of us if he could find younger men good enough. He talked of sending Sidgwick[417] an exposition of his views.

I cannot say how much relieved I have been since I got your letter, and I will not attempt to. Only thanking you once more.

I remain

Ever yours truly
James Ward

---

[414] [Vice-Chancellor of the University of Cambridge. Either Charles Smith (1844–1916), who held the position from 1895 to 1897, or Alexander Hill (1856–1929), who held it from 1897 to 1899.]

[415] [George Frederick (G.F.) Stout (1860–1944), St John's 1879; Classical Tripos, Part I, 1st Class, 1881; Part II, 1st Class, 1882; Moral Sciences Tripos, 1st Class, 1883; B.A. 1883; Fellow, 1884; University Lecturer in Moral Science, 1894–96; Wilde Reader in Mental Philosophy at Oxford, 1898–1903; Gifford Lecturer at Edinburgh University, 1919–21; Professor of Logic and Metaphysics at St. Andrews, 1903–36; editor of *Mind*, 1891–1920.]

[416] [**Alexander Bain** (1818–1903). Ward had travelled to Aberdeen where he delivered the Gifford Lectures, 1896–98, which were published as *Naturalism and Agnosticism*.]

[417] [**Henry Sidgwick** (1838–1900).]

114.
GCA C80/3
James Ward to John Venn
6 March 1897
Trinity College, Cambridge

My dear Venn,

I have just learnt from Keynes[418] – on finding your name absent from the lecture list – that you have resigned your lectureship at Caius and that you talk of ceasing to reside in Cambridge.[419] I need hardly say that I am very sorry for this on many grounds. Your place as a teacher, of course, cannot be filled; Advanced Logic at any rate has simply to fall out of our scheme for the present. But on personal grounds we shall all miss your presence among us; and I in particular shall have the painful feeling that I have been the unwitting occasion of all this. My position was really a most perplexing one. I knew you only contemplated holding the professorship for five years or so, and this for my own private means was what I wished. Till the last week before the election I believed and had every reason for believing your appointment a certainty. Then the situation I had to face was the familiar now or never: in fact that thought the vacant chair was, so to say, two chairs in one, only one of the two was to be filled, and that not the one to which your claims were unique. The fundamental mistake was made long ago, made that is by say [*sic*] in the days of the Studies Syndicate when our Board asked for such a single chair and not for two chairs.

I know from the kind and magnanimous way in which you wrote to me that you do not really blame me; but will my new position add bitterness to the regrets that I should in any case have felt most sincerely at your withdrawal.

Believe me ever

Your sincere friend and grateful pupil,
James Ward

115.
GCA C80/4
James Ward to John Venn
13 March 1897
Trinity College, Cambridge

My dear Venn,

Many thanks for your letter. It relieves my mind to one great point: I am glad to find that you do not blame me. But I confess I blame myself more or less: it seems

[418] [John Neville Keynes. See note **349**.]

[419] [Venn took his failure to obtain the professorship – his fourth and final attempt at a chair – rather hard. Among other things, acting upon a wish expressed in 1887–88 (see letters **79** and **85**), he resigned his post as Lecturer and turned away from logic completely. By the time that the Trinity men Alfred North Whitehead and Bertrand Russell revolutionized logic from Cambridge, Venn was devoting himself entirely to the affairs and history of his college. Furthermore, Venn briefly moved his family from Cambridge to Eastbourne before determining to make their final home in Cambridge in 1901.]

to me I was too vacillating. I said definitely that I should not go & allowed myself to be persuaded into giving in after all. My only excuse is the change of circumstances.

You refer to Sidgwick's[420] communications with me just before the election. He certainly never urged me to go in on the ground that the decision would probably be between Stout[421] and me. Stout in fact at that time was not a candidate & he too said that he did not mean to stand. His application like mine was posted in Aberdeen on the Friday, i.e. the day before the last day.

It was not Sidgwick who told me that the electors were likely to "handicap" Logic, nor did he speak of the election to me at all till I asked him whether he thought that I was free to become a candidate after I had said so definitely that I sh[d] not be one. Speaking of you, I remember him saying: I know Venn will & I am satisfied that he will consider you perfectly free to change your mind. You had better write & tell him frankly how the situation now appears to you.

Sidgwick was extremely anxious to see you elected & believed that you would be. I never asked him to advise me on the main question; but he said himself: I can as an elector only say that we want as many candidates as possible.

It is, I am sure, a mistake to suppose that Sidgwick or [John Neville] Keynes regarded your standing, as you say, as quite useless, or mine as more than a precaution against contingencies.

Pardon so much on a subject, of which I hearsay you are tired.

Ever yours sincerely
James Ward

116.
GCA C50/2
A.S. Lea[422] to John Venn
5 May 1899
11 Cavendish Road, Harringway, London N.

My dear Venn,

Many of us feel that our College should possess some pictorial memento of you, not merely for the pleasure it will give us but in order that our successors may know what you looked like as one of its most distinguished Fellows & its Historian. My object in writing is to know whether you can give the artist sittings for an oil portrait, Kitkat size, during August or September, in Cambridge if possible, but if not,

---

420 [**Henry Sidgwick**.]

421 [G.F. Stout. See note **315**.]

422 [Arthur Sheridan Lea (1853–1915), Trinity 1872; Natural Science Tripos, 1st Class, 1875; B.A. 1876; Sc.D., 1886; Fellow of Caius, 1885; tutor and bursar; demonstrator in physiology, 1883; University Lecturer in Physiology, 1883–96; collector of engravings and mezzotints of distinguished members of Caius.]

then wherever it suits your convenience.[423] We propose to ask M^r^. Brock[424] of Cambridge to make the portrait, since there is a general opinion that his painting of Prof. Cowell[425] in the Hall at Corpus is excellent. Brock has work on hand to occupy him to the end of July and can do nothing new before August. As soon as I hear from you I can then ask Brock to write and make all the necessary arrangements for your sittings to him. [...]

Yours very sincerely,

A.S. Lea

IV. Looking Back: 1900–1923

117.
TCL Add.Ms.c.95/175
John Venn to Henry Sidgwick
13 June 1900
Hotel Gr. Central, London

Dear Sidgwick,

I was so very sorry to hear from Bradbury[426] yesterday that you, like me, were lying ill here in London. It must have come upon us, very suddenly, after your pleasant little dinner at Newnham. I think it was only about 5 days after that I was suddenly knocked down by a desperate attack of pleurisy & pneumonia, which only just did not take me off.[427] Yours, I understand, was some internal complaint requiring an operation. I trust it has been successful.

I can't possibly be present at the Election[428] on the 30th, but it has set me thinking on the past: – '66, when you did not stand & I did (to my shame I afterwards felt

---

[423] [The seated, three-quarter length portrait of Venn, wearing scarlet robes, is held at Gonville and Caius College, Cambridge, accession number GC0085.]

[424] [Charles Edmund Brock (1870–1938), successful painter of members of the Royal Family and the aristocracy and book illustrator for Jonathan Swift, Jane Austen and several others as well as for several magazines.]

[425] [Edward Byles Cowell (1826–1903), Professor of Sanskrit, 1867–1903; Fellow of Corpus Christi, Cambridge, 1874–1903.]

[426] [A Cambridge doctor.]

[427] [Venn's account of his sudden illness has been preserved: 'Thermometric index of a desperate illness', CMS C44/1. Venn fell into an 'incessant delirium', full of 'grotesque visions', for three weeks; it was only after more than six weeks that he could return home. In addition to the visions, Venn later recalled the costs involved: £600.]

[428] [Early in 1900, **Henry Sidgwick** was forced by ill-health to resign the Knightbridge Professorship of Philosophy, which he had held since 1883. His successor was William Ritchie Sorley (1855–1935), previously Regius Professor of Moral Philosophy at the University of Aberdeen, 1894–1900; he is remembered today as a British Idealist and an opponent of the admission of women as students to the University of Cambridge.]

against the then candidates): to '72, when we both stood, & the cold thin face of [Thomas Rawson] Birks smiled down upon us; & to '83 when you stood & I did not. Well, the Tripos has not been what ardent young teachers of 1865 expected.[429]

Pray express my very earnest sympathy with Mrs Sidgwick,[430] as I know you will with Mrs Venn. She (M^rs^. V) bore up wonderfully for all the first 6 weeks,[431] but has now had to go back home, & leave me in charge of a sister. I hope to move home in a week.

I know not what may be allotted to us in the way of any future work, & I am too tired to write more. I think I shall pull through: in fact I feel pretty confident. I do trust the same for you too.

Well goodbye. Yours truly,

J. Venn

118.
GCA C72/1
Henry Sidgwick to John Venn
14 June 1900
18 Langham Street, Portland Place W.

My dear Venn

Many thanks for your kind and sympathetic letter.

I have known now for nearly six weeks that I have an internal complaint & incurable but of which it is hoped the fatal termination may be postponed for some time – they talk even of years – by the operation from which I am now recovering.[432] So far I have had little to endure – practically nothing in the way of pain and really very slight discomfort. I hope to be able to move about in a week or 10 days, but my career is over – the most I hope for is time to finish in some sort for publication one or more pieces of literary work in philosophy.[433] You I hope on the contrary when you have pulled through will return to full health and a normal expectation of life. As you say our work in Moral Sciences at Cambridge does not afford a triumphant

---

429 [On the development of the Moral Sciences Tripos see Palfrey, "The Moral Sciences Tripos at Cambridge".]

430 [Eleanor Mildred Sidgwick (née Balfour) (1845–1936), married to **Henry Sidgwick** in 1876. Known to her friends as Nora.]

431 [This was not entirely true. As Venn later recalled: 'Susie had herself by then [i.e. after 6,5 weeks] become so dangerously ill from anxiety & overwork that one of the nurses had to take her home'. See 'Thermometric index of a desperate illness', CMS C44/1. But even here Venn did not tell the entire truth, as the nurses were actually required to restrain his wife from throwing herself from the window of the London hotel. See Clewlow, "Intersecting Sets," p. 187.]

432 [**Sidgwick** died in August 1900, two months after writing this letter.]

433 [Henry Sidgwick, *Philosophy: Its Scope and Relations. An Introductory Course of Lectures* (London: Macmillan and Co., 1902). This book, posthumously published in 1902 and edited by James Ward, contained a course of eleven lectures in which Sidgwick defined the scope of philosophy and its relation to psychology, logic and history'. Sidgwick, *Philosophy*, on p. v.]

retrospect, but I endeavour to seek consolation out of even this, by reflecting that it was after all about time that mine shall come to an end.

I have never had a sanguine temperament and therefore I think the blow that has fallen on me is somewhat less of a shock than it would otherwise be, but I have found the world on the whole a very delightful world and a kind world and do not pretend to myself that I am not very sorry to leave it. I have been thinking much both of you and your wife since I heard the news about myself and about you, and should have written long before had I thought you were in a condition to take pleasure in letters.

I look forward to our meeting somehow and meanwhile I am ever your most sincerely

Henry Sidgwick

119.
GCA C12/3
Leslie Stephen to John Venn
30 October 1900
22 Hyde Park Gate, S.W.

My dear John,

I have been in a manner compelled to write an obituary notice of Sidgwick for 'Mind'.[434] Mrs [Nora] Sidgwick & others have given me some help. I should be glad if you could add anything. I fancy that you must have seen a great deal of him in early Cambridge years. The article must necessarily be short: but any little anecdote or fact wh would make that time clearer wd be valuable. I have fancied that he, like you, was a friend of John Grote[435] and may have been more or less stimulated by him. If you can tell me anything about that – or anything else – I should be grateful.

I heard constantly of your illness from my sister[436] & was pleased to hear from her the other day that you had completely recovered. I heartily congratulate you & your wife.

My own health has been shaky of late but I hope to do a little work still. I feel, however, that I am becoming survivor of a past generation. Sidgwick is a great loss to me: and the dust of Parker[437] of Peterhouse snapped another like of my old associations with Cambridge. The place is rather a sad one for me now – but I will not preach you a sermon.

Your aff^e^ cousin L. Stephen

---

[434] [Leslie Stephen, "Henry Sidgwick," *Mind* 10:37 (1901): pp. 1–17.]

[435] [Cf. pp. 127–128 of Venns *Annals* and letter **40**.]

[436] [Caroline Emelia Stephen (1834 –1909), philanthropist and writer on Quakerism who played an important role in the life of her niece Virginia Woolf. Venn's physical and his wife's mental illness in 1900 illustrates the continuing connection between the Venns, the Stephens and the Diceys. Henrietta Venn alerted her brother about it; a series of letters subsequently went around in a circle from Henry Venn's wife to Caroline Stephen, then to **Albert Venn Dicey** and back to Henry Venn. See Clewlow, "Intersecting Sets," p. 188.]

[437] [Caroline Emelia Stephen was known as Milly to her friends and family.]

120.
GCA 12/4
Leslie Stephen to John Venn
2 September 1902
Fritham, Lyndhurst

My dear John,

I must send you one word of sympathy.[438] Among my earliest memories are some of your father's links at Highgate & of playing there with his children. Much as we have drifted apart since, I have always, I hope, retained the feeling which I had then & in later years. Your father has always seemed to me to have been one of the best men I ever knew: and though I have seen very little of Henrietta in late years, I have always seemed to be in relation to her, especially through her friendship with Milly[439] – to my sister, the loss will be only less than to you & if only far that [*Illegible*], I should feel it strongly. Let me say too that I have a genuine affection for you - & could not be indifferent to any trouble of your's.

The old world to which we belonged is vanishing very rapidly: I [hear/have?] sufficient warnings that I shall not be a remnant of it for very long; & I have lost too much to feel any reluctance to follow. But while I live I am yr affectionate cousin, L. Stephen

121.
SGL
John Venn to Laura Forster[440]
4 December 1902
Vicarsbrook, Chaucer Road, Cambridge

Dear Miss Forster,

[…] I have been living mostly in the past of late, having been immersed in Family memoirs, which I have been putting together & hope some day to print.[441] It is a fascinating pursuit when one has got into it, & with me it is something of a duty as well as a pleasure, for my grandfather of Clapham began to enquiry, & my father was always extremely interested in it during the very few & short bits of leisure that he ever secured. They both kept everything they could, so the result is that I have a huge mass of letters, journals, account books, &c, from about 1750 onwards.[442] The difficulty is to select the one percent, or one tenth per cent, which can be got into a

---

[438] [**Leslie Stephen** was expressed his sympathies on the loss of Venn's sister, Henrietta, who had died in September 1902.]

[439] [Possibly Edward Parker (1821–1895), Peterhouse 1841; B.A., 1845; P.C. of Waddington, Yorks., 1862–95.]

[440] [**Laura Mary Forster** (1839–1924).]

[441] [Venn was then working on *Annals of a Clerical Family*, published in 1904.]

[442] [The Venn Family Papers are extensive, including many diaries, letters, and notebooks from the 1750s to the 1870s. They were bequeathed to the Church Missionary Society by Venn's son **John Archibald Venn** in 1958.]

small volume. I find the reading of these in one way rather melancholy work, as it serves to emphasize a very deep conviction that I have held for very many years, viz. the enormous superiority of those who have gone before me. It is a feeling that I sometimes felt almost crushing when I was with my father.

[…] You doubtless heard of my sister's death three months ago. She had had to face the end before her for just ten months. Many of the letters & diaries I have mentioned came from her house, which accounts for my being so full of them just now. My father kept almost everything, & my sister never parted with anything whatever, so the result was that when we had to clear the house all the insummerable [*sic*] little things which tell of ones childhood emerged to light, & had to be dealt with.

Believe me,

Yours very sincerely,
J. Venn

122.
SGL
John Venn to Laura Forster
14 December 1902
Vicarsbrook, Chaucer Road, Cambridge

Dear Miss Forster,

[…] You ask, by the way, if I am a recluse here. Not more than I can help for I keenly enjoy much of what life has to give. For some time we were in the full swim of it here, but now for many years we have been kicked about, by health conditions, like a football. Archie,[443] as a child, required incessant care; & this necessity first drove us to Bournemouth, &, after a short try here, again to Eastbourne. Mrs Venn also, for a number of years, has been able to go out but very little. Moreover, two or three years ago I had a desperate illness in London which it took a long time to get free from.

[…] As to Leslie Stephen you have probably heard that he had to undergo an operation two days ago. I hope it will give relief & lengthen life, but his sister here is not very hopeful. What you say about him interests me. I wish I could have seen more of him in early days, but in those days we did not fit in well together. I suppose he saw in me the sanctimonious pig, & I in him the athletic rough. I don't know which was furthest wrong.

[…] Believe me,

Yours very sincerely,
J. Venn

---

[443] [**John Archibald Venn** (1883–1958).]

123.
SGL
John Venn to Laura Forster
27 January 1903
Vicarsbrook, Chaucer Road, Cambridge

Dear Miss Forster,

[…] I am glad you liked my Commemoration address.[444] There is another 'impending' some days – as you said of one of your letters – but there is time to avert it. It is strange how one grows in time to appreciate the past. I suppose it is one of the signs of advancing age. Chronologically I belong to the period of the ardent young University reformers of the sixties, who started with the conviction that everything which had gone on here for more than 30 or 40 years must be bad. One of my objects in writing the History of the College[445] was to try to explain how those of the past tried to realize their ideals & what sort of problems they had before them.

[…] Believe me,

Yours very truly,
J. Venn

You read a great deal, I know. Have you seen William James' 'Varieties of Religious Experience'?[446] I have been much interested in it, & am reading it now for the second time. I never came on a book which covered the same ground. It is a very acute & sympathetic account of the [*illegible*] – so far as the human mind is concerned, of all the various creeds. It is difficult to say which of them all he supports himself: he really almost seems as if he would like to believe them all.

124.
SGL
John Venn to Laura Forster
15 February 1903
Vicarsbrook, Chaucer Road, Cambridge

Dear Miss Forster,

[…] My curate days are now far behind, but I am as you know, not one of those who have pulled off their surplice & stamped upon it, – or cut it up into towels & napkins. On the contrary, I have still strong sympathies with much in the clerical life. At this day there are few posts which I should more enjoy, in many ways, than that of a country vicar. But unfortunately there is a shut gate, with 39 iron spikes on

---

[444] [Venn probably sent John Venn, "Address Delivered in the College Chapel, After the Commemoration Service, Dec. 21 1894," *The Caian* IV (1895), pp. 188–198.]

[445] [Venn's *Biographical History of Gonville and Caius College*.]

[446] [William James, *The Varieties of Religious Experience: A Study in Human Nature, Being the Gifford Lectures on Natural Religion Delivered at Edinburgh in 1901–1902* (London & New York: Longmans, Green and Co., 1902).]

it, which bars the way.[447] You have been enough in Darwin society[448] to understand that 8 generations of one profession imply a good deal of a selection. So if you detect an obstinate disposition to step up into the pulpit, when I have had to give an address in our Chapel or Hall, you must begin by laying the blame on William Venn who became vicar of Otterton in 1599.[449]

I thought you must have met Dicey at the Farrers,[450] but I did not know he had ever been at your house. He is not strictly a cousin but a cousin's cousin (through the Stephens), but we were of the same age & much thrown together in childhood & ever since so that he is the most intimate friend I have amongst relations.

[...] Believe me,

Yours very truly,
J. Venn

125.
GCA C65
E.S. Roberts[451] to John Venn
18 February 1903
Gonville and Caius College, Cambridge

My dear Venn,

I hope you will do us the honour of accepting the historic office of President. I say "us" because, though I had no doubt about it myself, I wished to be able to make the offer not only in pursuance of my own wish but also in accordance with the strongly expressed wish of the Fellows; and of this I can now assure you.

Believe me

Yours sincerely,
E.S. Roberts

---

[447] [In 1883, 10 years after his father's death, Venn had made use of the provision of the Clerical Disabilities Act to resign Holy Orders. Unlike Leslie Stephen, for example, Venn never became agnostic: he would continue to attend church and chapel services and always retained a heartfelt affection for the clerical life.]

[448] [**Laura Forster** was a lifelong friend of Charles Darwin's eldest daughter Henrietta, with whom she conducted an extensive correspondence. Many of these letters have survived, including some which clearly convey her leaning towards Darwinism. At one point in 1879, when Charles Darwin travelled to London where he was presented with the Baly Medal, she let him stay in her house in Abinger, Surrey. See Francis Darwin, ed. *The Life and Letters of Charles Darwin. Volume III* (London: John Murray, 1887), p. 224.]

[449] [Venn's great-great-great-grandfather William Venn, Vicar of Otterton, 1600–1621.]

[450] [Probably the family of Thomas Farrer (1819–1899), civil servant and statistician educated at Oxford; for many years Secretary of the Board of Trade; President of the Royal Statistical Society from 1894 to 1896; he was the father of Emma Cecilia Farrer (1854–1946), who married Horace Darwin (1851–1928), son of Charles Darwin.]

[451] [Ernest Stewart Roberts (1847–1912), President, 1894–1903, and Master, 1903–1912, of Caius. See note **321**.]

126.
SGL
John Venn to Laura Forster
30 May 1903
Vicarsbrook, Chaucer Road, Cambridge

Dear Laura,

I rather like your notion that I belong to the 18th century. It is a period for which I have a great respect & liking. I know it was a poor enough time here, & in the Church history of England it is well used to hearing itself apologized for & explained. But it was a grand time in literature, & [John Robert] Seeley long ago taught me what a grand time it was in the political history of our 'Expansion'.[452]

[…] I do want to let those who come after me know how & why I abandoned clericalism. I think I told you that but for the Articles & Creeds & Liturgy (a tolerably big 'but'!') I should now wish for no other life than that of a country parson. And except for what I did not say probably few would notice anything amiss in my preaching. The old associations are very deep in me still. […]

Yours very truly,

J. Venn

127.
GCA C32/2
Francis Galton to John Venn
26 May 1904
42 Rutland Gate, S.W.

My dear Venn

It gave me very great pleasure indeed, to see your handwriting again, to read your letter, and to see that you have become the historiographer of your remarkable family.[453]

Even under the terms of circular, both Mrs Russell Gurney[454] & Miss Stephen[455] would be rated as "noteworthy". They both have written & the former was the subject of a (not over-good) biography.[456] But the fullness of their claims can only be known to their family or to their intimates.

---

[452] [Venn's Cambridge friend **John Robert Seeley** (1834–1895), In *The Expansion of England*, Seeley had argued that the British Empire had been founded largely by conquest and war, particularly its defeat of Louis XIV.]

[453] [Venn's *Annals of a Clerical Family* had then just been published. Galton had always been impressed by the eminence of the Venn family, which he entered into his work *Noteworthy Families* (1906).]

[454] [Venn's cousin **Emelia Russell Gurney** (1832–1896). She knew Galton through his wife Louis Jane (née Butler).]

[455] [Probably Caroline Emelia Stephen. See note **436**.]

[456] [Ellen Mary Gurney, ed., *Letters of Emelia Russell Gurney* (London: James Nisbet & Co., Ltd., 1902).]

I did suggest working "in concert with members of your family" but a mention of noteworthiness is very difficult indeed to obtain. Perhaps when the present inquiry is over, some material improvements will have suggested themselves, for use in an extension of the inquiry.

I wish you had told me something still more about yourself & yours, & about your son especially – I am getting older by uniform stages, but can still get about a good deal, in any way but walking. My walking powers are now my weakest faculty. [...]

I am too deaf to attend public meetings with profit, & keep much at home now in the evenings wondering when the end will come.[457] But the fog end of life is still pleasant to me.

With kind remembrances to your wife

Very sincerely yours
Francis Galton

128.
GCA C51
F.W. Maitland[458] to John Venn
21 July 1904
Downing College, Cambridge

Dear Venn,

Leslie Stephen left behind a sort of request that I, after some autobiographical papers, should put together a memoir of him.[459] In a way, I regret his choice, for I am never happy in so recast a time as the nineteenth century. However, I must do my best and am now looking about me for help. Having said this much, I need hardly add that I am going to appeal to your kindness. What I should especially like to have from you would be a paragraph or two about Stephen as a young man, as undergraduate and as don. I think that you were probably so placed that you could see certain sides of him that were not visible to by any means all of his contemporaries, and I know that anything you can bring yourself to give me will be – I could not say this of all contributors – absolutely accurate.

Miss [Caroline Emelia] Stephen has supplied me with some family papers which will enable me to perceive the main outlines of the boyhood. But I shall have some difficulty in understanding and explaining how this delicate child became the athlete. There is a parallel difficulty about mental development. And it is here that you may be able to give me some invaluable assistance.

---

[457] [**Francis Galton** would pass away some seven years later, on 17 January 1911.]

[458] [Venn's former pupil Frederick William Maitland (1850–1906), Trinity 1868; Moral Sciences Tripos, 1st Class, 1872; B.A. 1873; Downing Professor of the Laws of England at Cambridge, 1888–1906.]

[459] [**Leslie Stephen** had passed away on 22 February 1904. The memoir referred to would appear in 1906 as F.W. Maitland, ed., *The Life and Letters of Leslie Stephen* (London: Duckworth, 1906.]

I shall be best pleased if in the end (there is no hurry) you will write a little. Meanwhile would you talk a little, and if so would you appoint a time when I (not without a pencil) might call upon you in the hope of hearing you talk. Whether you grant all or any of my petitions, you will I trust forgive me for troubling you.

Believe me,

Yours very truly
F.W. Maitland

129.
GCA C64/1
William Ridgeway[460] to John Venn
4 June 1905
Fen Ditton, Cambridge

My dear Venn,

I hear that Swaine,[461] who is J.W. Clarke's[462] assistant at the Registry, is leaving, so that he will want another assistant. The duties are readily learned and are not arduous except a certain times each term. There is nothing that needs a man to have taken a high degree. It therefore occurs to me that it might be worth thinking about it for Archie.[463] Of course, as you know, I am the last person in Cambridge to be able to effect anything with J.W. except to rouse him to his great extremes of malice,[464] and I fear that the fact that Archie is engaged to my daughter will be regarded by the Registrary as a fatal objection to him. On the other hand you supported his library scheme[465] so that you are, I should say, more or less in favour, and I understand that at present you are begging down the list of Old Caians for his library Fund.

It may therefore be worth your while to talk to him, for the post would just be the kind of thing that we want for Archie.

Yours very truly
William Ridgeway

460 [William Ridgeway (1853–1926), B.A. from Trinity College, Dublin in 1875; Peterhouse 1877, migrated to Caius 1878; B.A. (5th Classic) 1880; Fellow, 1880–87 and 1893–1926; Professor of Greek at Queen's College, Cork, 1883–92; Disney Professor of Archeology at Cambridge, 1892–1926; vigorous opponent of the admission of women to Cambridge. Ridgeway was the father of Lucy Marion Ridgeway (1882–1958), who married Venn's son in 1906.]

461 [Not identified.]

462 [John Willis Clark (1833–1910), Trinity, 1851; B.A. (12th Classic) 1856; Fellow, 1858; President of the Cambridge Antiquarian Society, 1883–84; Registrary of the University, 1891–1910; Rede Lecturer, 1894.]

463 [Venn's son **John Archibald Venn**, who had taken his honours in the History Tripos of 1904–05.]

464 [The reason for their antagonism is unknown.]

465 [The details of the 'scheme' and 'fund' referred to have not been ascertained. There is no mention of a 'library scheme' or 'fund' in A.E. Shipley, ed., *"J." A Memoir of John Willis Clark. Registrary of the University of Cambridge and Sometime Fellow of Trinity College* (London, 1913).]

130.
GCA C28
F.Y. Edgeworth[466] to John Venn
23 June 1905
All Souls College, Oxford

My dear Venn

It gave me great pleasure to hear from you. It is too long since I have had an opportunity of taking counsel with you. The auspicious event which you mention reminds me how long a period separates the present from the not forgotten past. Nothing [*illegible*] the flight of time so strikingly as the maturity of one's friends' sons – except indeed that of their grandsons.

I should like to have had a contribution from you to the discussion of my paper. You will see that I still incline to a more favourable judgement on the work of the elder mathematical writers than you in your anxiety to correct exaggeration here expressed. I don't deny that there have been exaggeration.

Remember me kindly to Mrs Venn.

Yours very sincerely
F.Y. Edgeworth

131.
TCL Add.Ms.c.103/129
John Venn to Nora Sidgwick
28 February 1906
Vicarsbrook, Chaucer Road, Cambridge

Dear Mrs Sidgwick,

Many thanks for the 'Memoir'[467] which has come here, & which I have just begun to read. It is no idle praise to say that it seems to me excellent, – in the only way you would wish it to be praised, viz. in letting him speak for himself.

Every such life that I read, of the great & good men with whom I have come in contact here, leaves me with a certain bitterness of regret & self-reproach, for not having better used my opportunities while I had them. This is so, however much I admired them at the time, & in this case as you know, the admiration was great at the time.

I see that he wrote several letters in a similar tone to that of the one I received, – & greatly valued – in the beginning of June, 1900. I was just coming round from a most desperate illness in a London Hotel when Bradbury[468] happened to mention that he too was very ill. I managed to write a few lines, & got in reply the letter which I have.

Believe

Yours most truly
J. Venn

---

[466] [**Francis Ysidro Edgeworth** (1845–1926).]

[467] [*Henry Sidgwick. A Memoir by E.M.S and A.S.* (London: Macmillan and Co., 1906). 'A.S.' refers to Henry Sidgwick's cousin Alfred Sidgwick and 'E.M.S.' to Eleanor Mildred Sidgwick ('Nora').]

[468] [Sidgwick's and Venn's Cambridge doctor.]

132.
TCL Add.Ms.c.103/130
John Venn to Nora Sidgwick
20 March 1906
Vicarsbrook

Dear Mrs Sidgwick,

In the 'Memoir' (which I have been reading with increased interest & admiration) there is one passage in a letter which seems to have been written under a slight misapprehension. I refer to page 353, where your husband not unnaturally supposed that he had been summarily passed over for a junior, & former pupil. It was not quite so abrupt as this, the facts being as follows.[469]

The Vice-Chancellor (Dr Perowne)[470] in the first place offered the post to me. I told him that I could not accept it, as I considered that no one but Sidgwick ought to be appointed. He invited me to call upon him, & we had a long talk about the matter. His defense was that as he was choosing a deputy for Prof. [Thomas Rawson] Birks he could not fitly select any one whose opinions were so entirely opposed to his. Of course this did not alter my opinion, but I felt that there was something in what he said. He then urged me to accept the place; saying that as he should certainly not appoint Sidgwick I should not in the least be standing in his way. I could not see my way to this, & definitely declined. I think he then offered the post to [John Neville] Keynes, for I remember speaking to the latter on the subject, & saying that I thought he would probably feel on the subject as I did. The offer to [William] Cunningham was therefore not so abrupt as may have been supposed.

I have never mentioned these factors to any one before but there is no reason why you should not know them now.

Believe me
Yours very sincerely
J. Venn

---

[469] [Despite opposition from men like Sidgwick and Venn, who embraced a more secular approach to the moral sciences, in 1873 the Evangelical **Thomas Rawson Birks** had been appointed Knightbridge Professor of Moral Philosophy. The strength of the religious tradition at Cambridge was shown, however, when the Vice-Chancellor, the Evangelical Edward Perowne (1826–1906), offered Venn the position of deputy to Birks – who was then very ill – making it clear that he would not consider someone like Sidgwick for such a role given his unorthodox religious opinions. Venn declined out of respect for Sidgwick. Their former pupil, William Cunningham was chosen instead. In the letter, dated 18 May 1881, to which Venn refers, Sidgwick assumed that he had simply been passed over. See H. Sidgwick to F.W. Cornish, 18 May 1881, in Sidgwick and Sidgwick, *Memoir of Henry Sidgwick*, on p. 353.]

[470] [Edward Henry Perowne (1826–1906), Corpus Christi, 1846; B.A. (Senior Classic) 1850; Fellow, 1850; Tutor, 1858–79; Master, 1879–1906; Vice-Chancellor, 1879–81; Hulsean Lecturer, 1866; described as a sound disciplinarian and rigid conservative.]

133.
GCA C72/2
Nora Sidgwick to John Venn
25 March 1906
Newnham College, Cambridge

Dear Dr Venn

I am much interested in your letter about Henry [Sidgwick] and the deputy-professorship & it is very kind of you to write to tell me about the circumstances. Henry would of course have felt differently about it if he had known it had been offered to you and also if it had occurred to him that it was because the vice-chancellor regarded himself as bound to appoint some one whom he regarded as more or less in agreement with Prof. [Thomas Rawson] Birks – which is no doubt a tenable view, though he would probably still have thought it showed his election to the professorship itself to be impossible. May I say that it was very high minded of you not to accept when so pressed.

I am very glad you are interested in the book & like it.

I do not think I have written since in acknowledging the book you returned and the letter he wrote to you when you were ill in London. I remember well how much he was touched by your letter to him & the feeling with which he wrote to you.

Believe me

Yours sincerely
Eleanor Sidgwick

134.
SGL
John Venn to Laura Forster
3 June 1906
Vicarsbrook, Chaucer Road, Cambridge

Dear Laura,

I came back here all right, after my most pleasant two days stay with you, on Monday where I found things going on moderately well. [...]

I particularly enjoyed my ramble along the downs opposite you, as it vividly brought back to me many a delightful walk there about with two or three dear friends long since gone. Those of my standing naturally recall Tennyson's 'Valley of Cauteretz'[471] in those circumstances, & feel that it is indeed the voice of the dead that is a living voice to them. But I have no doubt the younger generation have their own prophets of the day who answer as well as to their various moods as Tennyson happened to do to those of my date. [...] I [suppose] that the present young generation have no one dominant poet who sings for all, any more than they have a dominant teacher (like Mill) who expounds for all.

---

[471] [Tennyson's *Valley of the Cauteretz*, penned in 1861 after revisiting the Cauteretz Valley in the Pyrenees that he had once visited with his friend, Arthur Hallam, who had died in 1833 at the age of twenty-two.]

I liked Shere church[472] very much. As I said, my position is a peculiar one, utterly indefensible in logic, & which leaves me out of sympathy with most of the men here. Emotionally all my nature clings to the faith of the past, or at least to the practices & forms founded on that faith, whilst intellectually I doubt & mistrust in every direction. And unfortunately there is no known means of reconciling differences between the different parts of our nature. So I wobble, & tend to become an object of pity or reproach to the robust who find the path clear on one side or the other. [...]

Well, I must say goodbye now.

Ever yours truly,
J. Venn

135.
SGL
John Venn to Laura Forster
25 December 1906
Vicarsbrook, Chaucer Road, Cambridge

Dear Laura,

[...] We have had a very great loss of Cambridge in Professor Maitland.[473] I have been seeing, & communicating with him a good deal lately as he was gathering early recollections of L. Stephen for the Biography[474] which he had only just completed when he went abroad. He must have died almost the day after his arrival at the Canaries. I dare say you have seen his life of L.S.[475] is marvelously well done, as was everything Maitland undertook. On the whole, as regards the subject of the biography, my impression is left as it was; but I realize more what an unusually fortune life, as regards all external circumstances, he enjoyed. Excellent health after boyhood, health on the part of all near to him (I hardly count the death of his wife in this connection: every marriage must be so cut short for one party or the other): no anxiety about means; & the universal respect & esteem, which however much a man may deserve, does so much to make life happy & successful. [...]

Of books to talk about, I don't think I have been reading much lately. Most of my time has been occupied on my old trade of Logic. I thought I had done with this, & taken to College & University history; but the publishers have asked for a new Edition of a book called Empirical Logic,[476] so I am at work brushing up half-forgotten acquirements & referring to long-neglected authorities.

---

[472] [St James's Shere Church, Surrey, near Abinger, where **Laura Forster** lived.]

[473] [Venn's former pupil and Sidgwick's biographer F.W. Maitland, who died on 19 December 1906. See note **458**.]

[474] [*Life and Letters of Leslie Stephen*, published in 1906. See note **459**.]

[475] [**Leslie Stephen**.]

[476] [The second edition of Venn's *Principles of Empirical or Inductive Logic*, first published in 1889, appeared in 1907.]

We are getting on fairly well on the whole. The young people seem flourishing, & I think Archie[477] is now developing a taste, – rather oddly, in the antiquarian direction. I say 'odly' because this is seldom a taste of early acquirement. Few people, for instance, begin to care for Family History until the last of their grand-aunts or those who could have told them all, are dead. And then they have to puzzle it out for themselves. A. is however beginning upon University records in the shape of the early matriculations, &c.[478] If he goes on, I hope it may lead eventually to some opening at the University Registrary.[479] [...]

Well good bye, & let me know how you are getting on.

Yours very truly,
J. Venn

136.
SGL
John Venn to Laura Forster
7 January 1907
Vicarsbrook, Chaucer Road, Cambridge

Dear Laura,

[...] I think our early life was a decidedly dreary one; & I am really coming to the opinion that the loss which is thus entailed in childhood is irreparable in after life. I do not mean that it leaves a permanent sense of injustice & bitterness; – I feel nothing of that: – but that in some respects it hampers the very power of enjoyment in the succeeding years of youth & early growth; & is apt to leave one in a morbid & solitary mood when one begins to come out into life.

[...] I am certainly not one who ought to preach about others but when I think how much I owe to those from whose teaching I have very widely departed, can't but wonder what will be the outcome of two or three generations such as we see all round us now. National characteristics do, no doubt, change, & peoples may suddenly develop unexpected enthusiasm. But an average British public, absolved from all religious feeling, & without even any second-hand tradition of devotion behind them, does not seem a promising agent for high achievements. The Japanese have shown us what marvelous self-sacrifice can spring from mere patriotism, without anything that we could call religion; but one disquieting element in our present position, to my thinking, is the display almost of anti-patriotism in what [comes?] out of the dominating elements of our modern society. However our grand-nephews & grand-nieces will see what comes of it all for themselves. [*Rest of letter missing.*]

---

[477] [Venn's son **John Archibald Venn.**]

[478] [*The Book of Matriculations and Degrees, 1544–1659. Compiled by John Venn and J.A. Venn, 1913* (Cambridge: Cambridge University Press, 1913.]

[479] [Cf. letter **129**.]

137.
SGL
John Venn to Laura Forster
24 December 1907
Vicarsbrook, Chaucer Road, Cambridge

Dear Laura,

[…] In the old days of Cambridge all who had a relation at College were thinking of the Mathematical Tripos. It began about Jan 1, & we just seized 2 or 3 days for a visit home at Christmas. I am sure there were many thoughts & wishes & prayers at Stisted vicarage for the success of one young Johnian of my year.[480] Almost the last time I saw your brother[481] to speak to was a day or two before the examination. He asked me to dine in his rooms to meet the expected senior wrangler (he entirely failed: – the S.W. I mean, not your brother) & a few of our other competitors. Private dinners in our own rooms were very rare occurrences in those simple days, & I have a vivid recollection of the evening, & of those whom I met. I think I am now the only survivor.

Naturally as time goes on, & the days are long indoors, reading & writing take up most of one's time. And my particular reading & writing just now is of such a special kind just now that it can hardly interest any but the specialist. It is mostly of the College & University antiquarian kind, &, just now, of the dullest species of this kind, viz. long lists of names & dates of matriculations & degrees.[482] But even these things can show an interesting side when looked at in the light of history, – churches, politics or social.

Archie is rather taking to this work, & has been preparing some statistical tables to show the rise & fluctuations of the University, & of the various colleges. As a rule, I think antiquarianism gives prospects which only please those who are on the downward slopes of life. For myself I cared for nothing of the kind until I was past middle life. [….]

I have not yet had an opportunity of reading Jebb's Life.[483] Just at present our book-procuring opportunities are rather poor. We did take in the Times for 2 years – newspaper, library & all – but have now given it up, & scramble as best we may. I pay an annual call on Lady Jebb,[484] & am putting it off from mouth to mouth till I

---

[480] [**Laura Forster**'s father, Charles Forster (1789–1871), was the Rector of Stisted, where she was born.]

[481] [Charles Thornton Forster (1836–1891), St John's, 1855, migrated to Jesus, 1856; B.A. (Classical Tripos, 1st Class) 1859; vicar of Hinxton, Cambs., 1865–91; chaplain to the Earl of Leven and Melville, 1875.]

[482] [Cf. note **478**.]

[483] [*Life and Letters of Sir Richard Claverhouse Jebb, by his Wife Caroline Jebb* (Cambridge: Cambridge University Press, 1907). **Laura Forster**'s father had also published a "life" of a Jebb, in 1836; that of Richard Jebb's uncle John Jebb, Bishop of Limerick.]

[484] [Caroline Jebb (née Reynolds) (1840–1930), American intellectual and socialite who married Richard Claverhouse Jebb in 1874. They lived in Glasgow, but moved to Cambridge in 1889, when Richard Jebb was appointed as Regius Professor of Greek there. She became Lady Jebb after her

can say I have read the book. She would soon find me out if I tried to discuss it from general outside knowledge.

[…] Yours ever truly,

J. Venn

138.
GCA C11/10
Henry Venn of Walmer to John Venn
6 June 1910
Canterbury

My dear John,

I send you back your "Memories"[485] with another instalment of my own of which I shall send you some more in a week or so.

I am very glad to have read your Apologia pro vitâ suâ or shall I say (as Augustine used the word) Confessio?

Do you remember how when you decided to give up your Orders you sent me about 3 lines announcing your purpose & then "This is not for discussion"? Of course I never spoke of it after that. Once – some 15 or 20 years ago – I was at Lincoln & went to lunch with the Mansells[486] & he asked me if I had any idea why you had taken the step but of course I could not tell him anything. He said he had asked you about it but to no purpose.

Your Confessiones make it all plain. It was a gradual process. And yet if – as you say – you had gone to a Parish where there was a great deal of active work & much less unpractical talk your life might have yielded much more of that sort of which Appleton[487] was but a single specimen so far as we know. With your training & given shrewdness you could not have gone to a worse man than John Thomas.[488]

In one of his sermons Phillips Brooks[489] says that if a man is by nature inclined to be "skeptical" he ought to occupy himself in active & hard work for others – not in preaching to them but in useful work for them.[490] His favourite text was "If any man willeth to do His will he shall know of the doctrine". This is pretty much what you say on p. 121.

Allow me to thank you sincerely in giving me the whole history of your decision.

husband was knighted in 1900. Her niece, Maud du Puy, was married to George Darwin, the Plumian Professor of Astronomy and Experimental Philosophy at Cambridge and son of Charles Darwin.]

485 [Venn's *Annals*, reproduced in Part I of this volume.]

486 [Not identified.]

487 [Not identified.]

488 [John Thomas Manley, vicar of Mortlake, Venn's second parish as curate.]

489 [Phillips Brooks (1835–1893), an American Episcopal clergyman and author; Rector of Boston's Trinity Church; received honorary degrees from Harvard (1877), Columbia (1887) and the Doctor of Divinity degree by the University of Oxford (1885).]

490 [Scepticism was a recurring theme in Phillips Brooks' work but the particular sermon referred to is probably 'The Safety and Helpfulness of Faith', in Phillips Brooks, *Twenty Sermons. Fourth Series* (New York: E.P. Dutton & Company, 1890).]

I am wondering if the next generation, & the one below it, will much care for our memories. […] I had thought of printing it for private distribution but I am doubtful if it is worth the expense. [Lorie?][491] does not think the grandchildren will appreciate it, and she would feel pained if they did not, when I had taken the trouble to print it for them.

Yours ever

HV

139.
GL Ms.Gen.508/21
John Venn to Albert Venn Dicey
10 March 1912
Vicarsbrook, Chaucer Road, Cambridge

Dear Albert,

I was glad to hear from you, that you got back comfortably & are, apparently, flourishing: – at least as much as men of our age & our sentiments, can be expected to flourish in these times.

I wish I could quite believe, with the current Spectator,[492] that the Manchester election[493] & the London vows[494] have put the finishing stroke to the prospects respectively of Home Rule[495] & female suffrage. No doubt they are both hard hit, but more than that, I fear, is needed to stamp the life out of such growths as these.[496] […]

We are all much as usual.

Yours ever,
J. Venn

---

[491] [Probably a family acronym for **Henry Venn of Walmer**'s second wife Louisa Jane Venn (née) (Wardlay-Ramsay) (1847–1927).

[492] [The weekly British magazine, founded in 1828. Its political outlook in its first 30 years or so had reflected the Liberal-Radical agenda, but it rallied behind the Conservative Robert Peel when he split the Tory Party by repealing the Corn Laws in 1846, which marked a decided shift toward free trade in Britain.]

[493] [In March 1912, the third Asquith government had lost the election in Manchester, which traditionally was a Liberal stronghold.]

[494] [Reference unknown.]

[495] [Between 1870 and the end of the World War I, the Irish Home Rule movement campaigned for self-government (or 'home rule') for Ireland within the United Kingdom of Great Britain. It had come close to success in 1886, when the Liberal Gladstone government introduced the First Home Rule Bill, which was defeated in the House of Commons. A Second Home Rule Bill was introduced in 1893; it was defeated in the House of Lords. The Third Home Rule Bill was introduced in 1912, leading to the Home Rule Crisis, with the Ulster Volunteers threatening to resist by force of arms the implementation of the Act. See, for instance, James Doherty, *Irish Liberty, British Democracy: The Third Home Rule Crisis, 1909–1914* (Cork: Cork University Press, 2019).]

[496] [Like many like-minded, educated liberals around him, such as **Ferrers**, **Seeley**, **Sidgwick**, **Dicey** and **Leslie Stephen**, Venn was a Liberal Unionist. Despite his (previous) support for education of women, he opposed female suffrage and thought of Home Rule as a constitutional threat. Unlike Dicey, among others, Venn did not take a public position on the subject.]

140.
GCA C25/17
Albert Venn Dicey to John Venn
30 April 1914
The Orchard, Oxford

My dear John,

[…] Somehow or other it seems to me that the best men of the present day – I mean of the generation younger than our own – have many virtues peculiar to themselves, especially a real care for the condition of the poor unknown to myself and most of my contemporaries, but they do seem to me to have lost, as compared with the generation before me, a certain kind of public spirit in common sense. I cannot say that anything ever strikes me as more futile than the belief that "Federalism"[497] is a solution of the Irish difficulty. To construct a federal government is in any circumstance the most difficult task which any constitution can set before himself. Federalism is opposed to the whole course of English history. […]

I am instructing the Nineteenth Century[498] people to send you the May number which will contain my article.[499] I hope on the main point you may agree with me, though very few people do. The King can cause an appeal to the nation by way of a dissolution. He ought to do it. His constitutional course is first to urge upon Asquith[500] to counsel this proceeding and carry it out. If Asquith will not do this the King should apply to Balfour, Bonar Law or Landsdowne.[501] […]

---

[497] [The 'Federal Solution' had not been seen as an option in the 1880–90s, as it was a little-known system of government. By the time of the Third Home Rule Bill of 1912, federal options gained greater consideration, especially after the passage of the 1911 Parliament Act, which had made it increasingly likely that some form of Home Rule would be implemented. Interestingly, while a staunch Unionist, it was Albert Venn Dicey himself who had done much to acquaint the English with the very possibility of a federal solution in his monumental *Law of the Constitution* (1885), which afterwards formed a major doctrinal influence upon English constitutional thought. See John Kendle, *Ireland and the Federal Solution. The Debate over the United Kingdom Constitution, 1870–1921* (Kingston & Montreal: McGill-Queen's University Press, 1989), chapter 7. On Dicey see Hugh Tullock, "A.V. Dicey and the Irish Question: 1870–1922," *Irish Jurist* 15:1 (1980): pp. 137–165.]

[498] [The *Nineteenth Century* was a popular British monthly literary magazine founded in 1877. It changed its name to *The Nineteenth Century and After*, ceasing publication in 1972.]

[499] [Albert Venn Dicey. "The Appeal to the Nation," *Nineteenth Century and After* 75 (May 1914): pp. 945–957. Dicey himself had made room for royal intervention in his *Law of the Constitution*, where he had written that 'No one really supposes that there is not a sphere, though a vaguely defined sphere, in which the personal will of the King has under the Constitution very considerable influence'. Albert Venn Dicey, *Lectures Introductory to the Study of the Law of Constitution* (London: Macmillan and Co., 1885), on p 388.]

[500] [Herbert Henry Asquith (1852–1928), British statesman and Liberal politician who served as Prime Minister of the United Kingdom from 1908 to 1918.]

[501] [Arthur James Balfour; see note **228**. Andrew Bonar Law (1858–1923), British Conservative politician. Henry Charles Keit Petty-fitzmaurice, 5th Marquess of Lansdowne (1845–1927), British statesman who served on both Liberal and Conservative Party governments. In 1914, a letter from Albert Venn Dicey was passed on to George V advising him that, 'while constitutionally the king's

You should read Balfour's last speech in to-day's Times. It is a very fine one. It is by far the most striking speech of his that I have ever read.

Your affectionate cousin,

A.V. Dicey

141.
GL Ms.Gen.508/22
John Venn to Albert Venn Dicey
3 May 1914
Vicarsbrook, Chaucer Road, Cambridge

Dear Albert,

Many thanks for the article you have sent me, & which we have, of course, all read. I think it splendid & agree with every word. I only wish it could have due effect.

On the particular point about the King I am at one with you. We were talking about it in the Combination Room last night & one or two earnest Unionists regretted the idea of, as the phrase is, 'dragging the King into politics'. I think that there are cases where this should be done, & that the present case is one of them. As I said once before, a sort of automatic King does not appeal to me. That there would be of course a radical outcry may be quite true, but I think it would only affect those who are already determined that way; would leave every Unionist unaffected; & would moreover, in my opinion, probably appeal to thousands who like even a King 'to stand up for himself'. But what is to come now?!

Ever yours aff^{y}

J. Venn

I will send the Article to Mrs Sidgwick[502]

---

right of veto remained intact, a dissolution of parliament set in motion by the king's seeking – and successfully seeking – a minister willing to form a new government would hopefully overcome the problem of the crown's interfering politically when it derived its essential strength from its political detachment'. Tulloch, "A.V. Dicey and the Irish Question," on pp. 158–159.]

[502] [Dicey's 'The Appeal to the Nation'. 'Mrs Sidgwick' refers to **Nora Sidgwick**.]

142.
GCA C63
Robert P. Richardson[503] to John Venn
9 November 1915
5010 Parkside Ave., Philadelphia, PA.

Dear sir,

I find that the copy of the Monist[504] containing the article[505] by myself and Professor Landis,[506] which I mentioned in my previous letter, does not seem to have been forwarded to you by the publishers. I am hence forwarding you one myself. Though from what you say in your note of Oct. 3rd I presume you will hardly care at the present time to more than glance at anything on such a topic,[507] yet I wish to put a copy of our article in your hands as a tribute to a man whose works have done so much for me.

Your *Symbolic Logic* and *Logic of Chance* have been a constant source of inspiration to me; I have benefitted, not merely by the knowledge I acquired from them, but by having before me subjects systematically worked up with a perfection I have never found elsewhere, either in Logic or in any other branch of science.

Even in John Stuart Mill, who has influenced me far more than any other author (not merely in Logic but in all the other fields he covers) I find here and there flaws, and the like is true of the other works I esteem most highly with the exception of these two of yours. I will not say that I might not be able to find one or two points with which I would raise an objection; it is now some time since I read the books, and on rereading anything even for the fourth or fifth time you nearly always get some new impression. But my previous perusals gave me an impression of perfection that I have never received from any other works and I have read rather widely both in Logic and elsewhere. I think I may say that Professor Landis has a like feeling; at all events he has been greatly impressed by your works and so has another friend of mine to whose attention I quite recently brought your *Symbolic Logic*. I have also read your Hulsean Lecture, *On Some Characteristics of Belief* and your *Empirical Logic* I have gone over but have not read throughout. From neither of these works did I gain as much as from your other two. I have not come across any other books or articles by you.

---

503 [Robert Porterfield Richardson (?–?), American mathematician affiliated to the Lehigh University, South Bethlehem, Pennsylvania. No biographical information has been found.]

504 [An academic journal devoted to philosophy, founded in 1890. It initially accepted papers from philosophers, scientists and mathematicians. It ceased publication in 1936, but was resumed in 1962 and has appeared continuously since then.]

505 [Robert P. Richardson and Edward H. Landis, "Numbers, Variables and Mr. Russell's Philosophy," *The Monist* 25 (1915): pp. 321–364. It was reproduced in book form in 1917.]

506 [Edward Horace Landis (?–?). No biographical information has been found.]

507 [The paper examined Russell's philosophy of mathematics. It focused solely on Russell's *The Principles of Mathematics* of 1903, neglecting the *Principia Mathematica* and Russell's journal articles published since then.]

It is incomprehensible to me that Cambridge University could have preferred any other candidate to you for a Professorship in which Logic was concerned, whether the chair covered Mental Philosophy or not. I knew, of course, that you were not a Professor, but I took it for granted that some personal reasons prevented you from accepting such a post; not that you had been a candidate and had not been chosen.

My eyes just now fell on my copy of "Lewis Carroll's" *Symbolic Logic, Part 1, Elementary*. This is a work which I esteem most highly and I have often wondered whether the fragments the author left of Part 2 and 3 will ever be published.[508] I have heard that some manuscript was left and I have always thought that you would be the only man to edit it properly. It really seems a pity that such matter should be lost to the world, and I have been hoping that some publisher might take the matter up and bring out the Advanced and Transcendental parts.

I have just gone over the last of the galley proofs of our book, *Fundamental Conceptions of Modern Mathematics, part 1, Variables and Quantities*.[509] It ought to soon be through the press and I shall then send you a copy. This work is really more philosophical than mathematical; it deals with Mathematics in the same sense as does your *Logic of Chance* – through assuredly I would not presume to compare it with your book.

Yours sincerely
Robert P. Richardson

143.
GL Ms.Gen.508/23
John Venn to Albert Venn Dicey
16 June 1916

Dear Albert,

I return H. Cox's article[510] with thanks. I have read it with much interest, & I am glad to think that I voted for him, – at least I believe I did so,[511] – when he stood for the University some years ago. His tone is excellent, & certainly marks the change

[508] [Charles L. Dodgson (Lewis Carroll) died in 1898, leaving the remaining parts of his *Symbolic Logic*, intended to be sub-titled 'advanced' and 'transcendental', unfinished. Most of this material is probably lost, if it had even been written. Some fragments of Part II survived, however, and were published in 1977 by W.W. Bartley, III, in a book which also reproduced Part I. W.W. Bartley, III, ed., *Lewis Carroll's Symbolic Logic* (Hassocks, Sussex: Harvester Press, 1977.]

[509] [Robert P. Richardson and Edward H. Landis, *Fundamental Conceptions of Modern Mathematics. Variables and Quantities. With a Discussion of the General Conception of Functional Relation* (Chicago & London: The Open Court Publishing Company, 1916). Many more part of the book were announced, but these do not seem to have materialized.]

[510] [Homersham Cox (1821–1897), Jesus, 1839; B.A, 1844; lawyer, judge, and historian; author of several works on English government. The article referred to has not been traced.]

[511] [Venn may have confused Homersham Cox, who died in 1897, with his son Homersham Cox (1857–1918) Trinity, 1975; B.A. (4th Wrangler) 1880; Fellow, 1881; Professor of Mathematics at the Muir Central College, Allahabad, 1891–1918. The position Cox stood for is unknown.]

of view since the days, some 10 years ago, when C. Bauerman[512] contemptuously called the whole question of Tariff-Reform, 'Foolery'.

I rather demur to your statement – 'H.C. says, & you deny that the policy of free imports is necessarily beneficial for the country'.[513] I do not think that I have gone beyond much doubting the fact, & saying that I have seen no proof of it. Something turns on the word 'beneficial'. If it merely means getting the commodity cheaper, it is very likely that they are right. But many other considerations come in: – the labour of our own people, the possibility of a state of war, &c &c.

I understand by 'Laissez Faire' the wide principle, which includes Free-Trade, viz. that of allowing complete individual liberty (short of doing injury to others: an exception which opens out a troublesome amount of explanation) on the ground that government control is sure to lead to waste & other mischief. Was not Herbert Spencer the great exponent of this?[514] If all nations acted alike, & no huge syndicates & combines existed, this condition might be tolerated. But as soon as one country (e.g. Germany) is organized from top to bottom in order to carry out its international schemes acquiescence is no longer possible. Is not the pacifist theory a parallel? <u>If</u> all nations did without any army all might get on. But the moment one of them started a permanent armed force all others would be compelled to do the same.

As to Malthus.[515] My own view is that he was probably quite right <u>in the long run</u>. But the run has been much longer than he & others expected. I can't but think that the population of the world, & of the European countries in particular, is quite large enough at present, & that a stationary state, as regards numbers, would be good all round. But here again the same difficulty arises. <u>If</u> one nation won't stop breeding fast, what about the others? Had France, & Germany changed places in this respect there would have been no war. If Germany continues to breed as they have done they <u>must</u> overflow somewhere, & it is hard to see how this can be without war.

---

[512] [Probably C.W. Bowerman (1851–1947), prominent trade unionist and Labour politician who long resisted Tariff Reform, the protectionist policy to protect British industry from foreign competition.]

[513] [The main alternative to Tariff Reform was Free Trade. Tariff Reform had split the Conservative Party and their government coalition allies in the Liberal Unionist Party, which played a key role in tits 1906 defeat to the Liberals who advocated Free Trade. Ironically, some wartime tariffs were introduced by the Liberal Chancellor Reginald McKenna in 1915. See, for instance, Antony Howe, *Free Trade and Liberal England, 1846–1946* (Oxford: The Clarendon Press, 1997).]

[514] [The prominent philosopher, biologist, anthropologist and sociologist Herbert Spencer (1820–1903), who coined the phrase 'survival of the fittest', was an advocate of radical individualism. He argued that the prospects for England were bad if the general public continued to believe that the general welfare could be improved by legislation and policies.]

[515] [Thomas Robert Malthus (1766–1834), author of *An Essay on the Principle of Population* which warned of future difficulties for society based on the idea that population growth is exponential while the growth of food production is linear, eventually leading to the want of food and famine, unless birth rates decreased.]

Authorities say that they, like most other peoples, have of late shown a falling birth-rate. Whether this is due to deliberate constraint, or to some unknown cause due to increased civilization, one does not know.

Ever yours,

J. Venn

144.
SGL
John Venn to Laura Forster
25 December 1916
Cambridge

Dear Laura,

[…] Archie[516] has been very well. He is in one of what are called the 'provisional battalions' – the 62th – whose function is to guard the coast.[517] So long as there is a powerful German fleet in existence there is always a chance of an attempt at invasion. Of actual work there is plenty, in the way of drill, instruction in trench – &c &c. Also, they have now a complete establishment of anti-zeppelin arrangements in the way of guns, search light &c, which are making them hope that the zepps will try again.[518]

As to my work here I find a certain satisfaction in a long, continuous monotonous job in which I get immersed. A few years ago Archie came into an arrangement with the Press here to undertake the publication of a complete (so far as possible) list of all known Cambridge men from the earliest times, – I commonly tell people that it covers the period from the creation of the world to the year A.D. 1900.[519] As he can't do it now, I am steadily plodding at it. Of course it is monotonous work, but, as I say, that is one good of it, in such anxious & perturbing times as these. And it has the further advantage – though it may sound odd to call it one, – that it can never be finished, at least not in my life-time. […]

Well: good bye & all good wishes from yours ever,

J. Venn

[516] [Venn's son **John Archibald Venn**.]

[517] [**John Archibald Venn** was lieutenant for 3 years in the Cambridgeshire Regiment. He afterwards served as a statistician in the Food Production Department.]

[518] [The main German campaign against Britain had started in January 1915 using airships for strategic bombing. These were generally referred to as 'Zeppelin raids'.]

[519] [The *Alumni Cantabrigienses: A Biographical List of All Known Students, Graduates and Holders of Office at the University of Cambridge, from the Earliest Times to 1900*, edited by John Venn and **John Archibald Venn** and published by Cambridge University Press in elven volumes between 1922 and 1954, covering over 130,000 names.]

145.
GL Ms. Gen.508/24
John Venn to Albert Venn Dicey
28 February 1917

Dear Albert,

[…] Now as to 'Free Trade', to use the accepted term. By this I understand, as I have said before, not what the doctrine has now been whittled down to, but the general principle of laissez faire. What Herb. Spencer as a theorist, & Cobden & Bright as practical men,[520] held was surely this: – that any collective or government interference in trade matters was sure to be corrupt or mischievous, & that therefore all dealing must be left to the play of free competition: the only checks being the usual legal ones against fraud & violence.

There was a time when some support could have been given to such a view but how any one can maintain it under modern conditions I cannot understand. I have just been reading Dr. Dillon's book 'Ourselves & Germany'.[521] It is a rather dismal one, for he is much of a pessimist as regards the present & the future, but, as to the past, he seems fully of true & useful information. His account of the gigantic system of organized effort, commercial & political, carried out by Germany in such countries as Russia & Italy & Turkey, is an eye-opener. Banks, consuls, newspapers, &c &c, are all combined, with lavish expense & bribes & secret information, to foster the German trade; – &, of course, to prepare the way for the civil & military power at a later date.

It is simply impossible for old-fashioned methods to stand up against this. […] Of course there are evils in government interference, as there are evils in a standing army, but the conduct of foreign nations has made both of them absolutely necessary. This is a very brief reply to your interesting letter: – I only wish, with you, that we could meet & talk it over. […]

Yours ever,

J. Venn

146.
GL Ms. Gen.508/25
John Venn to Albert Venn Dicey
20 January 1919

Dear Albert,

I was deeply interested in your letter, & naturally I did not want to reply without a little thought: – which must explain apparent delay. I think it is one of the advantages of getting old, &, still more, of living in such very different times

---

[520] [Herbert Spencer. See note **514**. Richard Cobden (1804–1865) and John Bright (1811–1889), British Radical and Liberal politicians who both advocated Free Trade by repeal of the Corn Laws, forming the Anti-Corn Law League in 1836. The Corn Laws were taxes on imported grain introduced in 1815.]

[521] [Emile Joseph Dillon, *Ourselves and Germany* (London: Chapman and Hall, 1916).]

& circumstances from those we can remember, that one can speak one's mind freely.

I can't but think that most of the difficulty about the 'infinite' knowledge, goodness, &c, of God, comes from our using the word in an absolute sense. As you say, very well, the theist recognizes a power far beyond anything he can conceive & naturally calls this power omnipotent…In my old mathematical days I had a good deal to do with the 'infinite', – lines, space, &c. What the modern speculators make of it now, I do not know, but as we commonly interpreted it then, there was no insurmountable difficulty. Mathematical figures, like the hyperbole, which had 'infinite' branches, simply meant that we could prolong to produce them without any limit whatever. As I understand it what we meant was, not that the curve was, in actual fact, an infinite thing at the present moment, but simply that we could go on, without coming upon any hindrance or stoppage, prolonging or producing it for ever. So, – I should say, – with infinite power or knowledge, I can suppose the most learned or the most powerful person being more so, without any limit. But if I regard him as already infinite I get into contradictions. Your suggestion as to what an infinite Being might try to achieve in the way of making a triangle which should also be a circle, is a case in point. You may remember a paper by Dr. Whewell as to what would result if an 'infinite' force encountered an 'immovable obstacle'?[522]

I do not find, as I get older, any change in my conviction that the course of the world is not a random or mechanical business, but that it is ruled & managed by some Being whom we may, in a conventional way, call 'infinite'. The popular objection, at the present day, that if there were such a being He could never have suffered such a war as this to take place, seems to me merely the illusion of over-estimating what strikes us at the moment.

Suppose that some one with the requisite power & information were to make an 'inventory' of all the suffering & guilt which has taken place, year by year, over the whole during the last 20 centuries, & were to display it in figures, or a diagram, do you suppose that we should notice any peculiar rise in the figures for 1914–8? I very much doubt it. I can't get over a similar conviction that the corresponding chart for the happiness & the good of the world would represent a slow & steady rise on the whole. But no proof is available.

I will write again after refreshing my mind reading [John Stuart] Mill.

Yours ever,
J. Venn

---

[522] [William Whewell (1794–1866), Trinity, 1812; B.A. 2nd Wrangler and Second Smith's Prizeman) 1816; Master, 1841–66; Vice-Chancellor, 1842–43; F.R.S, 1820; Professor of Mineralogy, 1828–32; Knightbridge Professor of Moral Philosophy, 1838–55; played a key role in founding the Moral Sciences Tripos. The paper referred to has not been traced. Venn's reference is probably to Whewell's *An Elementary Treatise on Mechanics*, first published in 1824.]

147.
GCA C25/26
Albert Venn Dicey to John Venn
16 March 1919

My dear John,

[…] Now I will turn to your letter.

We both I think must feel a great change in our own attitude towards Mill & also recognize an immense change in the influence exerted by his works since 1859 or [i.e.?] during the last 60 years.

His 'On Liberty'[523] was to me a sort of political & social gospel. Its influence was qualified I think by two things as far as I was concerned – an inherited respect for the Whigs & for Macaulay[524] & a belief wh. suited my position as a Union speaker in Universal Suffrage & Republicanism. I have till quite recent years felt it hard even to be fair to minority representation & now I greatly doubt whether in most cases it would produce half the good anticipated from it by its supporters.

At present I think Mill's great gifts in the statement of his arguments is underrated but I also believe that this literary skill in the statement of a case deceived both his disciples & even more himself.

[…] Personally I don't think Mill now a great or original thinker but I do think he rendered great services to thought & thus promoted the belief in the immense importance of truth & also of reasoning.

At one time, I should have entirely agreed with you. I even now completely understand your feeling of a future life is needed in order that the terribly wicked may in a way pay for the wrong they have done. But I feel doubtful how far I still share this feeling. Surely this would mean for most of us a terrible amount of punishment. I have had a far easier life & in many ways a less tempted life, than I imagine thousands of men & women. Ones 'sense of sin' to use a theological term wh. contains a much forgotten truth makes me feel that the cessation of personal existence might for many of us be the justest & in every way the best end of human life. […]

My dear fellow keep yourself alive as long as you can. Life is growing too solitary for me.[525] I know no one of the name Dicey. I doubt whether except you & Harry anybody is living who really knew me as a boy. I am the oldest member of All Souls.

Yours affec^{y}

A.V. Dicey

---

523 [**Mill**'s *On Liberty*, first published in 1859. It had a similar effect on Venn. See *Annals*, p. 96.]

524 [The historian and Whig politician Thomas Babington Macaulay (1800–1859.]

525 [By nature, **Dicey** was 'withdrawn, unworldly, ineffectual. An inborn muscular dystrophy cut him off from an active and involved existence, and increasing deafness and marriage to an invalid wife reinforced his sense of isolation'. Tulloch, "A.V. Dicey and the Irish Question," on p. 148.]

148.
GCA C25/27
Albert Venn Dicey to John Venn
23 March 1919

My dear John,
I want to tell you two things:

(1) I have read thro' your one vol. of Biographical Annals[526] with extreme interest from beginning to end
(2) I have ended it with disappointment & with some anxiety for the very obvious reason that whilst it purports on the title page to be written in 1903 it ends in fact on p. 147 […] at about apparently 1866 with the election of Maurice to the Moral Philosophy professorship & reflections thereon.

In the thoughts on it by the way you do yourself much injustice. It is so often true that to choose a man of established reputation is the right course, that I can't blame the electors. But his reputation should be in line of the professorship. I don't doubt that your election would have been of more benefit to the University.[527] In later life I have always preached in case of two good men choose the younger.

Surely there must be another vol. of the Autobiographical sketch in existence. I am heartily glad to have got it in any case, but it does not contain the very thing you told me it would explain or state your changes of opinion. All that is said is […] to my mind both perfectly convincing & explains what no one who knew you could ever doubt: the perfect honesty of all you ever thought, said & did.

I am a little anxious for a reply to this for I fear you may have thought you sent me some second vol I have not received.

[…] Your affectionate Cousin,

A.V.D.

149.
GCA C49
John Maynard Keynes[528] to John Venn
31 August 1921
King's College, Cambridge

Dear Dr Venn,

I have asked my publishers to send you a copy of my book A Treatise on Probability,[529] which will appear this week. I send it, if I may say so, in a spirit of

[526] [Venn's *Annals*, reproduced in Part I of the present volume.]

[527] [**Dicey**'s reference is to Venn's candidature for the Knightbridge Professorship of Moral Philosophy in 1862, when Venn was in his late twenties.]

[528] [John Maynard Keynes (1883–1946), King's College 1902; B.A. (12th Wrangler) 1905; Fellow of King's, 1908; famous economist who turned from philosophy to economics on Alfred Marshall's prodding; son of Venn's former pupil and colleague John Neville Keynes.]

[529] [John Maynard Keynes, *A Treatise on Probability* (London: Macmillan and Co. Ltd, 1921). The book was an enlarged edition of his fellowship dissertation at King's College, Cambridge.]

piety to the father of this subject in Cambridge. It is now no less than 55 years since the appearance of your first edition[530]; yet mine is the systematic treatise on the logic of the subject, next after yours, to be published from Cambridge; nor, so far as I know, has there been any treatise in the meantime in the English language. Yours was nearly the first book on the subject that I read; and its stimulus to my mind was of course very great. So, whilst you are probably much too wise to read any more logic (as I hope I shall be in my old age), I beg your acceptance of this volume, the latest link in the very continuous chain (in spirit of differences of opinion)[531] of Cambridge thought.

Yours sincerely,

J.M. Keynes

150.
GL Ms.Gen.508.31
John Venn to Albert Venn Dicey
28 September 1921
Vicarsbrook, Chaucer Road, Cambridge

Dear Albert,

[…] As you know, Cambridge has just lost one of its best known social institutions in Henry Jackson.[532] I have, of course, known him for many years; but we were never very intimate, as his particular type of University Radicalism was not quite to my taste.[533] I agreed with much of his earlier energies, but found myself disagreeing much more frequently in later days.

What a welter of a world we live in! I used to pray (1914–18) that I might live to see Right prevail over Wrong. To pray now that any of us may live to see Sanity triumph over Madness would be almost presumptuous.

Ever yours truly,

J. Venn

---

[530] [The first edition of Venn's *Logic of Chance* appeared in 1866.]

[531] [Keynes criticized Venn's entire outlook, which identified probability with long-run statistical frequency, arguing that other kinds of probability were subject to logical analysis too.]

[532] [Henry Jackson (1839–1921), Trinity 1858; B.A. (3rd Classic) 1862; became a Cambridge Apostle in 1863; Fellow, 1864–1921; Assistant Tutor, 1866; Praelector in Ancient Philosophy, 1875; Vice-Master, 1914–19; Regius Professor of Greek, 1906–21; F.B.A., 1903. Henry Jackson had died on 25 September 1921.]

[533] [Together with **Sidgwick**, Henry Jackson had been an all-round radical reformer; he contributed to the reform of the Triposes, the admission of women for university education (he voted for women's degrees), the abolition of tests.]

Biographical List of Names

As explained in the description of the editorial practices in the Introduction, this list provides short biographical sketches of all individuals whose names appear in bold in the annotations to Part I and Part II. Unless otherwise indicated, reference is to Cambridge on academic matters and Britain on general matters.

Bain, Alexander (1818–1903). Psychologist, logician and educationalist. Life-long friend of John Stuart Mill, whom he helped with the revision of the manuscript of *System of Logic*. Bain was appointed Professor of Mathematics and Natural Philosophy at Anderson's University in Glasgow in 1845; he resigned the post to dedicate himself to writing, being a frequent contributor to the *Westminister Review*. He was subsequently Examiner in Logic and Moral Philosophy at the University of London, 1857–62 and 1864–69, Regius Professor of Logic and Regius Professor of English Literature at the University of Aberdeen, 1860–1880. Author of several books on philosophy, linguistics, and psychology, for which he gained most recognition. Bain was the founder of the journal *Mind*, established in 1876.

Birks, Thomas Rawson (1810–1883). Clergyman and theologian. Trinity, 1829; B.A. (2nd Wrangler) 1834; Fellow, 1834; Tutor, 1834–36; Curate of Watton, Herts., Rector of Kelshall, 1844–66; Vicar of Holy Trinity, Cambridge, 1866–77; Knightbridge Professor of Moral Philosophy, 1872–83; popular preacher and controversialist.

Clayton, Charles (1813–1883). Prominent Evangelical. Caius, 1832; B.A. (21st Wrangler, Classical Tripos, 1st Class) 1836; Fellow, 1838–66; returned to Caius as Tutor after holding several curacies; Hebrew Lecturer, 1842–44; Greek Lecturer, 1842–46; Vicar of Holy Trinity, Cambridge, 1851–65; Rector of Stanhope, Durham, 1865–83.

Dicey, Albert Venn (1835–1922). Whig jurist and constitutional theorist. Educated at Balliol College, Oxford, graduating with First in Classical Moderations in 1856 and in Literae Humaniores in 1858; won a Fellowship at Trinity College, Oxford, in 1860, which he forfeited upon his marriage in 1872; appointed Vinerian

L. M. Verburgt, *John Venn: Unpublished Writings and Selected Correspondence*, Studies in History and Philosophy of Science 56,
https://doi.org/10.1007/978-3-030-79829-1

Professor of English Law at Oxford in 1882, a post he held until 1909; cousin of Fitzjames and Stephen Stephen and of John Venn; prominent Liberal Unionist; author of *Introduction to the Study of Law of the Constitution* (1885) in which he outlined the principles of parliamentary sovereignty.

Edgeworth, Francis Ysidro (1845–1926). Economist and statistician. Educated at Trinity College Dublin, studying classics, where he graduated in 1865, and at Balliol College, Oxford, where he graduated in 1869 with First in Literae Humaniores; qualified as an English barrister in 1877; turned to the study of economics in 1879, having briefly served as a teacher of Greek at Bedford College; an autodidact, Edgeworth became well-versed in mathematics and philosophy only after his graduation; became Lecturer on Logic at King's College, London, in 1880; became interested in (mathematical) statistics and probability due to Galton and Venn, writing many articles on the application of statistics to the moral or social sciences; appointed Professor of Political Economy at King's College, London, in 1881, and Drummond Professor of Political Economy at Oxford in 1891; founding editor of *The Economic Journal*; influential figure in the development of neo-classical economics through his work on utility theory.

Ferrers, Norman Macleod (1829–1903). Mathematician and university administrator. Caius, 1847; B.A. (Senior Wrangler and First Smith's Prizeman) 1851; Fellow, 1852–80; Dean, 1860–65; Tutor and Mathematical Lecturer, 1865–80; Master, 1880–1903; Vice-Chancellor, 1884–86; author of several mathematical works and co-editor, 1855–1891, with J.J. Sylvester, of *The Quarterly Journal of Pure and Applied Mathematics*.

Forster, Laura Mary (1839–1924). Philanthropist. Aunt of the novelist E.M. Forster (1879–1970) and niece of Marianne Thornton (1797–1887). A lifelong friend and correspondent of Charles Darwin's daughter Henrietta ("Etty") Darwin (1834–1927).

Galton, Francis (1822–1911). Polymath: statistician, human geneticist, eugenicist, tropical explorer, geographer, inventor. Studied at King's College London Medical School before going up to Cambridge in 1840; like his half-cousin Charles Darwin, Galton took a "poll" B.A. degree in 1844. Hon. Sc.D., 1895; Hon. Fellow, 1902; F.R.S., 1856; General Secretary of the British Association, 1863–67; founder of the Anthropometric Laboratory at the International Health Exhibition, 1884–85; founder of the Eugenic Laboratory, 1904, and the journal *Biometrika*. Galton published more than 300 books and papers on a wide variety of topics, including fingerprints and the laws of heredity.

Grote, John (1813–1866). Philosopher. Trinity, 1831; B.A. (Classical Tripos, 1st Class) 1835; Fellow, 1837–66; succeeded William Whewell as Knightbridge Professor of Moral, Philosophy, 1855–66; vicar of Trumpington, 1847–66. Author of *Exploratio Philosophica* (1865); his *Examination of Utilitarian Philosophy* and *Treatise on Moral Ideas* were published posthumously. Founder of the 'Grote Club', through which he influenced Venn, Sidgwick and others involved in the Moral Sciences Tripos.

Gurney (née Batten), Emelia Russell (1823–1896). Prominent campaigner for women's higher education, patron and benefactor. After her marriage to the lawyer and Conservative Party politician Russell Gurney (1804–1898) known as Mrs. Russell Gurney. Venn's cousin, she was the daughter of Venn's aunt Caroline Venn.

Hopkins, William (1793–1866). Mathematician, geologist and famous "Wrangler-maker", with William Thomson (Lord Kelvin), George Gabriel Stokes, James Clerk Maxwell, Francis Galton and Isaac Todhunter among his pupils. Peterhouse, 1823; B.A. (7th Wrangler) 1827; disbarred from a Fellowship due to his (re)marriage; F.R.S., 1837; President of the Geological Society, 1851–53; President of the British Association, 1853; author of several mathematical books.

Jevons, William Stanley (1835–1882). Economist and philosopher of science. Educated at University College, London, and granted B.A. and M.A. degrees from the University of London. After having been a Tutor there, he was appointed Professor of Logic and Mental and Moral Philosophy and Cobden Professor of Political Economy at Owens College, Manchester in 1866. In 1876, he moved to London to become Professor of Political Economy in University College, London, a post he resigned in 1880. He drowned in 1882. A prolific writer who contributed many books and papers the fields of logic, political economy and practical economics, for which he first received general recognition, for instance for his *The Coal Question* (1865).

Maurice, Frederick Denison (F.D.) (1805–1872). Clergyman and theologian. Trinity, 1823, migrated to Trinity Hall, 1825; Civil Law Classes, 1st Class, 1826–27; entered Exeter College, Oxford, 1829; B.A. (Oxford) 1831; Professor of Modern History and English Literature at King's College, London, 1840–46; Professor of Ecclesiastical History, 1846–53; Knightbridge Professor of Moral Philosophy, Cambridge, 1866–72. One of the original members of the Cambridge Apostles, Maurice was as controversial (and idiosyncratic) as he was influential as a liberal Anglican theologian. He was accused of heterodoxy in 1852; and in 1853 he was asked to retire by the Council of King's College after the publication of his *Theological Essays*.

Mayor, Joseph Bickersteth (1828–1916). Philosopher and classical scholar. St. John's, 1847; B.A. (2nd Classic) 1851; Fellow, 1852–64; Tutor, 1861–63; appointed College Lecturer in Moral Sciences as St. John's in 1861; Headmaster of Kensington Proprietary School, 1863–69; Professor of Classical Literature at King's College, London, 1870–79, and Professor of Moral Philosophy, 1879–83. Mayor served as the first editor of the *Classical Review*, and published and edited several volumes on theology, the classics and ancient philosophy. He edited the works of John Grote (his wife Alexandrina Jessie Grote's uncle), including *Exploratio Philosophica*.

Mill, John Stuart (1806–1873). Philosopher, economist, advocate of women's rights and Member of Parliament. Eldest son of the utilitarian philosopher and Benthamite reformer James Mill (1773–1836). A child prodigy, John Stuart Mill was educated at home by his father and private tutors. After a mental crisis, Mill set out to qualify and humanize his inherited Benthamism on the basis of Romanticism, which had come to him mainly via F.D. Maurice. A major advocate of utilitarianism, Mill contributed to many fields of study and wrote several books, including *System of Logic* (1843), *Principles of Political Economy* (1848) and *On Liberty*

(1859), that dominated nineteenth-century thought and that are still widely read today.

Monro, Cecil James (1833–1882). Trinity, 1851; B.A. (38th Wrangler and 8th Classic) 1855; Fellow, 1856. Due to ill health, he was forced to live largely abroad, devoting himself to the work of letter-writing and criticism of proof-sheets for friends, such as James Clerk Maxwell, on physics, Augustus De Morgan, on pure mathematics and probability, and William Stanley Jevons and Venn, on algebraic logic. Brother of Charles Henry Monro, Venn's College friend.

Monro, Charles Henry (1835–1908). Caius, 1853; B.A. (8th Classic) 1857; Fellow, 1857–1908; College Law Lecturer, 1872–96; left his estate to Caius, which founded a Monro Memorial Fellowship and Celtic Lectureship. Venn's College friend at Caius.

Owen, Richard (1804–1892). Biologist, comparative anatomist and palaeontologist. A controversial figure, and an outspoken critic of Darwin's theory of evolution by natural selection, Owen was generally considered to be an outstanding naturalist and expert on interpreting fossils. He achieved success with a campaign that led to the establishment, in 1881, of the Natural History Museum in London.

Pearson, Karl (1857–1936). Statistician and eugenicist. King's, 1875; B.A. (3rd Wrangler) 1879; Fellow, 1880; Hon. Fellow, 1903. Professor of Applied Mathematics at University College, London, 1885–1911; Gresham Professor of Geometry, 1892–94; Galton Professor of Eugenics in the University of London, and Director of the Laboratory for National Eugenics, 1911; F.R.S, 1896. A protégé and biographer of Francis Galton, and supporter of the program of social Darwinism and eugenics. Pearson, was one of the founders of modern mathematical statistics, and established the world's first university Department of Statistics at University College, London, in 1911.

Ransome, Arthur (1834–1922). Physician. Caius, 1853; Natural Sciences Tripos, 1st Class, 1856; B.A., 1857; M.B., 1859; M.D., 1869; Hon. Fellow, 1892–1923; studied medicine at Manchester, Paris and St. George's Hospital, London; F.R.S., 1885. A leading authority on the nature and treatment of consumption, Ransome was Professor of Hygiene at Owens College, Manchester, 1880–95. Venn's College friend.

Roberts, Ernest Stewart (1847–1912). Classicist and academic administrator. Caius, 1865; B.A. (6th Wrangler) 1869; Fellow, 1870–1903; Junior Proctor, 1875–76; Tutor, 1876; Senior Proctor, 1884–85; Senior Tutor, 1885; President, 1894–1903; Master, 190 –1912; Vice-Chancellor, 1906–08.

Seeley, John Robert (1834–1895). Historian. Christ's, 1852; B.A. (Senior Classic) 1857; Chancellor's (Classical) Medal, 1857; Fellow, 1858–69; Professor of Latin at University College, London, 1859–69; Regius Professor of Modern History at Cambridge, 1869–95; Fellow of Caius, 1882–95. Seeley is best known as the author of *Ecce Homo: A Survey in the Life and Work of Jesus Christ* (1865), controversial for its focus on Jesus's moral and political life, and *The Expansion of England* (1883).]

Sidgwick, Henry (1838–1900). Philosopher. Trinity, 1855; B.A. (33rd Wrangler and Senior Classic) 1859; Fellow, 1859–81, and 1883–1900; Assistant Tutor,

1859–68; Lecturer in Moral Sciences, 1868–75; Praelector in Moral and Political Philosophy, 1875–83; Knightbridge Professor of Moral Philosophy, 1883–1900. Sidgwick, a member of the Cambridge Apostles, played a key role in the early development of the Moral Sciences Tripos at Cambridge.

Sidgwick (née Balfour), Eleanor Mildred ('Nora') (1845–1936). Physics researcher, activist for the higher education of women and Principal of Newnham College, Cambridge. Sidgwick came from a prominent political family: her father was a Conservative Member of Parliament, and her brother, Arthur Balfour, a former pupil of Venn, became Prime Minister, 1902–05. In 1880, she became Vice-Principal of Newnham College under the founding Principal Anne Clough, who she succeeded as Principal in 1892. She was elected President of the Society for Psychical Research in 1908, in which field she was a leading figure.

Stephen, James Fitzjames, first baronet (1829–1894). Judge and writer. Son of Sir James Stephen and Jane Catherine Stephen (née Venn), brother of Leslie Stephen, cousin of Albert Venn Dicey and Venn. He was educated at Eton College, and for two years at King's College, London. He entered Trinity in 1846; B.A. 1852; Hon. Fellow, 1885. Member of the Cambridge Apostles. When he was called to the bar in 1854, he decided to read for an LL.B. from the University of London. Pursuing a legal career, Stephen became Legal member of Council of Viceroy of India, 1869–72 University Counsel, 1872–79, Professor of Common Law at the Inns of Court, 1875–79, and Judge of the Exchequer Division of the High Court, 1879–91. Author of several books, including *Liberty, Equality, Fraternity* (1873–74), a protest against John Stuart Mill's utilitarianism.

Stephen, Leslie (1832–1904). Author, literary critic, historian and first editor of the Dictionary of National Biography. Father of Virginia Woolf and Vanessa Bell, Leslie Stephen was Venn's cousin, being the son of Sir James Stephen and (Lady) Jane Catherine Stephen (née Venn). Like Venn and Albert Venn Dicey, another cousin, Leslie Stephen's family had belonged to the Clapham Sect, the early 19th century group of Evangelical Christian social reformers. He was educated at Trinity Hall, Cambridge; B.A. (20th Wrangler) 1854; Fellow, 1854–67; Tutor, 1856–62; Clark Lecturer at Trinity College, 1885. Leslie Stephen relinquished Holy Orders in 1875, becoming a leader of the 'agnostic school'. A regular contributor to the *Saturday Review* and *Pall Mall Gazette*, he was a critic and editor to the *Cornhill Magazine*, 1871–82, and served as the first editor of the Dictionary of National Biography, 1882–91.

Todhunter, Isaac (1820–1884). Mathematician and historian of mathematics. St. John's, 1841; B.A. (Senior Wrangler and First Smith's Prizeman) 1848; Fellow, 1849–64; F.R.S., 1862; Sc.D., 1883. After graduating, Todhunter remained in Cambridge as College Lecturer and as a successful mathematical coach for the rest of his life, also writing several mathematical textbooks, which went through many editions and were translated into many languages.

Venn, John Archibald (1883–1958). Historian and Master of Queen's College, Cambridge. Trinity, 1902; History Tripos, 1905; Fellow of Queen's College, 1927; Lecturer in the History and Economics of Agriculture, Cambridge, 1921–49; President of Queen's College from 1932 and Vice-Chancellor of the University of

Cambridge, 1941–43. Son of John Venn and Susanna Venn (née Edmonstone), he collaborated with his father on the *Alumni Cantabrigienses*, 8 vols., 1922–53, taking charge of all but the first two volumes.

Venn, Henry Venn (of C.M.S.) (1796–1873). Honorary Secretary of the Church Missionary Society. Queen's, 1814; B.A. (19th Wrangler) 1818; Fellow, 1819–29; close friend of Charles Simeon; Curate of St. Dunstan, 1821–24; returned to Cambridge, 1842; Dean and College Lecturer, Junior Proctor, 1825–26; Evening Lecturer at Great St. Mary's, Cambridge, 1826; Rector of Drypool Hull, 1827–34; Rector of St. John's, Holloway, 1834–47; Honorary Secretary of the C.M.S., of which his father, John Venn of Clapham, was one of the founders. A leader of the Evangelical body of the Church of England. Member of the Royal Commission on Clerical Subscription and Ritual Reform. Father of John Venn and Henry Venn of Walmer.

Venn, Henry (of Walmer) (1838–1923). Clergyman and Assistant Secretary of the C.M.S. Caius, 1857; B.A., 1861; Assistant Secretary of the C.M.S., 1861–67; Vicar of Walmer, 1893–1908; Hon. Canon of Canterbury, 1900–23.

Venn, Susanna Carnegie (née Edmonstone) (1844–1931). Author. Born in Marlboro, the eldest daughter of Rev. Charles Welland Edmonstone and the granddaughter of Neil Benjamin Edmonstone, a distinguished civil servant in the East India Company. She married Venn in 1867, and gave birth to their son John Archibald Venn in 1883. Author of five novels, of which *The Dailys of Sodden Fen* (1884) was considered her best. With her husband, she also co-edited *Admissions to Gonville and Caius College* (1887).

Ward, James (1834–1925). Philosopher and psychologist. Studied in Berlin and Göttingen, 1870–72. Before entering Cambridge as a non-collegiate student in 1872, he worked as a pastor for a year at Emmanuel Church, Cambridge, which he resigned because of changed theological views. Moral Sciences Tripos, 1st Class, 1874; B.A. 1875; Sc.D., 1887; Fellow of Trinity, 1875–1925; University Teacher in the Theory of Education, 1880–88; Professor of Mental Philosophy and Logic, 1897–1925. Gifford Lecturer at Aberdeen, 1895–97, and at St. Andrews, 1908. F.B.A., 1902. Author of several books, including *Naturalism and Agnosticism* and *Psychological Principles*. Together with Venn, Ward lobbied the University for a research laboratory for the study of psychophysical phenomena.

Daniel Wilson (1778–1858). Bishop of Calcutta. An Oxford graduate and prominent Evangelical, Wilson was associated with the Clapham Sect. He was vicar of St. Mary's Church, Islington, 1824–1932, before being consecrated Bishop of Calcutta and first Metropolitan of India and Ceylon in 1832. Henry Venn of C.M.S.'s living at St. John's, Holloway, had been in the gift of Wilson. He had also been President of the Islington Proprietary School, attended by Venn.

Acknowledgments and Permissions

My sincere thanks go to Christopher Stray and Jonathan Smith, for their support and help with some of the letters, and to Amirouche Moktefi, for sharing his knowledge on nineteenth-century British logic, especially regarding Lewis Carroll. I also wish to thank the archivists at the Cadbury Research Library, University of Birmingham, Cambridge University Library, Gonville and Caius College, Cambridge, John Rylands University Library of Manchester, Society of Genealogists Library, London, Trinity College Library, Cambridge and the Special Collections of University College London Library. I am especially grateful to James Cox, archivist at Gonville and Caius College, Cambridge, for making my visits both very pleasant and productive.

My thanks are due to the following for allowing me to reproduce material for which they hold the copyright: Church Missionary Society Unofficial Papers (Accession 81: Venn Papers), Cadbury Research Library: Special Collections, University of Birmingham; the Syndics of Cambridge University Library; the Master and Fellows of Gonville and Caius College, Cambridge; the John Rylands University Library, University of Manchester; the Society of Genealogists, London; the Master and Fellows of Trinity College, Cambridge; the University College of London Library; Richard Lethbridge and Christopher Lethbridge; University of Glasgow Library, Archives and Special Collections. Every afford has been made to secure permission for reproduction. If I have failed in any case to trace a copyright holder, I apologize for any apparent negligence.

This research was supported by a Veni grant from the Netherlands Organisation for Scientific Research (NWO), grant number 275-20-066.

L. M. Verburgt, *John Venn: Unpublished Writings and Selected Correspondence*, Studies in History and Philosophy of Science 56,
https://doi.org/10.1007/978-3-030-79829-1

Bibliography

Books and articles by John Venn

For a complete list of Venn's books and articles see the bibliography in Verburgt, John Venn: A Life in Logic.

Venn, John. [Anon]. "Science of History," *Fraser's Magazine* (May 1862): pp. 651–660.

Venn, John. "The Theology of the Nineteenth Century," *Christian Observer* (May 1865): pp. 371–387.

Venn, John. *The Logic of Chance. An Essay on the Foundations and Province of the Theory of Probability, with Especial Reference to Its Application to Moral and Social Science* (London & Cambridge: Macmillan and Co., 1866).

Venn, John. *On Some of the Characteristics of Belief, Scientific and Religious. Being the Hulsean Lectures for 1869* (London & Cambridge: Macmillan and Co., 1870a).

Venn, John. "Dr. Newman's Grammar of Assent," *Christian Observer* (October 1870b): pp. 727–739.

Venn, John. "Bishop Butler reviewed in the light of modern thought," *Christian Observer* (September 1871): pp. 681–692.

Venn, John. "Review of *The Principles of Science: A Treatise on Logic and Scientific Method.* By W. Stanley Jevons," *The Academy* (October 1874): pp. 381–384.

Venn, John. "Consistency and Real Inference," *Mind* 1:1 (1876a): pp. 43–52.

Venn, John. "Boole's Logical System," *Mind* 1:4 (1876b): pp. 479–491.

Venn, John. "The Foundations of Chance," *Princeton Review* (September 1878): pp. 471–510.

Venn, John. "The Difficulties of Material Logic," *Mind* 4:13 (1879): pp. 35–47.

Venn, John. "On the Diagrammatic and Mechanical Representation of Propositions and Reasonings," *The London, Edinburgh, and Dublin Philosophical Magazine* 10:59 (1880a): pp. 1–18.

Venn, John. "On the Forms of Logical Proposition," *Mind* 5:19 (1880b): pp. 336–349.

Venn, John. "Symbolical Logic," *Nature* 24 (1881a): pp. 140–141.

Venn, John. *Symbolic Logic* (London: Macmillan and Co., 1881b).

Venn, John. "Cambridge Anthropometry," *Journal of the Anthropological Institute of Great Britain and Ireland* 18 (1889a): pp. 140–154.

Venn, John. *The Principles of Empirical or Inductive Logic* (London: Macmillan and Co., 1889b).

L. M. Verburgt, *John Venn: Unpublished Writings and Selected Correspondence*, Studies in History and Philosophy of Science 56,
https://doi.org/10.1007/978-3-030-79829-1

Venn, John. "Natural Inheritance," *Mind* 14:55 (1889c); pp. 414–420.
Venn, John. "On the Nature and Uses of Averages," *Journal of the Royal Statistical Society* 54:3 (1891a): pp. 429–456.
Venn, John. "A Chapter in College History. (Address Given in the Large Lecture Room.)," *The Caian* 1:1 (Easter Term, 1891b): pp. 13–39.
Venn, John. "Results of Anthropometry at Cambridge," *Transactions of the Seventh International Congress of Hygiene and Demography. London, August 10th–17th, 1891. Volume X. Division II. Demography* (London, 1892): pp. 308–314.
Venn, John. *Symbolic Logic*, 2nd ed. (London: Macmillan, 1894).
Venn, John. "Some Personal Reminiscences of J.R. Seeley," *The Caian* 9:3 (Lent Term, 1895a): pp. 164–170.
Venn, John. "Address Delivered in the College Chapel, After the Commemoration Service, Dec. 21 1894," *The Caian* 4:3 (Lent Term 1895b): pp. 188–198.
Venn, John. *Biographical History of Gonville and Caius College, 1349–1897*. 3 vols. (Cambridge: Cambridge University Press, 1897–1901).
Venn, John. *Annals of a Clerical Family* (London: Macmillan and Co., 1904).
Venn, John. "College life and ways sixty years ago," in John Venn, ed., *Early Collegiate Life* (Cambridge: W. Heffer and Sons Ltd., 1913), pp. 253–282, on p. 256
Venn, John, and Francis Galton. "Cambridge Anthropometry," *Nature* 41 (1890): pp. 450–454.
Venn, John, and John Archibald Venn. *The Book of Matriculations and Degrees, 1544–1659*. (Cambridge: Cambridge University Press, 1913.
Venn, John, and John Archibald Venn. *Alumni Cantabrigienses: A Biographical List of All Known Students, Graduates and Holders of Office at the University of Cambridge, from the Earliest Times to 1900*. 11 vols. (Cambridge: Cambridge University Press, 1922–1954).
Venn, John, and Susanna C. Venn. *Admissions to Gonville and Caius College in the University of Cambridge, March 1558–9 to Jan. 1678–9* (London: C.J. Clay and Sons., 1887).

Other Authors

Allender, Tim. *Learning Femininity in Colonial India, 1820–1932* (Manchester: Manchester University Press, 2016).
Altholz, Joseph L. "The Mind of Victorian Orthodoxy: Anglican Responses to 'Essays and Reviews', 1860–1864,' in Gerald Parsons, ed., *Religion in Victorian Britain. IV: Interpretations* (Manchester & New York: Manchester University Press, 1988), pp. 28–40.]
Altholz, Joseph L. *Anatomy of a Controversy: The Debate over* Essays and Reviews *1860–1864* (Aldershot: Scolar Press, 1994).
Arnold, Guy. *Historical Dictionary of the Crimean War* (Lanham, MD: Scarecrow, 2002).
Bartley, W.W., III., ed. *Lewis Carroll's Symbolic Logic* (Hassocks, Sussex: Harvester Press, 1977.
Becher, Harvey. "Radicals, Whigs and Conservatives: The Middle and Lower Classes in the Analytical Revolution at Cambridge in the Age of Aristocracy," *The British Journal for the History of Science* 28:4 (1995): pp. 405–426.
Bennett, Joshua. "A History of 'Rationalism' in Victorian Britain," *Modern Intellectual History*, 15: 1 (2018): 63–91
Boole, George. *An Investigation of the Laws of Thought* (London; Macmillan and Co., 1854).
Brooke, Christopher. *A History of Gonville and Caius College* (Woodbridge: The Boydell Press, 1985).
Brooks, Phillips. *Twenty Sermons. Fourth Series* (New York: E.P. Dutton & Company, 1890).
Buckle, Henry Thomas. *History of English Civilization. Volume 1* (London: John W. Parker and Son, 1857).

Butts, Dennis and Pat Garrett, eds., *From The Dairyman's Daughter to Worrals of the WAAF: The Religious Tract Society, Lutterworth Press and Children's Literature* (Cambridge: Lutterworth Press, 2006).

Cambridge Examination Papers*: Being a Supplement to the University Calendar for the Year 1857* (Cambridge: Deighton, Bell and Co., 1857).

Carroll, Lewis. "A Logical Paradox," *Mind* 3:11 (1894): pp. 436–448.

Catalogue of a Collection of Books on Logic Presented to the Library by John Venn, Sc.D. (Cambridge, 1889).

Clewlow, Michelle. "Intersecting Sets: John Venn, Church and University, 1834–1923," unpublished PhD thesis, The Open University, 2007.

Cockburn, J.S., et.al., eds. "School: Highgate School," in *A History of the County of Middlesex*, vol. 1 (London: University of London Press), pp. 302–304.

Colaiaco, James A. *James Fitzjames Stephen and the Crisis of Victorian Thought* (London: Palgrave Macmillan, 1983).

Cook, Simon J. *The Intellectual Foundations of Alfred Marshall's Economic Science. A Rounded Globe of Knowledge* (Cambridge: Cambridge University Press, 2009).

Craik, A.D.D. *Mr Hopkins' Men: Cambridge Reform and British Mathematics in the 19th Century* (London: Springer-Verlag, 2008).

Crampton, Colin. "The Cambridge School: The Life, Work and Influence of James Ward, W.H.R. Rivers, C.S. Myers and Sir Frederic Bartlett," unpublished PhD thesis, University of Edinburgh, 1978.

Crelle, August Leopold. *Tables for Facilitating the Operations of Multiplication, Division, and Evolution. Abridged from Dr. A.L. Crelle's* Rechentafeln. By Lieut. J.A. Borris. (Washington: Government Printing Office, 1885).

Darwin, Charles. *On The Origin of Species by Means of Natural Selection, or the Preservation of Favoured Races in the Struggle for Life* (London: John Murray, 1859).

Davidson, Randall Thomas, and William Benham. *Life of Archibald Campbell Tait. Archbishop of Canterbury* (London: Macmillan and Co., 1891).

Day, Henry. "A Matterhorn Postscript. John and Henry Venn's Visit to Zermatt in 1865," available from alpinejournal.org.uk.

De Morgan, Augustus. *An Essay on Probabilities, and Their Application to Life Contingencies and Insurance Offices* (London, 1838).

Dicey, Albert Venn. *Lectures Introductory to the Study of the Law of Constitution* (London: Macmillan and Co., 1885).

Dicey, Albert Venn. "The Appeal to the Nation," *Nineteenth Century and After* 75 (May 1914): pp. 945–957.

Doherty, James. *Irish Liberty, British Democracy: The Third Home Rule Crisis, 1909–1914* (Cork: Cork University Press, 2019).

Essays and Reviews (London: John W. Parker and Son, 1860).

'F.M.T.'. "Head Growth in Students at the University of Cambridge," *Nature* 40 (1889): pp. 317–318.

Foster, Joseph. *Men-at-the-Bar: A Biographical Hand-List of the Members of the Various Inns of Court* (printed for the author by Hazell, Watson, and Viney, London, 1885).

Fry, Herbert, ed. *Our Schools and Colleges* (London: Robert Hardwicke, 1868).

Galton, Francis. "Typical Laws of Heredity," *Proceedings of the Royal Institution* 8 (1877): pp. 282–201.

Galton, Francis. "On Head Growth in Students at the University of Cambridge," *Nature* 38 (1888): pp. 14–15.

Galton, Francis. *Natural Inheritance* (London: Macmillan and Co., 1889a).

Galton, Francis. "Head Measures at Cambridge," *Nature* 40 (1889b).

Galton, Francis. "Method of Indexing Finger-Marks," *Proceedings of the Royal Society* 54 (1891): pp. 303–311.

Garden, Francis. *An Outline of Logic, for the Use of Students* (London, 1867).

Gaskell, Elizabeth Cleghorn. *The Life of Charlotte Brontë*, 2 vols. (London, 1857).

Gibson, Susannah. *The Spirit of Inquiry: How One Extraordinary Society Shaped Modern Science* (Oxford: Oxford University Press, 2019).

Gillham, Nicholas W. *A Life of Sir Francis Galton. From African Exploration to the Birth of Eugenics* (Oxford: Oxford University Press, 2001).

Glaisher, James W.L. "On the Law of Facility of Errors and Observations, and on the Method of Least Squares," *Memoirs of the Royal Astronomical Society* 39 (1872): pp. 75–124.

Groenewegen, Peter. *Alfred Marshall: Economist, 1842–1924* (Basingstoke: Palgrave Macmillan, 2007).

Gurney, Ellen Mary ed. *Letters of Emelia Russell Gurney* (London: James Nisbet & Co., Ltd., 1902).

Hauser, Gertrud. "Galton and the Study of Fingerprints," in Milo Keynes, ed. *Sir Francis Galton, FRS* (London: Palgrave Macmillan, 1993), pp. 144–157.]

Hennell, Michael. *John Venn and the Clapham Sect* (London: Lutterworth Press, 1958).

Geison, Gerald L. *Michael Foster and the Cambridge School of Physiology. The Scientific Enterprise in Late Victorian Society* (Princeton: Princeton University Press, 1978).]

Gibbins, John R. *John Grote, Cambridge University and the Development of Victorian Thought* (Exeter: Imprint Academic, 2007).

Grattan-Guinness, Ivor. "The Correspondence Between George Boole and Stanley Jevons, 1863–1864," *History and Philosophy of Logic* 12:1 (1991): pp. 15–35.

Harrison, Frederic. "Neo-Christianity," *Westminster Review* 18 (October 1860): 293–332.

Harvey, Adrian. *Football: The First Hundred Years. The Untold Story* (London: Routledge, 2005).

H.J.P. "Anthropometric Measurements at Cambridge," *Nature* 40 (1889): pp. 593–594.

Howe, Antony. *Free Trade and Liberal England, 1846–1946* (Oxford: The Clarendon Press, 1997).

Hylson-Smith, Kenneth. *Evangelicals in the Church of England, 1734–1984* (Edinburgh: T. & T. Clark, 1988).

James, William. *The Varieties of Religious Experience: A Study in Human Nature, Being the Gifford Lectures on Natural Religion Delivered at Edinburgh in 1901–1902* (London & New York: Longmans, Green and Co., 1902).

Jebb, Caroline. *Life and Letters of Sir Richard Claverhouse Jebb* (Cambridge: Cambridge University Press, 1907).

Jevons, William Stanley. *Pure Logic: Or, The Logic of Quality Apart from Quantity: With Remarks on Boole's System and on the Relation of Logic and Mathematics* (London: Edward Stanford, 1864).

Jevons, William Stanley. *The Substitution of Similars* (London: Macmillan and Co., 1869).

Jevons, William Stanley. *Elementary Lessons in Logic: Deductive and Inductive* (London: Macmillan, 1870a).

Jevons, William Stanley. "On the Mechanical Performance of Logical Inference", *Proceedings of the Royal Society of London* 18 (1870b): 166–169.

Jevons, William Stanley. *The Principles of Science: A Treatise on Logic and Scientific Method* (London: Macmillan, 1874).

Jevons, William Stanley. *Studies in Deductive Logic: A Manuel for Students* (London: Macmillan and Co., 1880).

Jevons, William Stanley. "Symbolic Logic," *Nature* 24 (1881): pp. 233–234.

Johnson, W.E. "Review of *The Logical of Chance*. By John Venn," *Mind* 50 (1888): pp. 268–280.

Kendle, John. *Ireland and the Federal Solution. The Debate over the United Kingdom Constitution, 1870–1921* (Kingston & Montreal: McGill-Queen's University Press, 1989).

Keynes, John Maynard. *A Treatise on Probability* (London: Macmillan and Co. Ltd, 1921).

Kinzer, Bruce L., Ann P. Robson and John M. Robson. *A Moralist In and Out of Parliament. John Stuart Mill at Westminster, 1865–1868* (Toronto: University of Toronto Press, 1992).

Knight, William ed. *Memoir of Henry Venn, B.D. Prebendary of St. Pauls, and Honorary Secretary of the Church Missionary Society,* with an introductory biographical chapter and a notice of West African commerce by his sons, the Rev. John Venn, M.A., and the Rev. Henry Venn, M.A. (London: Longmans, Green, and Co., 1880).]

Kuper, Adam. *Incest & Influence. The Private Life of Bourgeois England* (Cambridge, MA: Harvard University Press, 2009).

Laita, L.M. "Influences on Boole's Logic: The Controversy Between William Hamilton and Augustus De Morgan," *Annals of Science* 36:1 (1979): pp. 45–65.

Lenzen, Wolfgang. "Ploucquet's "Refutation" of the Traditional Square of Opposition," *Logica Universalis* 2 (2008): pp. 43–58.

Levine, Philippa. *The Amateur and the Professional. Antiquarians, Historians and Archaeologists in Victorian England, 1838–1886* (Cambridge: Cambridge University Press, 1986).

Loizides, Antis ed. *Mill's* A System of Logic. *Critical Appraisals* (London: Routledge, 2014).

Maas, Harro. *William Stanley Jevons and the Making of Modern Economics* (Cambridge: Cambridge University Press, 2005).

MacColl, Hugh. "Symbolical Logic," *Nature* 24 (1881): pp. 124–126.

Maitland, Frederick William, ed. *The Life and Letters of Leslie Stephen* (London: Duckworth, 1906).

Mangan, J.A. *Athleticism in the Victorian and Edwardian Public School.* With a Foreword by Sheldon Rothblatt and an Introduction by Jeffrey Richards (Abingdon: Routledge, 2000).

Klein, Alexander. "How American is Pragmastism?" *Philosophy of Science*, forthcoming.

Kremer-Marietti, Angèle. "Introduction: Comte and Mill – the Philosophical Encounter," in *The Correspondence of John Stuart Mill and Auguste Comte.* Translated from the French and edited by Oscar A. Haac (New Brunswick & London: Transaction Publishers, 1995), pp. 1–32.

Martin, G. "The Cambridge Lectureship of 1866: A False Start in American Studies," *Journal of American Studies*, 7:1 (1973): 17–29

Michie, Michael. *An Enlightenment Tory in Victorian Scotland. The Career of Sir Archibald Alison* (Montreal & Kingston: McGill-Queen's University Press, 1997).

Mill, John Stuart. *System of Logic.* 7th ed. (London: Longmans, Green, Reader, and Dyer, 1868).

Minto, Wiliam. *Logic, Inductive and Deductive* (London: John Murray, 1893).

Misak, Cheryl. *Cambridge Pragmatism: From Peirce and James to Ramsey and Wittgenstein* (Oxford: Oxford University Press, 2016).

Moktefi, Amirouche. "Lewis Carroll and the British Nineteenth-Century Logicians on the Barber Shop Problem," *Proceedings of the Canadian Society for the History and Philosophy of Mathematics* 20 (2007): pp. 189–199.]

Moktefi, Amirouche, and Francine F. Abeles. "The Making of "What the Tortoise Said to Achilles": Lewis Carroll's Investigations Toward a Workable Theory of Hypotheticals,' *The Carrollian: The Lewis Carroll Journal* 28 (2016): 14–47.

Moktefi, Amirouche, and Sun-Joo Shin. "A History of Logic Diagrams," in Dov. Gabbay, Francis Jeffry Pelletier, and John Woods, eds., *Handbook of the History of Logic. Volume 11. Logic: A History of Its Central Concepts* (Amsterdam: North-Holland, 2012), pp. 611–682.

Monro, C.J. "Review of *Symbolic Logic*, by John Venn," *Mind* 6:24 (1881): pp. 574–581.

Niiniluoto, Ilkka. "From Possibility to Probability: British Discussion on Modality in the Nineteenth Century," in Simo Knuuttila, ed. *Modern Modalities. Studies of the History of Modal Theories from medieval Nominalism to Logical Positivism* (Dordrecht: Kluwer Academic Publishers, 1988), pp. 239–274.

Olaniyan, Richard. "British Imperial Politics and Diplomacy in West Africa, 1860–1865," *The Historian* 34: 4 (1972): pp. 650–669

Perowne, J.J. Stewart. *Immortality. Four Sermons. Preached before the University of Cambridge, being the Hulsean Lectures for 1868* (Cambridge: Deighton, Bell, and Co., 1869).

Porter, Theodore M. *Genetics in the Madhouse: The Unknown History of Human Heredity* (Princeton & Oxford: Princeton University Press, 2019).

Pearson, Karl, ed. *The Life and Letters of Francis Galton. Volume I: Birth 1822 to Marriage 1853* (Cambridge: Cambridge University Press, 1914).

Rigg, James H. *Modern Anglican Theology* (London, 1859).

Richardson, Robert P., and Edward H. Landis. "Numbers, Variables and Mr. Russell's Philosophy," *The Monist* 25 (1915): pp. 321–364.

Richardson, Robert P., and Edward H. Landis. *Fundamental Conceptions of Modern Mathematics. Variables and Quantities. With a Discussion of the General Conception of Functional Relation* (Chicago & London: The Open Court Publishing Company, 1916).

Riso, Mary. *The Narrative of the Good Death. The Evangelical Deathbed in Victorian England* (London & New York: Routledge, 2015).

Rüger, Jan. *The Great Naval Game: Britain and Germany in the Age of Empire* (Cambridge: Cambridge University Press, 2007).

Palfrey, David. "The Moral Sciences Tripos at Cambridge University, 1848–1860," unpublished PhD thesis, University of Cambridge, 2002.

Parsons, Gerald. "Victorian Roman Catholicism: Emancipation, Expansion and Achievement," in Gerald Parsons, ed. *Religion in Victorian Britain* (Manchester & New York: Manchester University Press, 1988), pp. 146–183.

Rahman, Shahid, and Juan Redmond. "Hugh MacColl and the Birth of Logical Pluralism," in Dov. M. Gabbay and John Woods, eds., *Handbook of the History of Logic. Volume 4: British Logic in the Nineteenth Century* (Amsterdam & Boston: North Holland, 2008), pp. 533–604.

Rice, Adrian. "'Everybody Makes Errors': The Intersection of De Morgan's Logic and Probability, 1837–41," *History and Philosophy of Logic* 24:4 (2003): pp. 289–305.

Rupke, Nicolaas A. *Richard Owen: Biology Without Darwin* (Chicago & London: The University of Chicago Press, 2009 [1994]).

Searby, Peter. *A History of the University of Cambridge. Volume III: 1750-1870* (Cambridge: Cambridge University Press, 1997).

Schultz, Bart. *Henry Sidgwick. Eye of the Universe. An Intellectual Biography* (Cambridge: Cambridge University Press, 2004).

Shea, Victor, and William Whitla. *Essays and Reviews: The 1860 Text and Its Reading* (Charlottesville & London: University Press of Virginia, 2000).

Shenk, Wilbert R. *Henry Venn Missionary Statesman* (Eugene, OR: Wipf & Stock Publishers, 1983).]

Shipley, A.E., ed. *"J." A Memoir of John Willis Clark. Registrary of the University of Cambridge and Sometime Fellow of Trinity College* (London, 1913).

Sidgwick, Henry. *Philosophy: Its Scope and Relations. An Introductory Course of Lectures* (London: Macmillan and Co., 1902).

Sidgwick, Eleanor Mildred, and Alfred Sidgwick. *Henry Sidgwick. A Memoir* (London: Macmillan and Co., 1906).

Sokal, Michael M. "Psychology at Victorian Cambridge: The Unofficial Laboratory of 1887–1888," *Proceedings of the American Philosophical Society* 116:2 (1972): pp. 145–147.

Stanley, A.P. "Theology of the Nineteenth Century," *Fraser's Magazine* 71 (February 1865): pp. 252–268.

Stephen, Fitzjames. "Buckle's *History*," *Edinburgh Review* 107 (1858): pp. 465–512.

Stephen, Fitzjames. [Anon.] "The Study of History," *Cornhill Magazine* III (June 1861a): pp. 666–680.

Stephen, Fitzjames. [Anon.] "The Study of History," *Cornhill Magazine* IV (July 1861b): pp. 25–41.

Stephen, Fitzjames. "English Jurisprudence," *Edinburgh Review* 114 (1861c), pp. 56–86.

Stephen, Leslie. "Henry Sidgwick," *Mind* 10:37 (1901): pp. 1–17.

Stigler, Stephen M. *The History of Statistics: The Measurement of Uncertainty Before 1900* (Cambridge: Belknap Press, 1986).

Stigler, Stephen M. "Francis Galton's Account of the Invention of Correlation," *Statistical Science* 4:2 (1989): pp. 73–79.

Stigler, Stephen M. *Statistics on the Table* (Cambridge, MA: Harvard University Press, 1999).

Testimonials in Favour of G. Croom Robertson, M.A., Candidate for the Examinership in Logic and Moral Philosophy in the University of London. Privately printed.

Tomlinson, Charles. *Experimental Essays* (London: Virtue Brothers & Co., 1863).

Tullock, Hugh. "A.V. Dicey and the Irish Question: 1870–1922," *Irish Jurist* 15:1 (1980): pp. 137–165.

Van Evra, James. "Richard Whately and the Rise of Modern Logic," *History and Philosophy of Logic* 5:1 (1984): pp. 1–18.

Veitch, John. *Institutes of Logic* (Edinburgh & London: William Blackwood and Sons, 1885).

Venn, Henry (of C.M.S.). *West-African Colonies. Notices of the British Colonies on the West Coast of Africa* (London: 1854).

Venn, Henry (of C.M.S.). "The Proper Interpretation of the Baptismal Service of the Church of England," *Christian Observer* (September 1870c): pp. 643–653.

Venn, John (of Hereford). *Assertions of a Roman Catholic Priest Examined and Exposed; or the Correspondence Between the Rev. John Venn and the Rev. James Waterworth, Respecting Certain Assertions Made by the Latter at the "Hereford Discussion"* (Hereford: Charles Anthony, 1845).

Verburgt, Lukas M. "Pragmatism at Cambridge, England before 1900," *British Journal for the History of Philosophy* 29:1 (2021a): 84–105.

Verburgt, Lukas M. "The Venn-MacColl dispute in *Nature*," *History and Philosophy of Logic* 41:3 (2020): pp. 244–251.

Verburgt, Lukas M. *John Venn: A Life in Logic* (forthcoming at The University of Chicago Press, 2022).

Vogeler, Martha S. "More Light on *Essays and Reviews*: The Role of Frederic Harrison," *Victorian Periodicals Review* 12 (1979): 105–116.

Wall, Byron E. "John Venn, James Ward, and the Chair of Mental Philosophy and Logic at the University of Cambridge," *Journal of the History of Ideas* 68:1 (2007): pp. 131–155.

Warwick, Andrew. *Masters of Theory. Cambridge and the Rise of Mathematical Physics* (Chicago & London: The University of Chicago Press, 2003).

Watson, Henry William, and Francis Galton. "On the Probability of the Extinction of Families," *The Journal of the Anthropological Institute of Great Britain and Ireland* 4 (1875): pp. 138–144.

Wilson, Robin, and Amirouche Moktefi, eds. *The Mathematical World of Charles L. Dodgson (Lewis Carroll)* (Oxford: Oxford University Press, 2019).

Yates, Nigel. *Anglican Ritualism in Victorian Britain, 1830–1890* (Oxford: Oxford University Press, 1999).

Index

L. M. Verburgt, *John Venn: Unpublished Writings and Selected Correspondence*, Studies in History and Philosophy of Science 56,
https://doi.org/10.1007/978-3-030-79829-1

J

K

L

M

N

O

P

CPSIA information can be obtained
at www.ICGtesting.com
Printed in the USA
BVHW010221011221
622954BV00002B/53

9 783030 798284